ACKNOWLEDGEMENTS

The publishers gratefully acknowledge the permissions granted to reproduce copyright material in this book. Every effort has been made to contact the holders of copyright material, but if any have been inadvertently overlooked, the Publisher will be pleased to make the necessary arrangements at the first opportunity.

Chapter 1
pg. 12 Chaiwuth Wichitdho/Shuttestock, stavklem/Shutterstock, Mikhail Zahranichny/Shutterstock, Corepics VOF/Shutterstock; pg. 13 Ufuk ZIVANA/Shutterstock, Olga Popova/Shutterstock; pg. 14 ANDREW LAMBERT PHOTOGRAPHY/SCIENCE PHOTO LIBRARY; pg. 16 Ufuk ZIVANA/Shutterstock; pg. 17 PROF. PETER FOWLER/SCIENCE PHOTO LIBRARY; pg. 18 GRAHAM J. HILLS/SCIENCE PHOTO LIBRARY; pg. 20 Syda Productions/Shutterstock; pg. 22 SSSCCC/Shutterstock.

Chapter 2
pg. 32 Swapan Photography/Shutterstock, ggw1962/Shutterstock, thieury/Shutterstock, grey color/Shutterstock, Arsenis Spyros/Shutterstock; pg. 33 Pressmaster/Shutterstock, piximage/Shutterstock, Triff/Shutterstock; pg. 34 Lakeview Images/Shutterstock; pg. 35 ANDREW LAMBERT PHOTOGRAPHY/SCIENCE PHOTO LIBRARY; pg. 38 brumhildich/Shutterstock, Yenyu Shih/Shutterstock; pg. 40 STILLFX/Shutterstock; pg. 41 Andre Adams/Shutterstock; pg. 48 demarcomedia/Shuttestock, Jeffrey B. Banke, Triff/Shutterstock, Gail Johnson/Shutterstock; pg. 49 Anneka/Shutterstock, Mona Makela/Shutterstock, Asmus Koefoed/Shutterstock, Popartic/Shutterstock; pg. 54 Shi Yali/Shutterstock, Lori Werhane/Shutterstock, izzzy71/Shutterstock; pg. 57 YuriyK/Shutterstock; pg. 66 iceink/Shutterstock; pg. 69 Jens Ottoson/Shutterstock; pg. 72 Tatiana Popova/Shutterstock, science photo/Shutterstock; pg. 73 Timur Djafarov/Shuttestock, Flegere/Shutterstock; pg. 78 Boris15/Shutterstock; pg. 79 ppl/Shutterstock; pg. 80 Yeko Photo Studio/Shutterstock;pg. 81 Georgi Roshkov/Shutterstock; pg. 82 Vladimir A Veljanovski/Shutterstock; pg. 83 Aumm graphixphoto/Shutterstock; pg. 84 AlexanderAIUS/Wikipedia, nobeastsofierce/Shutterstock; pg. 86 Konstantin L/Shutterstock; pg. 88 Syda Productions/Shutterstock.

Chapter 3
pg. 96 IIsshaya/Shutterstock; Dziewul/Shutterstock, Sabine Kappel/Shutterstock, BeautifulChemistry.net/SCIENCE PHOTO LIBRARY; pg. 97 CHARLES D. WINTERS/SCIENCE PHOTO LIBRARY; pg. 98 Fablok/Shutterstock, motorolka/Shutterstock, ANDREW LAMBERT PHOTOGRAPHY/SCIENCE PHOTO LIBRARY; pg. 99 Smith1972/Shutterstock; pg. 100 Evgeny Karandaev/Shutterstock; pg. 106 SCIENCE PHOTO LIBRARY/Shutterstock; pg. 107 GIPhotoStock/SCIENCE PHOTO LIBRARY; pg. 108 li jianbing/Shutterstock; pg. 112 Corepics VOF/Shutterstock; pg. 114 Andrey_Popov/Shutterstock; pg. 118 Albert Russ/Shutterstock, Eugene Sergeev/Shutterstock; pp. 122 GIPhotoStock/SCIENCE PHOTO LIBRARY; pp. 126 NagyDodo/Shutterstock; pp. 134 Dionisvera/Shutterstock, Ivaschenko Roman/Shutterstock, igor.stevanovic/Shutterstock, Slavoljub Pantelic/Shutterstock.

Chapter 4
pg. 156 pan_kung/Shutterstock, BeautifulChemistry.net/SCIENCE PHOTO LIBRARY, MARTYN F. CHILLMAID/SCIENCE PHOTO LIBRARY, Yuriyk/Shutterstock, anyaivanova/Shutterstock; pg. 157 HUANSHENG XU/ Shutterstock, Italianvideophotoagency/Shutterstock, ANDREW LAMBERT PHOTOGRAPHY/SCIENCE PHOTO LIBRARY, Decha Thapanya/Shutterstock; pg. 158 DenisNata/Shutterstock; pg. 159 tonyz20/Shutterstock; pg. 160 tarog/Shutterstock, ANDREW LAMBERT PHOTOGRAPHY/SCIENCE PHOTO LIBRARY, ANDREW LAMBERT PHOTOGRAPHY/SCIENCE PHOTO LIBRARY; pg. 162 SCIENCE PHOTO LIBRARY; pg. 163 ANDREW LAMBERT PHOTOGRAPHY/SCIENCE PHOTO LIBRARY; pg. 165 Andraž Cerar/Shutterstock, HUANSHENG XU/Shutterstock; pg. 168 TREVOR CLIFFORD PHOTOGRAPHY/SCIENCE PHOTO LIBRARY, Filip Fuxa/Shutterstock; pg. 174 ANDREW LAMBERT PHOTOGRAPHY/SCIENCE PHOTO LIBRARY; pg. 175 MARTYN F. CHILLMAID/SCIENCE PHOTO LIBRARY, CHARLES D. WINTERS/SCIENCE PHOTO LIBRARY, Charles D. Winters/SCIENCE PHOTO LIBRARY; pg. 176 Magdalena Kowalik/Shutterstock; pg. 177 pryzmat/Shutterstock; pg. 178 tanewpix/Shutterstock; pg. 179 extender_01/Shutterstock.

Chapter 5
pg. 188 eldar nurkovic/Shutterstock, yurok/Shutterstock, CHARLES D. WINTERS/SCIENCE PHOTO LIBRARY, jmarkow/sShutterstock, ggw1962/Shutterstock; pg. 189 EM Karuna/Shutterstock, the palms/Shutterstock; pg. 196 Michal Kowalski/Shutterstock; pg. 197 Lindsey Moore/Shutterstock; pg. 198, Ming-Hsiang Chuang/Shutterstock, motorolka/Shutterstock, Martin M303/Shutterstock; pg. 199 Dong liu/Shutterstock; pg. 205, MARTYN F. CHILLMAID/SCIENCE PHOTO LIBRARY; pg. 206 iladm/Shutterstock, RUI FERREIRA/Shutterstock; pg. 218 ANDREW LAMBERT PHOTOGRAPHY/SCIENCE PHOTO LIBRARY, MARTYN F. CHILLMAID/SCIENCE PHOTO LIBRARY; pg. 222 Dennis Sabo/Shutterstock.

Chapter 6
pg. 236 jordache/Shutterstock, Matee Nuserm/Shutterstock, PHOTO FUN/Shutterstock, Ammit Jack/Shutterstock, Hung Chung Chih/Shutterstock; pg. 237 jordache/shutterstock, Alessandro Colle/Shutterstock, Charles Knowles/Shutterstock, Khoroshunova Olga/Shutterstock, Filip Fuxa/Shutterstock; pg. 238 Jiri Vaclavek/Shutterstock, Nastya22/Shutterstock; pg. 239 jordache/Shutterstock; pg. 240 CHOKCHAI POOMICHAIYA/Shutterstock; pg. 242 Mark Schwettmann/Shutterstock, LianeM/Shutterstock, MAXIMILIAN STOCK LTD/SCIENCE PHOTO LIBRARY; pg. 243 Jose Arcos Aguilar/Shutterstock; pg. 244 Stockr/Shutterstock; pg. 245 hacohob/Shutterstock; pg. 246 Josef Hanus/Shutterstock; pg. 249 wavebreakmedia/Shutterstock, xshot/Shutterstock; pg. 250 Pavel L Photo and Video/Shutterstock, freedomnaruk/Shutterstock; pg. 251 Therina Groenewald/Shuttestock, Huguette Roe/Shutterstock; pg. 252 marekusz/Shutterstock, Africa Studio/Shutterstock; pg. 253 I love photo/Shutterstock, stockcreations/Shutterstock; pg. 254 sima/Shutterstock, Dmitry Kalinovsky/Shutterstock; pg. 255 dani3315/Shutterstock; pg. 256 iurii/Shutterstock; pg. 262 Anneka/Shutterstock; pg. 263 Balakir Alla/Shutterstock; pg. 264 Jiri Slama/Shutterstock; pg. 256 CREATISTA/Shutterstock, pilipphoto/Shutterstock, Anatol Tyshkevich/Shutterstock; pg. 257 Maxisport/Shutterstock, padu_foto/Shutterstock; pg. 260 Myimagine/Shutterstock, underverse/Shutterstock; pg. 266 Elena Schweitzer/Shutterstock; pg. 270 iurii/Shutterstock; pg. 272 Ulga/Shutterstock; pg. 273 pixinoo/Shutterstock; pg. 276 Darren Brode/Shutterstock, petrmalinak/Shutterstock; pg. 284 Ammit Jack/Shutterstock, James Steidl/Shutterstock; pg. 285 Khoroshunova Olga/Shutterstock; pg. 286 Unicus/Shutterstock, attem/Shutterstock; pg. 287 Ekaterina Pokrovsky/Shutterstock; pg. 291 Dudarev Mikhail/Shutterstock, Huguette Roe/Shutterstock; pg. 292 Sohel Parvez Haque/Shutterstock, Silken Photography/Shutterstock, Saikat Paul/Shutterstock; pg. 293 Matty Symons/Shutterstock; pg. 294 Filip Fuxa/Shutterstock, martin33/Shutterstock, Tomasz Darul/Shutterstock; pg. 298 donikz/Shutterstock; pg. 299 Warren Price Photography/Shutterstock; pg. 300 Alexander Raths/Shutterstock, MikeDotta/Shutterstock; pg. 301 Petr Vopenka/Shutterstock, J. Helgason/Shutterstock, Laurence Gough/Shutterstock; pg. 302 ventdusud/Shutterstock, muratart/Shutterstock; pg. 303 goodcat/Shutterstock; pg. 304, Reddogs/Shutterstock; pg. 305 Vaclav Volrab/Shutterstock.

Contents

You can use this book if you are studying Combined Science:

you will need to master all of the ideas and concepts on these pages.

OCR Gateway GCSE (9–1)

Chemistry

Student Book

Ann Daniels
Series editor: Ed Walsh

William Collins' dream of knowledge for all began with the publication of his first book in 1819.

A self-educated mill worker, he not only enriched millions of lives, but also founded a flourishing publishing house. Today, staying true to this spirit, Collins books are packed with inspiration, innovation and practical expertise. They place you at the centre of a world of possibility and give you exactly what you need to explore it.

Collins. Freedom to teach

HarperCollins Publishers
1 London Bridge Street
London SE1 9GF

Browse the complete Collins catalogue at
www.collins.co.uk

First edition 2016

10 9 8 7 6 5 4 3 2 1

ISBN 978-0-00-815095-2

www.collins.co.uk

A catalogue record for this book is available from the British Library

Commissioned by Lucy Rowland, Lizzie Catford and Joanna Ramsay

Edited by Hamish Baxter

Project managed by Elektra Media Ltd

Development edited by Tim Jackson and Rich Cutler

Copy edited by Dr Sarah Ryan and Lynette Woodward

Proofread by Laurice Suess and Ali Craig

Typeset by Jouve India and Ken Vail Graphic Design

Cover design by We are Laura

Printed by Bell and Bain Ltd, Glasgow

Cover images © Shutterstock/Komsan Loonprom, Shutterstock/Everett Historical

OCR Endorsement statement

This resource is endorsed by OCR for use with specification J248 GCSE Gateway Science Suite – Chemistry A (9–1). In order to gain OCR endorsement, this resource has undergone an independent quality check. Any references to assessment and/or assessment preparation are the publisher's interpretation of the specification requirements and are not endorsed by OCR. OCR recommends that a range of teaching and learning resources are used in preparing learners for assessment. OCR has not paid for the production of this resource, nor does OCR receive any royalties from its sale. For more information about the endorsement process, please visit the OCR website, www.ocr.org.uk.

How to use this book

Remember! to cover all the content of the OCR Chemistry Specification you should study the text and attempt the End of Chapter Questions.

Learning objectives which are Higher tier only appear in a purple background box.

These tell you what you will be learning about in the lesson and are linked to the OCR specification.

This introduces the topic and puts the science into an interesting context.

Each topic is divided into three sections. The level of challenge gets harder with each section.

Chemistry – The particle model (C1.1)

Three states of matter

Learning objectives:

- use data to predict the states of substances
- explain the changes of state
- use state symbols in chemical equations
- explain the limitations of the particle model.

KEY WORDS

changes of state
condensing
limitations
particle

What happens to ice in a cold drink? How can we describe what is happening as the solid becomes a liquid? A particle model can be used to describe the three states of matter – solid, liquid and gas – and used to help us to visualise what happens during changes of state.

The states of matter

The three states of matter are solid (s), liquid (l) and gas (g).

- Melting and freezing take place at the melting point.
- Boiling and **condensing** take place at the boiling point.

Gas
Sublimation
Boiling
Condensation
Freezing
Solid
Liquid
Melting

Figure 1.1 The three states of matter

Substance	Melting point (°C)	Boiling point (°C)	State at room temperature
W	–18	42	liquid
X	150	875	
Y	–190	–84	
Z	–56	16	

Substance W in the table will have melted at –18 °C but at room temperature, which is around 25 °C, it will not have boiled. So it is a liquid.

1 What are the states of substances X, Y and Z at room temperature?

The three states of matter can be represented by a simple model. In this model, particles are represented by small solid spheres.

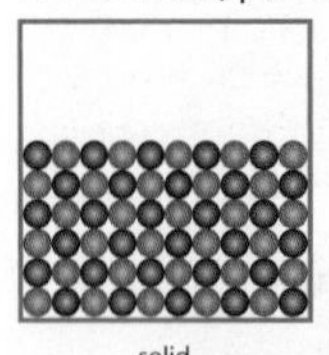
solid

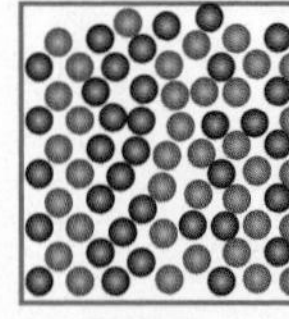
liquid

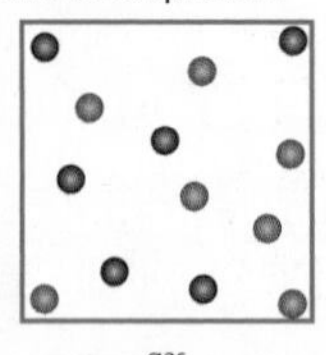
gas

Figure 1.3 Particle model diagrams of solid, liquid and gas

Look again at substance W in the table. At 25 °C it will be a liquid. However, a few particles will have enough energy to escape the surface of the liquid and evaporation will be taking place.

DID YOU KNOW?

Iodine can turn straight from a solid to a gas. This is called sublimation.

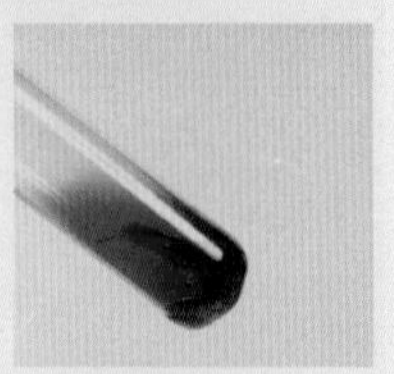

Figure 1.2 Iodine changing from a solid to a gas

14 OCR Gateway GCSE Chemistry: Student Book

Important scientific vocabulary is highlighted. You can check the meanings in the Glossary at the end of the book.

Changes of state

1.1

Particle theory can help to explain the changes of state. Melting, boiling, freezing and condensing are processes that depend on changing the forces *between* the particles.

In melting and boiling the strength of the forces between particles becomes less. The distance between particles increases and the arrangement becomes more random. The particles move more so more energy is needed from the surroundings.	In freezing and condensing the strength of the forces remains the same. The distance between particles decreases and the arrangement become less random. The particles move less so less energy is needed.

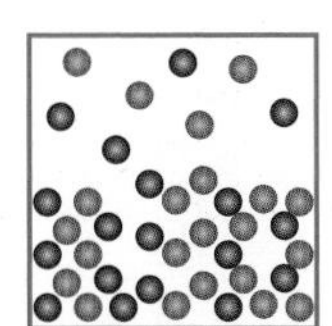

Figure 1.4 Particle model of evaporation

The *amount* of energy needed to change state from solid to liquid and from liquid to gas depends on the strength of the forces between the particles of the substance.

The strength of the forces between particles depends on the nature of the particles involved, on the type of bonding and the structure of the substance.

The stronger the forces between the particles the higher the melting point and boiling point of the substance.

2. **The data in the table on the previous page for substance W shows its melting point is –18 °C. Describe what is happening to the particles and the forces between them at these temperatures: –28 °C, –18 °C, –14 °C**
3. **Look at the boiling point data for substance W in the table. Describe what is happening to the particles and the forces between them at these temperatures: 25 °C, 38 °C, 42 °C, 46 °C**
4. **Suggest why iron has high melting and boiling points.**
5. **Ethanol has a boiling point of 78.4 °C. Propane has a boiling point of –42 °C. Suggest why.**

KEY INFORMATION

These diagrams only show a model. For example, in the model of a solid the individual spheres represent particles that are not, themselves, solid. It is only when lots of these particles are arranged closely in a regular pattern that they together represent a solid.

Each section has level-appropriate questions, so you can check and apply your knowledge.

DID YOU KNOW?

In chemical equations, the three states of matter are shown as (s), (l) and (g), with (aq) for aqueous solutions.

Each topic has some fascinating additional background information.

HIGHER TIER ONLY

Limitations of this model

This simple model is limited because:

- there are no forces represented between the spheres
- all the particles are represented as spheres
- the spheres represented are solid and inelastic.

This means that the changes in forces and collisions between particles cannot be represented fully. However, it is a useful model to show spatial arrangement, both regular and random.

6. **Explain the limitations of the particle theory when considering the process of condensing.**

KEY INFORMATION

We see that when the particles move more rapidly the 'state' of the bulk of the matter changes from solid to liquid to gas. These changes are physical changes. If particles of matter from different substances collide, react and join to make new substances, this change is a chemical change. It can happen in any state.

For Foundation tier, you do not need to understand the content in the Higher tier only boxes. For Higher tier you should aim to understand the other sections, as well as the content in the Higher tier only boxes.

Google search: 'melting office graph' 15

Look for the internet search terms at the bottom of the page.

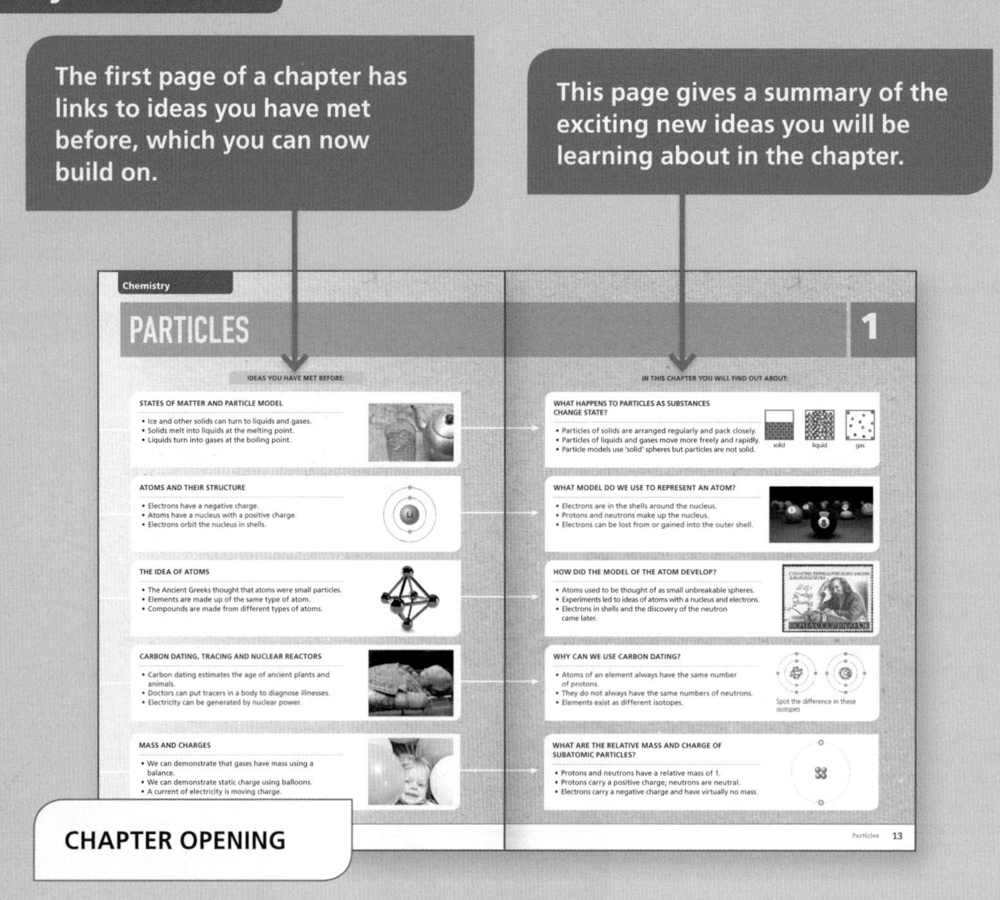

The Key Concept pages focus on a core ideas. Once you have understood the key concept in a chapter, it should develop your understanding of the whole topic.

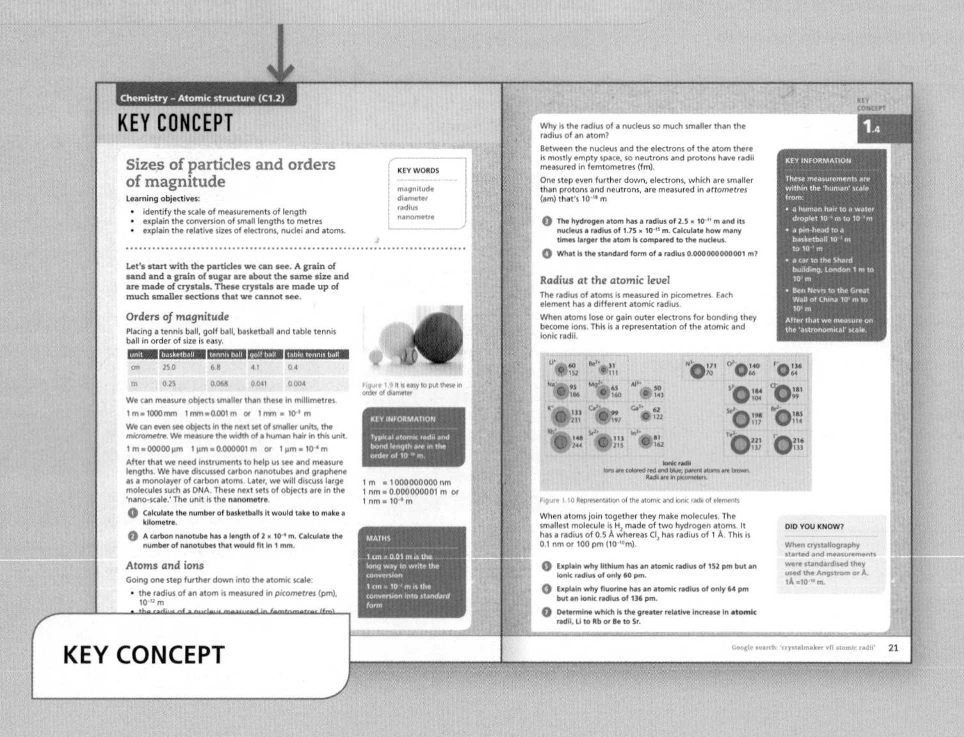

There are dedicated pages for practicals. They help you to analyse the practical and to answer questions about it.

The tasks – which get a bit more difficult as you go through – challenge you to apply your science skills and knowledge to the new context.

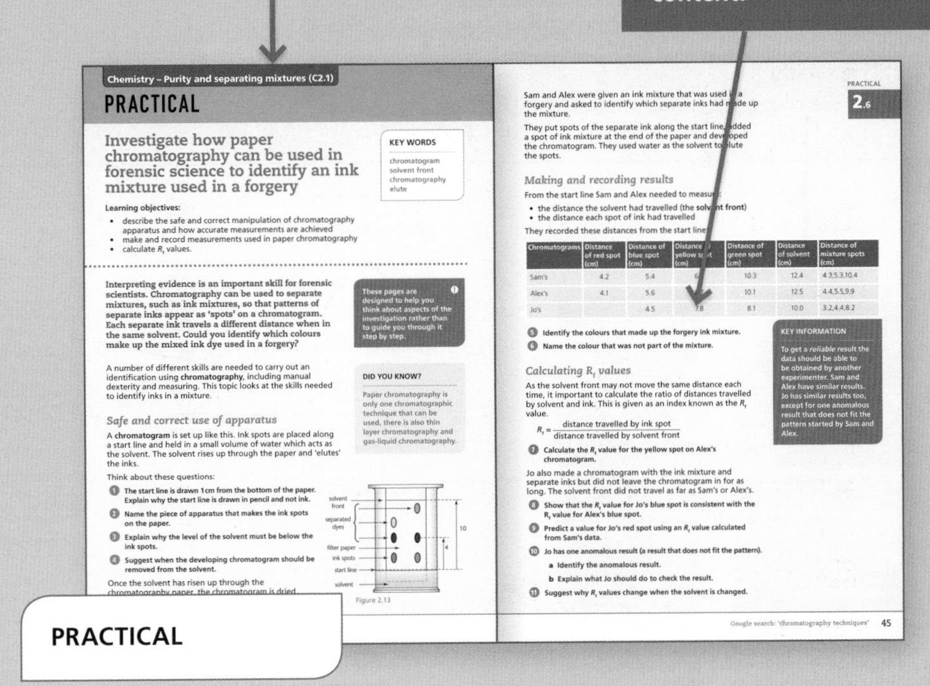

Chemistry – Purity and separating mixtures (C2.1)

PRACTICAL

Investigate how paper chromatography can be used in forensic science to identify an ink mixture used in a forgery

KEY WORDS
chromatogram
solvent front
chromatography
elute

Learning objectives:

- describe the safe and correct manipulation of chromatography apparatus and how accurate measurements are achieved
- make and record measurements used in paper chromatography
- calculate R_f values.

Interpreting evidence is an important skill for forensic scientists. Chromatography can be used to separate mixtures, such as ink mixtures, so that patterns of separate inks appear as 'spots' on a chromatogram. Each separate ink travels a different distance when in the same solvent. Could you identify which colours make up the mixed ink dye used in a forgery?

These pages are designed to help you think about aspects of the investigation rather than to guide you through it step by step.

A number of different skills are needed to carry out an identification using chromatography, including manual dexterity and measuring. This topic looks at the skills needed to identify inks in a mixture.

DID YOU KNOW?
Paper chromatography is only one chromatographic technique that can be used, there is also thin layer chromatography and gas-liquid chromatography.

Safe and correct use of apparatus

A chromatogram is set up like this. Ink spots are placed along a start line and held in a small volume of water which acts as the solvent. The solvent rises up through the paper and 'elutes' the inks.

Think about these questions:

1. The start line is drawn 1 cm from the bottom of the paper. Explain why the start line is drawn in pencil and not ink.
2. Name the piece of apparatus that makes the ink spots on the paper.
3. Explain why the level of the solvent must be below the ink spots.
4. Suggest when the developing chromatogram should be removed from the solvent.

Once the solvent has risen up through the chromatography paper, the chromatogram is dried...

Figure 2.13

PRACTICAL

Sam and Alex were given an ink mixture that was used in a forgery and asked to identify which separate inks had made up the mixture.

They put spots of the separate ink along the start line, added a spot of ink mixture at the end of the paper and developed the chromatogram. They used water as the solvent to elute the spots.

PRACTICAL 2.6

Making and recording results

From the start line Sam and Alex needed to measure:

- the distance the solvent had travelled (the solvent front)
- the distance each spot of ink had travelled

They recorded these distances from the start line:

Chromatograms	Distance of red spot (cm)	Distance of blue spot (cm)	Distance of yellow spot (cm)	Distance of green spot (cm)	Distance of solvent (cm)	Distance of mixture spots (cm)
Sam's	4.2	5.4	6	10.3	12.4	4.3,5.3,10.4
Alex's	4.1	5.6		10.1	12.5	4.4,5.5,9.9
Jo's		4.5	7.8	8.1	10.0	3.2,4.4,8.2

5. Identify the colours that made up the forgery ink mixture.
6. Name the colour that was not part of the mixture.

KEY INFORMATION
To get a *reliable* result the data should be able to be obtained by another experimenter. Sam and Alex have similar results. Jo has similar results too, except for one anomalous result that does not fit the pattern started by Sam and Alex.

Calculating R_f values

As the solvent front may not move the same distance each time, it important to calculate the ratio of distances travelled by solvent and ink. This is given as an index known as the R_f value.

$$R_f = \frac{\text{distance travelled by ink spot}}{\text{distance travelled by solvent front}}$$

7. Calculate the R_f value for the yellow spot on Alex's chromatogram.

Jo also made a chromatogram with the ink mixture and separate inks but did not leave the chromatogram in for as long. The solvent front did not travel as far as Sam's or Alex's.

8. Show that the R_f value for Jo's blue spot is consistent with the R_f value for Alex's blue spot.
9. Predict a value for Jo's red spot using an R_f value calculated from Sam's data.
10. Jo has one anomalous result (a result that does not fit the pattern).
    a Identify the anomalous result.
    b Explain what Jo should do to check the result.
11. Suggest why R_f values change when the solvent is changed.

Google search: 'chromatography techniques' 45

The Maths Skills pages focus on the maths requirements in the OCR specification, explaining concepts and providing opportunities to practise.

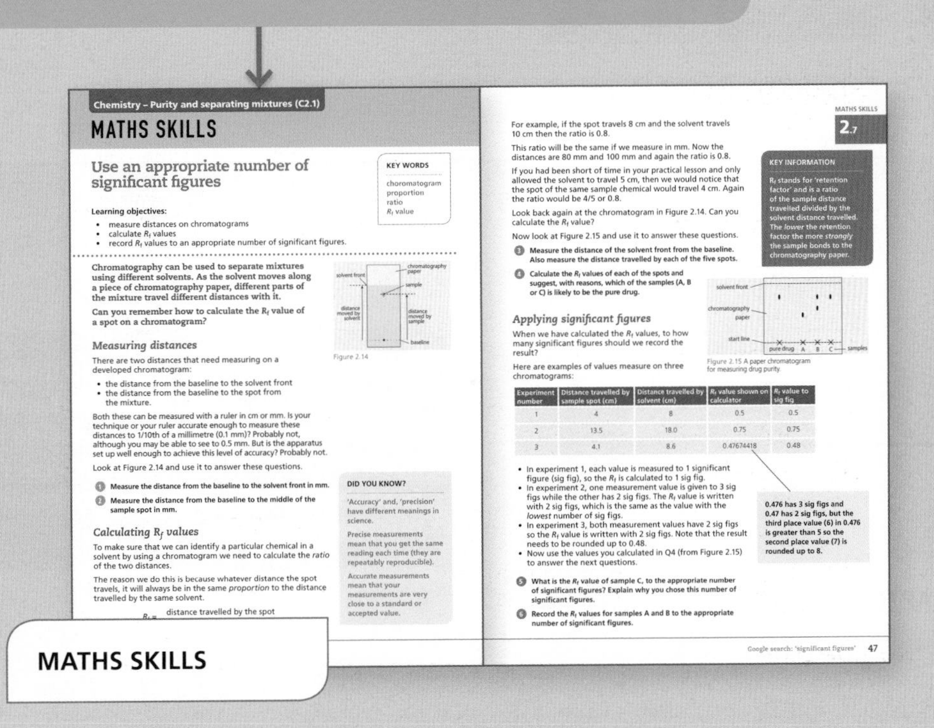

Chemistry – Purity and separating mixtures (C2.1)

MATHS SKILLS

Use an appropriate number of significant figures

KEY WORDS
chromatogram
proportion
ratio
R_f value

Learning objectives:

- measure distances on chromatograms
- calculate R_f values
- record R_f values to an appropriate number of significant figures.

Chromatography can be used to separate mixtures using different solvents. As the solvent moves along a piece of chromatography paper, different parts of the mixture travel different distances with it.

Can you remember how to calculate the R_f value of a spot on a chromatogram?

Figure 2.14

Measuring distances

There are two distances that need measuring on a developed chromatogram:

- the distance from the baseline to the solvent front
- the distance from the baseline to the spot from the mixture.

Both these can be measured with a ruler in cm or mm. Is your technique or your ruler accurate enough to measure these distances to 1/10th of a millimetre (0.1 mm)? Probably not, although you may be able to see to 0.5 mm. But is the apparatus set up well enough to achieve this level of accuracy? Probably not.

Look at Figure 2.14 and use it to answer these questions.

1. Measure the distance from the baseline to the solvent front in mm.
2. Measure the distance from the baseline to the middle of the sample spot in mm.

DID YOU KNOW?
'Accuracy' and, 'precision' have different meanings in science.
Precise measurements mean that you get the same reading each time (they are repeatably reproducible).
Accurate measurements mean that your measurements are very close to a standard or accepted value.

Calculating R_f values

To make sure that we can identify a particular chemical in a solvent by using a chromatogram we need to calculate the *ratio* of the two distances.

The reason we do this is because whatever distance the spot travels, it will always be in the same *proportion* to the distance travelled by the same solvent.

R_f = distance travelled by the spot

MATHS SKILLS

For example, if the spot travels 8 cm and the solvent travels 10 cm then the ratio is 0.8.

This ratio will be the same if we measure in mm. Now the distances are 80 mm and 100 mm and again the ratio is 0.8.

If you had been short of time in your practical lesson and only allowed the solvent to travel 5 cm, then we would notice that the spot of the same sample chemical would travel 4 cm. Again the ratio would be 4/5 or 0.8.

Look back again at the chromatogram in Figure 2.14. Can you calculate the R_f value?

Now look at Figure 2.15 and use it to answer these questions.

3. Measure the distance of the solvent front from the baseline. Also measure the distance travelled by each of the five spots.
4. Calculate the R_f values of each of the spots and suggest, with reasons, which of the samples (A, B or C) is likely to be the pure drug.

MATHS SKILLS 2.7

KEY INFORMATION
R_f stands for 'retention factor' and is a ratio of the sample distance travelled divided by the solvent distance travelled. The lower the retention factor the more *strongly* the sample bonds to the chromatography paper.

Figure 2.15 A paper chromatogram for measuring drug purity

Applying significant figures

When we have calculated the R_f values, to how many significant figures should we record the result?

Here are examples of values measure on three chromatograms:

Experiment number	Distance travelled by sample spot (cm)	Distance travelled by solvent (cm)	R_f value shown on calculator	R_f value to sig fig
1	4	8	0.5	0.5
2	13.5	18.0	0.75	0.75
3	4.1	8.6	0.47674418	0.48

- In experiment 1, each value is measured to 1 significant figure (sig fig), so the R_f is calculated to 1 sig fig.
- In experiment 2, one measurement value is given to 3 sig figs while the other has 2 sig figs. The R_f value is written with 2 sig figs, which is the same as the value with the *lowest* number of sig figs.
- In experiment 3, both measurement values have 2 sig figs so the R_f value is written with 2 sig figs. Note that the result needs to be rounded up to 0.48.
- Now use the values you calculated in Q4 (from Figure 2.15) to answer the next questions.

0.476 has 3 sig figs and 0.47 has 2 sig figs, but the third place value (6) in 0.476 is greater than 5 so the second place value (7) is rounded up to 8.

5. What is the R_f value of sample C, to the appropriate number of significant figures? Explain why you chose this number of significant figures.
6. Record the R_f values for samples A and B to the appropriate number of significant figures.

Google search: 'significant figures' 47

These lists at the end of a chapter act as a checklist of the key ideas of the chapter. In each row, the green box gives the ideas or skills that you should master first. Then you can aim to master the ideas and skills in the blue box. Once you have achieved those you can move on to those in the red box.

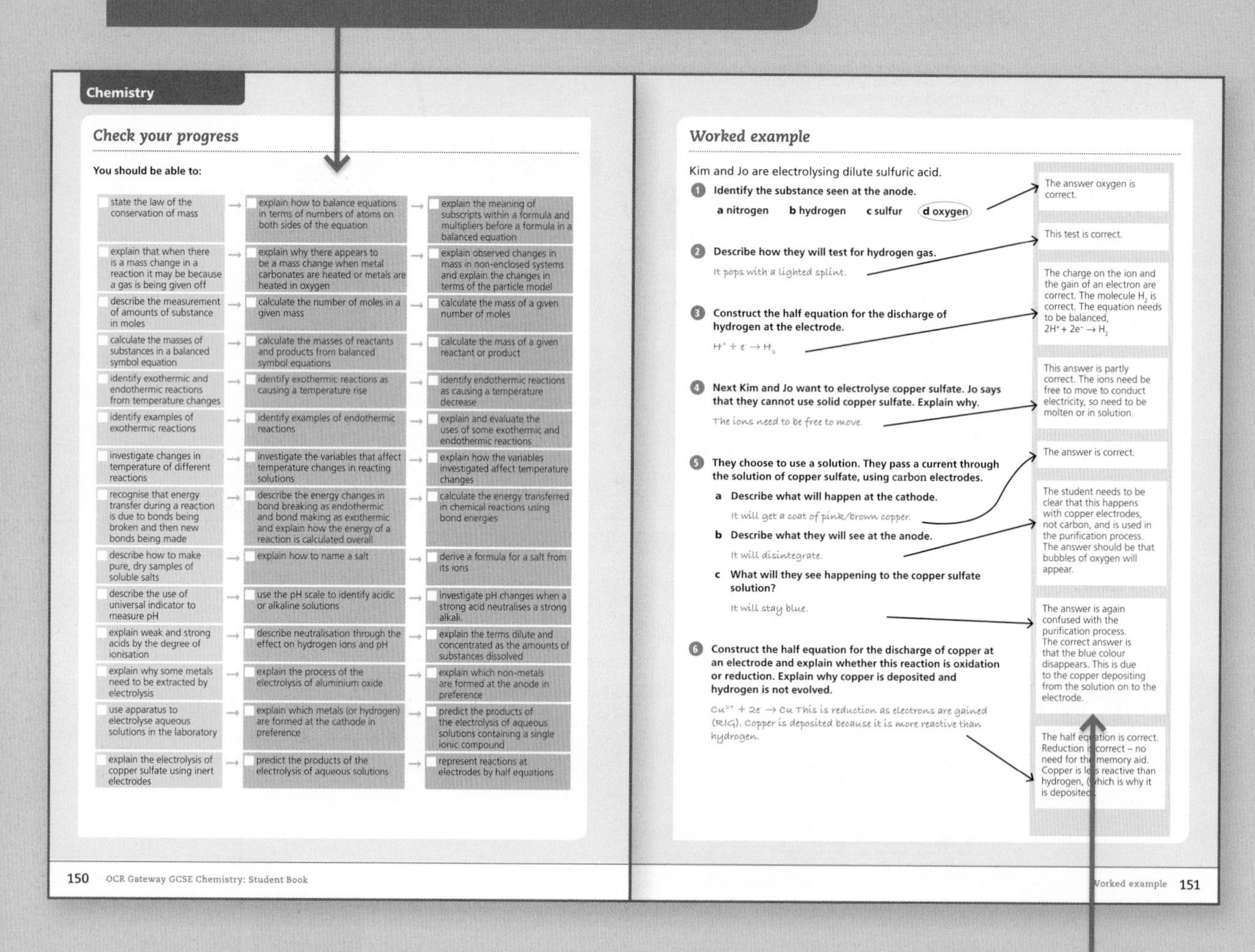

Chemistry

Check your progress

You should be able to:

state the law of the conservation of mass →	explain how to balance equations in terms of numbers of atoms on both sides of the equation →	explain the meaning of subscripts within a formula and multipliers before a formula in a balanced equation
explain that when there is a mass change in a reaction it may be because a gas is being given off →	explain why there appears to be a mass change when metal carbonates are heated or metals are heated in oxygen →	explain observed changes in mass in non-enclosed systems and explain the changes in terms of the particle model
describe the measurement of amounts of substance in moles →	calculate the number of moles in a given mass →	calculate the mass of a given number of moles
calculate the masses of substances in a balanced symbol equation →	calculate the masses of reactants and products from balanced symbol equations →	calculate the mass of a given reactant or product
identify exothermic and endothermic reactions from temperature changes →	identify exothermic reactions as causing a temperature rise →	identify endothermic reactions as causing a temperature decrease
identify examples of exothermic reactions →	identify examples of endothermic reactions →	explain and evaluate the uses of some exothermic and endothermic reactions
investigate changes in temperature of different reactions →	investigate the variables that affect temperature changes in reacting solutions →	explain how the variables investigated affect temperature changes
recognise that energy transfer during a reaction is due to bonds being broken and then new bonds being made →	describe the energy changes in bond breaking as endothermic and bond making as exothermic and explain how the energy of a reaction is calculated overall →	calculate the energy transferred in chemical reactions using bond energies
describe how to make pure, dry samples of soluble salts →	explain how to name a salt →	derive a formula for a salt from its ions
describe the use of universal indicator to measure pH →	use the pH scale to identify acidic or alkaline solutions →	investigate pH changes when a strong acid neutralises a strong alkali.
explain weak and strong acids by the degree of ionisation →	describe neutralisation through the effect on hydrogen ions and pH →	explain the terms dilute and concentrated as the amounts of substances dissolved
explain why some metals need to be extracted by electrolysis →	explain the process of the electrolysis of aluminium oxide →	explain which non-metals are formed at the anode in preference
use apparatus to electrolyse aqueous solutions in the laboratory →	explain which metals (or hydrogen) are formed at the cathode in preference →	predict the products of the electrolysis of aqueous solutions containing a single ionic compound
explain the electrolysis of copper sulfate using inert electrodes →	predict the products of the electrolysis of aqueous solutions →	represent reactions at electrodes by half equations

150 OCR Gateway GCSE Chemistry: Student Book

Worked example

Kim and Jo are electrolysing dilute sulfuric acid.

1 **Identify the substance seen at the anode.**

a nitrogen **b** hydrogen **c** sulfur **d** oxygen (circled)

The answer oxygen is correct.

2 **Describe how they will test for hydrogen gas.**

It pops with a lighted splint.

This test is correct.

3 **Construct the half equation for the discharge of hydrogen at the electrode.**

$H^+ + e^- \rightarrow H_2$

The charge on the ion and the gain of an electron are correct. The molecule H_2 is correct. The equation needs to be balanced, $2H^+ + 2e^- \rightarrow H_2$

4 **Next Kim and Jo want to electrolyse copper sulfate. Jo says that they cannot use solid copper sulfate. Explain why.**

The ions need to be free to move.

This answer is partly correct. The ions need be free to move to conduct electricity, so need to be molten or in solution.

5 **They choose to use a solution. They pass a current through the solution of copper sulfate, using carbon electrodes.**

a Describe what will happen at the cathode.

It will get a coat of pink/brown copper.

The answer is correct.

b Describe what they will see at the anode.

It will disintegrate.

The student needs to be clear that this happens with copper electrodes, not carbon, and is used in the purification process. The answer should be that bubbles of oxygen will appear.

c What will they see happening to the copper sulfate solution?

It will stay blue.

The answer is again confused with the purification process. The correct answer is that the blue colour disappears. This is due to the copper depositing from the solution on to the electrode.

6 **Construct the half equation for the discharge of copper at an electrode and explain whether this reaction is oxidation or reduction. Explain why copper is deposited and hydrogen is not evolved.**

$Cu^{2+} + 2e^- \rightarrow Cu$ This is reduction as electrons are gained (RIG). Copper is deposited because it is more reactive than hydrogen.

The half equation is correct. Reduction is correct – no need for the memory aid. Copper is less reactive than hydrogen, (which is why it is deposited.

Worked example 151

Use the comments to help you understand how to answer questions. Read each question and answer. Try to decide if, and how, the answer can be improved. Finally, read the comments and try to answer the questions yourself.

END OF CHAPTER

The End of Chapter Questions allow you and your teacher to check that you have understood the ideas in the chapter, can apply these to new situations, and can explain new science using the skills and knowledge you have gained. The questions start off easier and get harder. If you are taking Foundation tier try to answer all the questions in the Getting started and Going further sections. If you are taking Higher Tier try to answer all the questions in the Going further, More challenging and Most demanding sections.

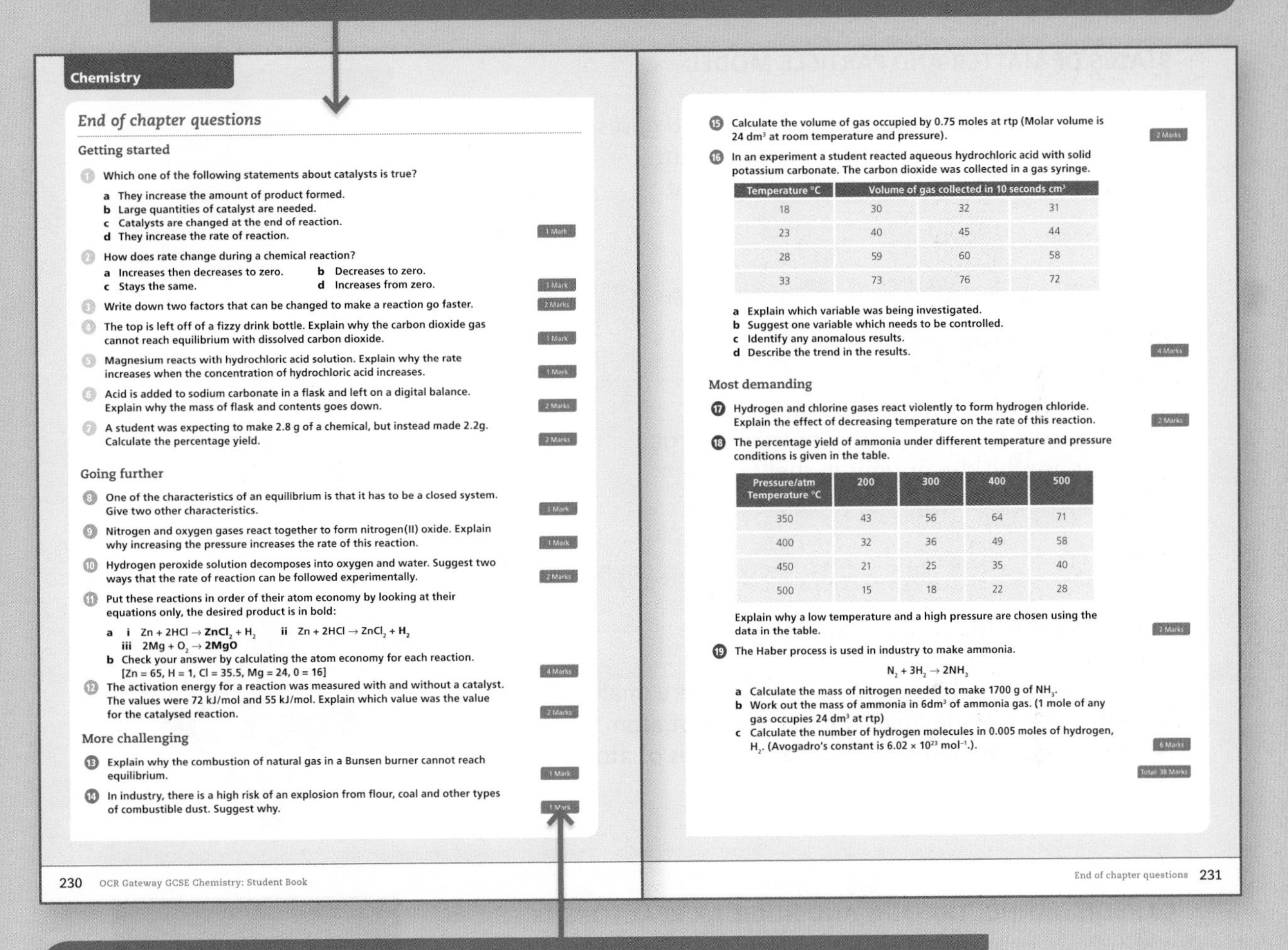

Chemistry

End of chapter questions

Getting started

1. Which one of the following statements about catalysts is true?
   - a They increase the amount of product formed.
   - b Large quantities of catalyst are needed.
   - c Catalysts are changed at the end of reaction.
   - d They increase the rate of reaction. (1 Mark)
2. How does rate change during a chemical reaction?
   - a Increases then decreases to zero.
   - b Decreases to zero.
   - c Stays the same.
   - d Increases from zero. (1 Mark)
3. Write down two factors that can be changed to make a reaction go faster. (2 Marks)
4. The top is left off of a fizzy drink bottle. Explain why the carbon dioxide gas cannot reach equilibrium with dissolved carbon dioxide. (1 Mark)
5. Magnesium reacts with hydrochloric acid solution. Explain why the rate increases when the concentration of hydrochloric acid increases. (1 Mark)
6. Acid is added to sodium carbonate in a flask and left on a digital balance. Explain why the mass of flask and contents goes down. (2 Marks)
7. A student was expecting to make 2.8 g of a chemical, but instead made 2.2g. Calculate the percentage yield. (2 Marks)

Going further

8. One of the characteristics of an equilibrium is that it has to be a closed system. Give two other characteristics. (1 Mark)
9. Nitrogen and oxygen gases react together to form nitrogen(II) oxide. Explain why increasing the pressure increases the rate of this reaction. (1 Mark)
10. Hydrogen peroxide solution decomposes into oxygen and water. Suggest two ways that the rate of reaction can be followed experimentally. (2 Marks)
11. Put these reactions in order of their atom economy by looking at their equations only, the desired product is in bold:
    - a i $Zn + 2HCl \rightarrow \mathbf{ZnCl_2} + H_2$ ii $Zn + 2HCl \rightarrow ZnCl_2 + \mathbf{H_2}$
      iii $2Mg + O_2 \rightarrow \mathbf{2MgO}$
    - b Check your answer by calculating the atom economy for each reaction. [Zn = 65, H = 1, Cl = 35.5, Mg = 24, 0 = 16] (4 Marks)
12. The activation energy for a reaction was measured with and without a catalyst. The values were 72 kJ/mol and 55 kJ/mol. Explain which value was the value for the catalysed reaction. (2 Marks)

More challenging

13. Explain why the combustion of natural gas in a Bunsen burner cannot reach equilibrium. (1 Mark)
14. In industry, there is a high risk of an explosion from flour, coal and other types of combustible dust. Suggest why. (1 Mark)

230 OCR Gateway GCSE Chemistry: Student Book

15. Calculate the volume of gas occupied by 0.75 moles at rtp (Molar volume is 24 dm^3 at room temperature and pressure). (2 Marks)
16. In an experiment a student reacted aqueous hydrochloric acid with solid potassium carbonate. The carbon dioxide was collected in a gas syringe.

Temperature °C	Volume of gas collected in 10 seconds cm^3		
18	30	32	31
23	40	45	44
28	59	60	58
33	73	76	72

   - a Explain which variable was being investigated.
   - b Suggest one variable which needs to be controlled.
   - c Identify any anomalous results.
   - d Describe the trend in the results. (4 Marks)

Most demanding

17. Hydrogen and chlorine gases react violently to form hydrogen chloride. Explain the effect of decreasing temperature on the rate of this reaction. (2 Marks)
18. The percentage yield of ammonia under different temperature and pressure conditions is given in the table.

Pressure/atm Temperature °C	200	300	400	500
350	43	56	64	71
400	32	36	49	58
450	21	25	35	40
500	15	18	22	28

   Explain why a low temperature and a high pressure are chosen using the data in the table. (2 Marks)
19. The Haber process is used in industry to make ammonia.

$$N_2 + 3H_2 \rightarrow 2NH_3$$

   - a Calculate the mass of nitrogen needed to make 1700 g of NH_3.
   - b Work out the mass of ammonia in 6dm^3 of ammonia gas. (1 mole of any gas occupies 24 dm^3 at rtp)
   - c Calculate the number of hydrogen molecules in 0.005 moles of hydrogen, H_2. (Avogadro's constant is 6.02×10^{23} mol^{-1}.). (6 Marks)

Total 38 Marks

End of chapter questions 231

There are questions for each assessment objective (AO) from the final exams. These help you to develop the thinking skills you need to answer each type of question.

AO1 – to answer these questions you should aim to **demonstrate** your knowledge and understanding of scientific ideas, techniques and procedures.

AO2 – to answer these questions you should aim to **apply** your knowledge and understanding of scientific ideas and scientific enquiry, techniques and procedures.

AO3 – to answer these questions you should aim to **analyse** information and ideas to: interpret and evaluate, make judgements and draw conclusions, develop and improve experimental procedures.

PARTICLES

IDEAS YOU HAVE MET BEFORE:

STATES OF MATTER AND PARTICLE MODEL

- Ice and other solids can turn to liquids and gases.
- Solids melt into liquids at the melting point.
- Liquids turn into gases at the boiling point.

ATOMS AND THEIR STRUCTURE

- Electrons have a negative charge.
- Atoms have a nucleus with a positive charge.
- Electrons orbit the nucleus in shells.

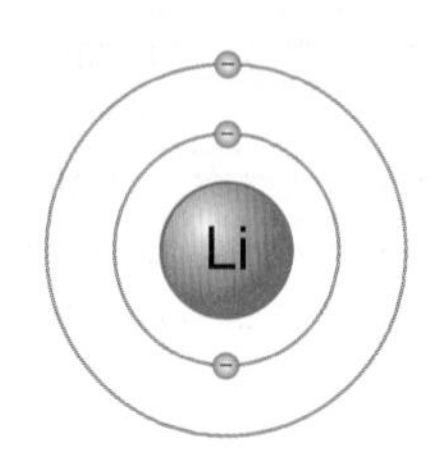

THE IDEA OF ATOMS

- The Ancient Greeks thought that atoms were small particles.
- Elements are made up of the same type of atom.
- Compounds are made from different types of atoms.

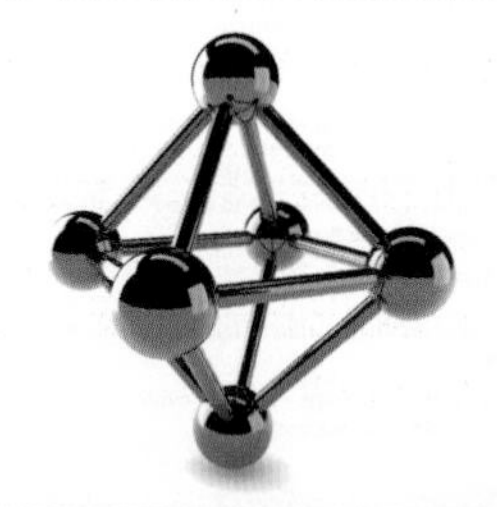

CARBON DATING, TRACING AND NUCLEAR REACTORS

- Carbon dating estimates the age of ancient plants and animals.
- Doctors can put tracers in a body to diagnose illnesses.
- Electricity can be generated by nuclear power.

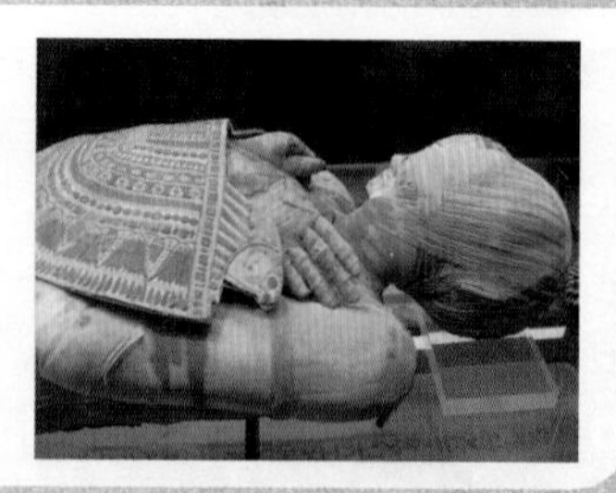

MASS AND CHARGES

- We can demonstrate that gases have mass using a balance.
- We can demonstrate static charge using balloons.
- A current of electricity is moving charge.

1

IN THIS CHAPTER YOU WILL FIND OUT ABOUT:

WHAT HAPPENS TO PARTICLES AS SUBSTANCES CHANGE STATE?

- Particles of solids are arranged regularly and pack closely.
- Particles of liquids and gases move more freely and rapidly.
- Particle models use 'solid' spheres but particles are not solid.

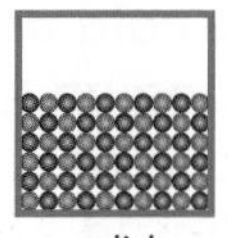
solid

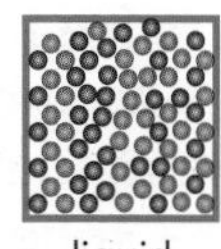
liquid

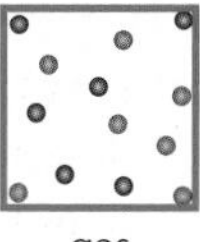
gas

WHAT MODEL DO WE USE TO REPRESENT AN ATOM?

- Electrons are in the shells around the nucleus.
- Protons and neutrons make up the nucleus.
- Electrons can be lost from or gained into the outer shell.

HOW DID THE MODEL OF THE ATOM DEVELOP?

- Atoms used to be thought of as small unbreakable spheres.
- Experiments led to ideas of atoms with a nucleus and electrons.
- Electrons in shells and the discovery of the neutron came later.

WHY CAN WE USE CARBON DATING?

- Atoms of an element always have the same number of protons.
- They do not always have the same numbers of neutrons.
- Elements exist as different isotopes.

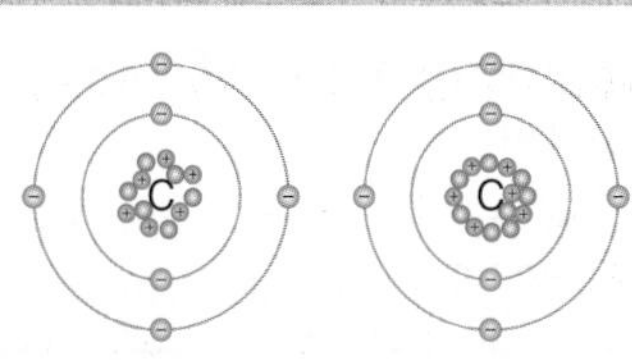

Spot the difference in these *isotopes*

WHAT ARE THE RELATIVE MASS AND CHARGE OF SUBATOMIC PARTICLES?

- Protons and neutrons have a relative mass of 1.
- Protons carry a positive charge; neutrons are neutral.
- Electrons carry a negative charge and have virtually no mass.

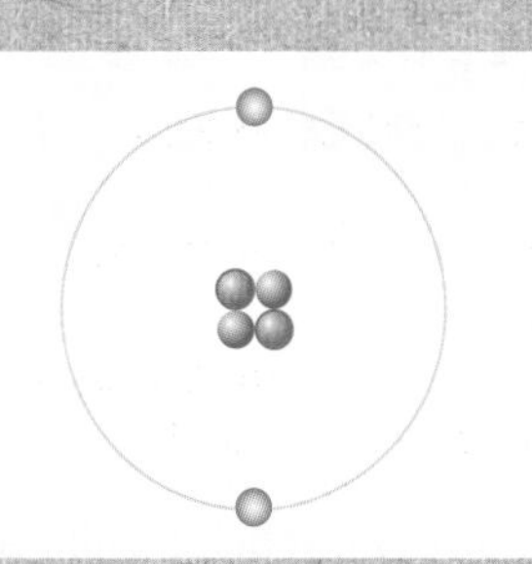

Three states of matter

Learning objectives:

- use data to predict the states of substances
- explain the changes of state
- use state symbols in chemical equations
- explain the limitations of the particle model.

KEY WORDS

changes of state
condensing
limitations
particle

What happens to ice in a cold drink? How can we describe what is happening as the solid becomes a liquid? A particle model can be used to describe the three states of matter – solid, liquid and gas – and used to help us to visualise what happens during changes of state.

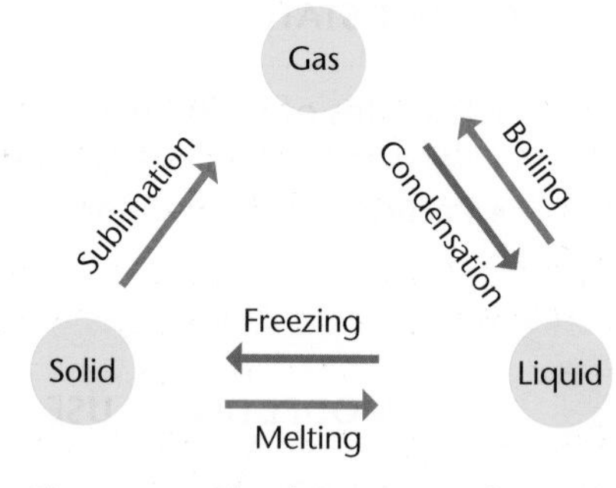

Figure 1.1 The three states of matter

The states of matter

The three states of matter are solid (s), liquid (l) and gas (g).

- Melting and freezing take place at the melting point.
- Boiling and **condensing** take place at the boiling point.

Substance	Melting point (°C)	Boiling point (°C)	State at room temperature
W	–18	42	liquid
X	150	875	
Y	–190	–84	
Z	–56	16	

Substance W in the table will have melted at –18 °C but at room temperature, which is around 25 °C, it will not have boiled. So it is a liquid.

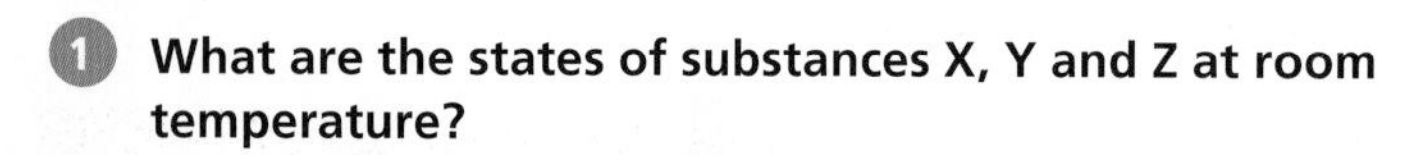

1 What are the states of substances X, Y and Z at room temperature?

The three states of matter can be represented by a simple model. In this model, particles are represented by small solid spheres.

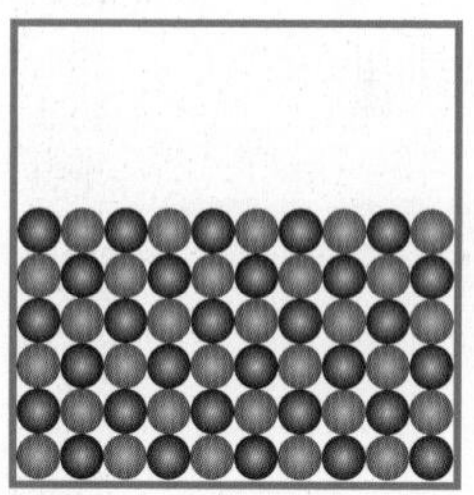

solid

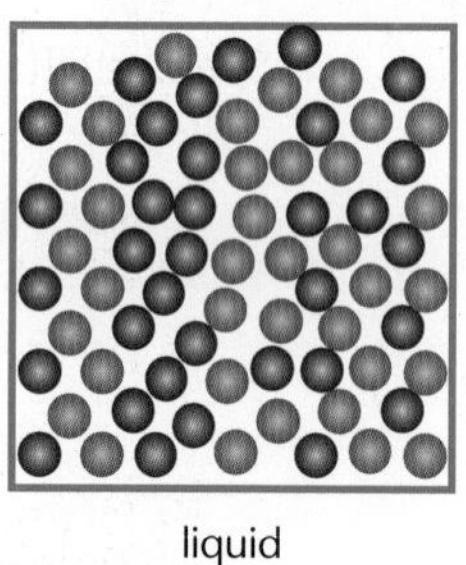

liquid

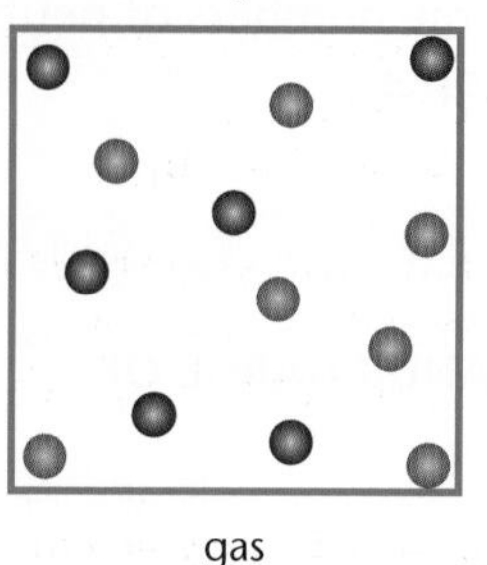

gas

Figure 1.3 Particle model diagrams of solid, liquid and gas

Look again at substance W in the table. At 25 °C it will be a liquid. However, a few particles will have enough energy to escape the surface of the liquid and evaporation will be taking place.

DID YOU KNOW?

Iodine can turn straight from a solid to a gas. This is called sublimation.

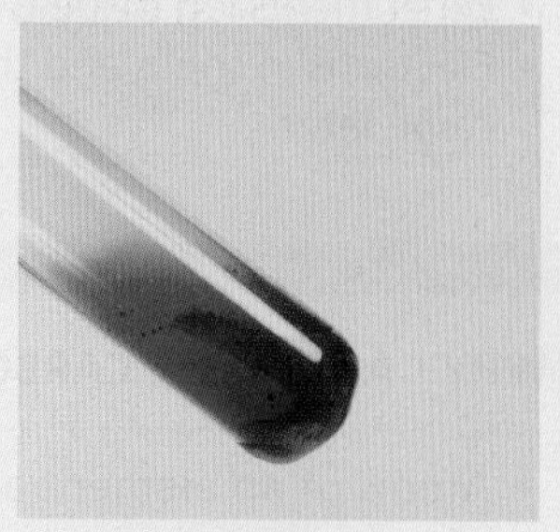

Figure 1.2 Iodine changing from a solid to a gas

Changes of state

Particle theory can help to explain the changes of state. Melting, boiling, freezing and condensing are processes that depend on changing the forces *between* the particles.

In melting and boiling the strength of the forces between particles becomes less. The distance between particles increases and the arrangement becomes more random. The particles move more so more energy is needed from the surroundings.	In freezing and condensing the strength of the forces remains the same. The distance between particles decreases and the arrangement become less random. The particles move less so less energy is needed.

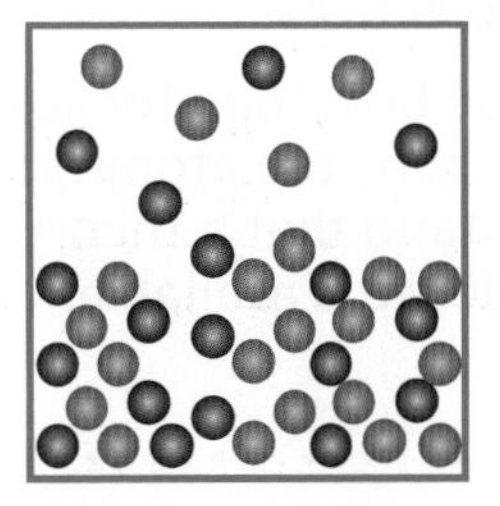

Figure 1.4 Particle model of evaporation

The *amount* of energy needed to change state from solid to liquid and from liquid to gas depends on the strength of the forces between the particles of the substance.

The strength of the forces between particles depends on the nature of the particles involved, on the type of bonding and the structure of the substance.

The stronger the forces between the particles the higher the melting point and boiling point of the substance.

2 The data in the table on the previous page for substance W shows its melting point is –18 °C. Describe what is happening to the particles and the forces between them at these temperatures: –28 °C, –18 °C, –14 °C

3 Look at the boiling point data for substance W in the table. Describe what is happening to the particles and the forces between them at these temperatures: 25 °C, 38 °C, 42 °C, 46 °C

4 Suggest why iron has high melting and boiling points.

5 Ethanol has a boiling point of 78.4 °C. Propane has a boiling point of –42 °C. Suggest why.

HIGHER TIER ONLY

Limitations of this model

This simple model is limited because:

- there are no forces represented between the spheres
- all the particles are represented as spheres
- the spheres represented are solid and inelastic.

This means that the changes in forces and collisions between particles cannot be represented fully. However, it is a useful model to show spatial arrangement, both regular and random.

6 Explain the limitations of the particle theory when considering the process of condensing.

KEY INFORMATION

These diagrams only show a model. For example, in the model of a solid the individual spheres represent particles that are not, themselves, solid. It is only when lots of these particles are arranged closely in a regular pattern that they together represent a solid.

DID YOU KNOW?

In chemical equations, the three states of matter are shown as (s), (l) and (g), with (aq) for aqueous solutions.

KEY INFORMATION

We see that when the particles move more rapidly the 'state' of the bulk of the matter changes from solid to liquid to gas. These changes are physical changes. If particles of matter from different substances collide, react and join to make new substances, this change is a chemical change. It can happen in any state.

Changing ideas about atoms

Learning objectives:

- describe how the atomic model has changed over time
- explain why the atomic model has changed over time
- understand that a theory is provisional until the next piece of evidence is available.

KEY WORDS

electron shell
Ernest Rutherford
Geiger and Marsden experiment
J. J. Thompson
James Chadwick
John Dalton
Niels Bohr

The idea of atoms has changed hugely over the years. At the moment, scientists believe atoms are very small, have a very small mass and are made of protons, electrons and neutrons. Our current theories were developed by imagination, evidence and advances in technology, with each new idea being built on the ideas of earlier scientists.

Developing the atomic theory

Explanations about atoms began about 400 BC, when the Greek philosopher Democritus described materials as being made of small particles. He called these particles 'atoms'. However, he had no evidence. It was just an idea.

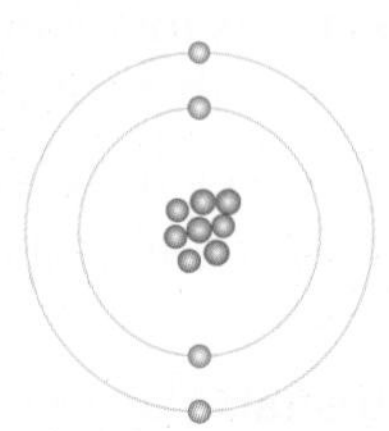

A current simple model of an atom with protons and neutrons in the nucleus surrounded by electrons in shells around it. This model was developed over time.

Little more was suggested for more than 2000 years, but in 1803 the British scientist **John Dalton** used his observations to describe the atom in more detail. His model described an atom as a 'billiard ball'.

Figure 1.5 Dalton's idea of atoms: they were like tiny billiard balls

Dalton's model was then changed as new evidence was found.

In 1897, 94 years later, **J. J. Thomson** discovered the electron. Thomson developed the way that the atom was thought of by using a 'plum pudding' model to describe atoms. Negative electrons were thought to be embedded in a ball of positive charge, rather like the fruit (the electrons) are part of a pudding (the ball of positive charge).

KEY INFORMATION

At each stage, the explanations of atomic theory were provisional until more convincing evidence was found to make the model better.

1. **Suggest why Dalton's atomic model did not include positive and negative charge.**
2. **Explain why the discovery of the electron changed the Dalton model of the atom.**

Changing theories

Sometimes ideas can develop rapidly because of unexpected results.

In 1909 **Geiger** and **Marsden** had really surprising results in their experiment with gold leaf and alpha particles. These results led Geiger, Marsden and **Rutherford** to propose a new idea that an atom has a nucleus. In 1911, Rutherford suggested the atom had a positively charged nucleus and much of the atom was empty space. This was the nuclear model of the atom.

Figure 1.6 Rutherford and Geiger in their lab in Manchester, UK

In 1913, **Niels Bohr** used theoretical calculations that agreed with experimental evidence to adapt the nuclear model. He explained that the electrons orbited the nucleus in definite orbits at specific distances from the nucleus. He explained that a fixed amount of energy (a *quantum* of energy) is needed for an electron to move from one orbit to the next. Electrons only exist in these orbits.

3. **Suggest why Bohr proposed that electrons orbited the nucleus in shells.**
4. **What is meant by the phrase 'quantum of energy'?**

DID YOU KNOW?

As a challenge you can find out about the Geiger and Marsden experiment that changed the theory from a 'plum-pudding' atom to a nuclear atom, it is a famous turning point in the understanding of atoms.

Further development of atomic theory

Later experiments gradually led to the idea that the positive charge of any nucleus can be sub-divided into a whole number of smaller particles. Each of these particles had the same amount of positive charge. In 1920 the term 'proton' was first used in print for these particles.

In 1932, **James Chadwick** discovered the neutron. Again this discovery involved experimental evidence and mathematical analysis.

5. **Draw a timeline of the discoveries that led to our present understanding of the atomic theory.**
6. **Suggest why it was twelve years between finding protons and finding neutrons.**

DID YOU KNOW?

The idea of atoms as small particles is not new. However, our ideas about the theory of atoms are still developing. Search on 'CERN LHC' to find out more.

Modelling the atom

Learning objectives:

- describe the atom as a positively charged nucleus surrounded by negatively charged electrons
- explain that most of the mass of an atom is in the nucleus
- explain that the nuclear radius is much smaller than that of the atom and with most of the mass in the nucleus.

KEY WORDS

charge
electron
nucleus
electron shell

Atoms are the building blocks of all matter, both living and non-living, simple and complex. Atoms join together in millions of different ways to make *all* the materials around us. We can explain how *everything*, including ourselves, is made by using ideas and models of atoms.

Figure 1.7 Magnified image of gold atoms

Atoms

Individual atoms are very small. There are about ten million million atoms in this full stop. i.e., 10^7 atoms.

An atom is made up of a **nucleus** that is surrounded by electrons.

- The nucleus carries the **positive charge**.
- **Electrons**, which surround the nucleus, each carry a **negative charge.**

It depends how it is measured but the diameter of an atom is about 10^{-10} m. That's 0.000 000 01 cm. If we imagine that an atom is blown up to the size of a football stadium the nucleus would be the size of a peanut placed on the centre spot.

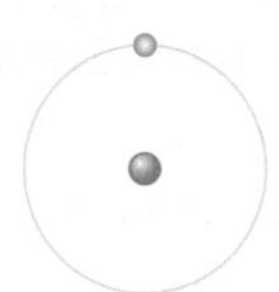

Figure 1.8 The structure of a hydrogen atom. What charge does an electron carry?

1. **What is the type of charge in the nucleus?**
2. **Helium has two positive charges in the nucleus. Predict the number of electrons in a helium atom.**

More on atoms

Electrons occupy the space around the nucleus in 'shells'. The space between the nucleus and the **electron shells** is completely empty.

The nucleus contains most of the mass of the atom and the electrons contribute very little. On the other hand, the radius of the atom, where the electrons are orbiting, is much larger than the radius of the nucleus in the centre.

When we are talking about these differences we are talking about small sizes. Atoms are *very* small. A typical atomic radius is about 0.1 nm (1×10^{-10} m). The radius of a nucleus is less than one ten-thousandth of the radius of an atom (about 1×10^{-14} m).

Typical atomic radius	Typical radius of a nucleus
1×10^{-10} m	1×10^{-14} m

The radius of an atom is measured in many different ways. This is because the outer electron shell is not a fixed boundary, and so its position can only be measured approximately.

3 **Most of the atom is empty space. What does this suggest about the size of an electron?**

4 **Explain why the radius of the nucleus is much smaller than the radius of the whole atom.**

HIGHER TIER ONLY

Atoms are very small. A typical atomic radius is about 0.1 nm (1×10^{-10} m). However, the radius of an atom increases down a group of elements in the periodic table.

For example the atomic radii of Li, Na and K increase as more electrons are 'added' to the atom.

5 **Suggest why the radius of potassium is larger than the radius of lithium.**

6 **The positive charge on a lithium nucleus is 3. The positive charge on a neon nucleus is 10. As more negative electrons are added one by one to atoms from Li up to Ne the radius gets smaller, not bigger. Suggest why. Use ideas about opposite charges.**

DID YOU KNOW?

An atom of gold has a mass of about 3.3×10^{-22} g and a radius of about 1.4×10^{-10} m. Most of the mass of the atom is in the middle, in the nucleus.

KEY INFORMATION

Remember that the typical radius of a nucleus is less than 1/10 000th of the typical radius of an atom.

KEY CONCEPT

Sizes of particles and orders of magnitude

Learning objectives:

- identify the scale of measurements of length
- explain the conversion of small lengths to metres
- explain the relative sizes of electrons, nuclei and atoms.

KEY WORDS

magnitude
diameter
radius
nanometre

Let's start with the particles we can see. A grain of sand and a grain of sugar are about the same size and are made of crystals. These crystals are made up of much smaller sections that we cannot see.

Orders of magnitude

Placing a tennis ball, golf ball, basketball and table tennis ball in order of size is easy.

unit	basketball	tennis ball	golf ball	table tennis ball
cm	25.0	6.8	4.1	0.4
m	0.25	0.068	0.041	0.004

Figure 1.9 It is easy to put these in order of diameter

We can measure objects smaller than these in millimetres.

1 m = 1000 mm   1 mm = 0.001 m   or   1 mm = 10^{-3} m

We can even see objects in the next set of smaller units, the *micrometre*. We measure the width of a human hair in this unit.

1 m = 00000 µm   1 µm = 0.000001 m   or   1 µm = 10^{-6} m

After that we need instruments to help us see and measure lengths. We have discussed carbon nanotubes and graphene as a monolayer of carbon atoms. Later, we will discuss large molecules such as DNA. These next sets of objects are in the 'nano-scale.' The unit is the **nanometre**.

KEY INFORMATION

Typical atomic radii and bond length are in the order of 10^{-10} m.

1 m = 1000000000 nm
1 nm = 0.000000001 m or
1 nm = 10^{-9} m

1. **Calculate the number of basketballs it would take to make a kilometre.**
2. **A carbon nanotube has a length of 2×10^{-9} m. Calculate the number of nanotubes that would fit in 1 mm.**

Atoms and ions

Going one step further down into the atomic scale:

- the radius of an atom is measured in *picometres* (pm), 10^{-12} m
- the radius of a nucleus measured in *femtometres* (fm), 10^{-15} m

MATHS

1 cm = 0.01 m is the long way to write the conversion

1 cm = 10^{-2} m is the conversion into *standard form*

Why is the radius of a nucleus so much smaller than the radius of an atom?

Between the nucleus and the electrons of the atom there is mostly empty space, so neutrons and protons have radii measured in femtometres (fm).

One step even further down, electrons, which are smaller than protons and neutrons, are measured in *attometres* (am) that's 10^{-18} m

3 **The hydrogen atom has a radius of 2.5×10^{-11} m and its nucleus a radius of 1.75×10^{-15} m. Calculate how many times larger the atom is compared to the nucleus.**

4 **What is the standard form of a radius 0.000000000001 m?**

Radius at the atomic level

The radius of atoms is measured in picometres. Each element has a different atomic radius.

When atoms lose or gain outer electrons for bonding they become ions. This is a representation of the atomic and ionic radii.

KEY INFORMATION

These measurements are within the 'human' scale from:

- a human hair to a water droplet 10^{-6} m to 10^{-3} m
- a pin-head to a basketball 10^{-3} m to 10^{-1} m
- a car to the Shard building, London 1 m to 10^{3} m
- Ben Nevis to the Great Wall of China 10^{3} m to 10^{6} m

After that we measure on the 'astronomical' scale.

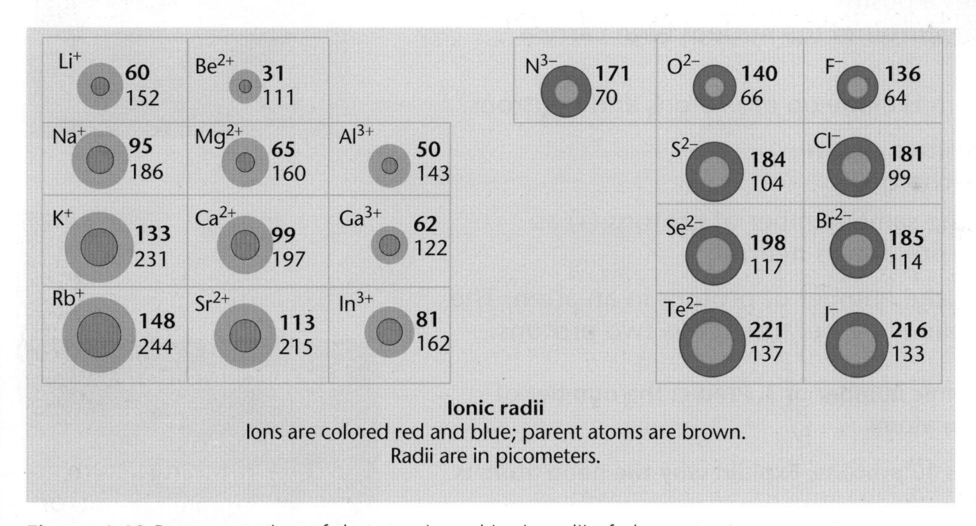

Figure 1.10 Representation of the atomic and ionic radii of elements

When atoms join together they make molecules. The smallest molecule is H_2 made of two hydrogen atoms. It has a radius of 0.5 Å whereas Cl_2 has radius of 1 Å. This is 0.1 nm or 100 pm (10^{-10}m).

DID YOU KNOW?

When crystallography started and measurements were standardised they used the Angstrom or Å. 1Å $=10^{-10}$ m.

5 **Explain why lithium has an atomic radius of 152 pm but an ionic radius of only 60 pm.**

6 **Explain why fluorine has an atomic radius of only 64 pm but an ionic radius of 136 pm.**

7 **Determine which is the greater relative increase in atomic radii, Li to Rb or Be to Sr.**

Google search: 'crystalmaker vfi atomic radii'

Relating charges and masses

Learning objectives:

- describe the structure of atoms
- recall the relative masses and charges of protons, neutrons and electrons
- explain why atoms are neutral.

KEY WORDS

atomic number
electron
neutral
neutron
proton
symbol

We have seen how ideas about atoms have changed over the years. Currently, scientists believe atoms are made of three important particles – protons, electrons and neutrons.

- **The numbers of protons and neutrons are important in nuclear reactions.**
- **The numbers of protons and electrons are important in chemical reactions.**

DID YOU KNOW?

Even these particles can be broken down further in huge particle accelerators, such as the one built deep underneath Switzerland by a joint team of scientists and engineers from many European countries.

Structure of atoms

An atom is made up of a nucleus that is surrounded by electrons.

- The nucleus carries a positive charge.
- The electrons that surround the nucleus each carry a negative charge.

The nucleus of an atom is made up of protons and neutrons.

- Protons have a positive charge.
- Neutrons have no charge.

An atom always has the same number of protons (+) as electrons (–) so atoms are always **neutral**.

The **atomic number** is the number of protons in an atom. The atomic number for helium is 2 because it has two protons.

1. **Lithium has an atomic number of 3. Predict the number of electrons in lithium atoms.**
2. **The neon atom has 10 protons. Explain why the neon atom is neutral.**
3. **Use the periodic table to identify the element with 3 protons.**
4. **Determine the number of protons in an atom of calcium, Ca.**

KEY INFORMATION

It is because a helium atom has two protons that it has an atomic number of 2.

Masses and charges

The nucleus of an atom is made up of particles (protons and neutrons) that are much heavier than electrons. The relative masses and charges of electrons, protons and neutrons are shown in the table.

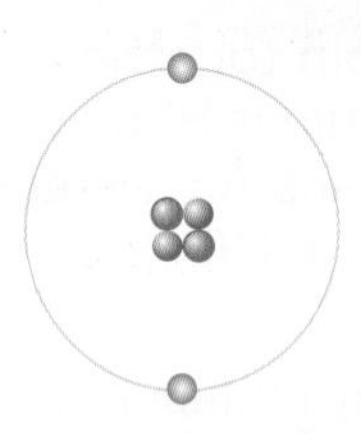

Figure 1.11 The structure of a helium atom. There are the same number of protons and electrons.

	Relative charge	Relative mass
Electron	–1	0.0005
Proton	+1	1
Neutron	0	1

DID YOU KNOW?

Electrons have such a small relative mass that it is usually treated as zero.

5 **A fluorine atom has 9 positive charges, 9 negative charges and a mass of 19. Describe the structure of its atom.**

6 **A chlorine atom has 17 electrons and a mass of 35. Describe the structure of its atom.**

Losing electrons

If an atom has an atomic number of 3 and a neutral charge, it must be a lithium atom. It has a neutral charge because the atom has three protons (+) and three electrons (–).

If the lithium atom loses one negatively charged electron it then becomes a *charged* particle with one positive charge that is not balanced out by a negative charge.

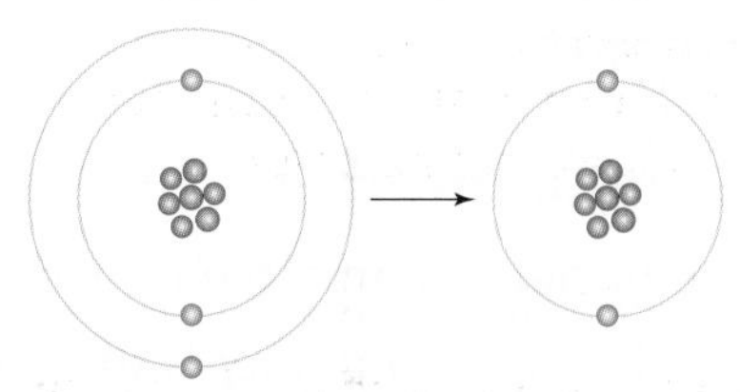

Figure 1.12 A neutral lithium atom loses an electron and becomes charged

	Atomic number	Number of protons	Number of electrons	Charge
Lithium atom	3	3	3	0
Lithium charged particle	3	3	2	+1

If an atom loses electrons and becomes charged, this charged particle is called a positive *ion.*

7 **If a magnesium atom loses 2 electrons it becomes a charged particle. It still has a mass of 24. Write out the atomic number, number of protons, number of electrons, number of neutrons and the charge of:**

a **a magnesium atom**

b **a magnesium ion.**

8 **Explain why a magnesium atom is neutral but a magnesium ion is charged.**

9 **Nitride ions have a 3– charge. Work out the number of electrons in a nitride ion, given that the atomic number of nitrogen is 7.**

Subatomic particles

Learning objectives:

- use the definitions of atomic number and mass number
- calculate the number of protons, neutrons and electrons in *atoms*
- calculate the number of subatomic particles in isotopes and ions.

KEY WORDS

atomic mass
isotope
neutrons
protons

Smoke detectors, archaeological dating and bone imaging all use isotopes. Some elements have more than one type of atom. These different types of atom have different numbers of neutrons and are called isotopes.

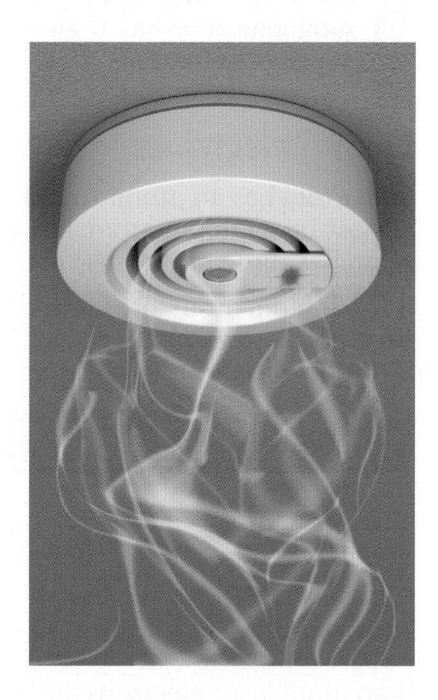

Atomic number and mass number

The nucleus of an atom is made up of **protons** and **neutrons**.

- The *atomic number* is the number of protons in an atom.
- The *mass number* of an atom is the total number of protons and neutrons in the atom.

If a particle has an atomic number of 11, a mass number of 23 and a neutral charge, it must have:

- 11 protons, because it has an atomic number of 11.
- 11 electrons, because there are 11 protons and the atom is neutral.
- 12 neutrons, because the mass number is 23 and there are already 11 protons (23 – 11 = 12).

Here are some more examples.

	Atomic number	Mass number	Number of protons	Number of electrons	Number of neutrons
Carbon	6	12	6	6	6
Fluorine	9	19	9	9	10
Sodium	11	23	11	11	12
Aluminium	13	27			

1. **Complete the row for an atom of aluminium, Al.**
2. **Work out the number of protons, electrons and neutrons in an atom with an atomic number of 15 and a mass number of 31.**

Isotopes

All atoms of carbon have 6 protons, so its atomic number is 6. Most carbon atoms have 6 neutrons, so the mass number is 6 + 6 = 12. This form of the carbon atom is written as $^{12}_{6}C$.

Another form of carbon, $^{14}_{6}C$, has an atomic number of 6 (6 protons) and a mass number of 14. It must therefore have 8 neutrons (14 – 6). $^{14}_{6}C$ is sometimes written as carbon-14. $^{12}_{6}C$ and $^{14}_{6}C$ are **isotopes** of carbon.

3 **Write the isotope symbol for an atom that has 17 protons and 18 neutrons.**

4 **Identify all the subatomic particles in an atom of carbon-13.**

Relative abundance of isotopes

Most elements have two or more isotopes. For example, hydrogen has three common isotopes.

Isotope	Electrons	Protons	Neutrons	Mass number
$^{1}_{1}H$	1	1	0	1
$^{2}_{1}H$	1	1	1	2
$^{3}_{1}H$	1	1	2	3

The relative atomic mass of an element is the average mass of the different *isotopes* of an element. Chlorine's A_r of 35.5 is an average of the masses of the different isotopes of chlorine.

There are two main isotopes $^{35}_{17}Cl$ and $^{37}_{17}Cl$. If there were 50% of each of the isotopes what would be the average mass? The answer is 36. But there are less of the $^{37}_{17}Cl$ isotopes. So we need a *relative abundance* calculation:

$$A_r = \frac{\begin{pmatrix}\text{mass of first isotope}\\ \times\text{\% of first isotope}\end{pmatrix} \times \begin{pmatrix}\text{mass of second isotope}\\ \text{\% of second isotope}\end{pmatrix}}{100}$$

For example, for chlorine the abundance values are:

75% $^{35}_{17}Cl$ and 25% $^{37}_{17}Cl$

Therefore:

$$A_r = \frac{(75 \times 35) + (25 \times 37)}{100}$$
$$= \frac{2625 + 925}{100}$$
$$= 35.5$$

5 **Explain the similarities and differences between the three isotopes of hydrogen.**

6 **Element X has two isotopes, mass 27 and 29. Calculate the relative atomic mass of X if the first isotope has abundance of 65% and the second isotope has 35% abundance.**

KEY INFORMATION

In the symbol $^{12}_{6}C$, the smaller number (6) is the atomic number and the larger number (12) is the mass number.

DID YOU KNOW?

The mass numbers of the isotopes of hydrogen are 1, 2 and 3. However, there are unequal proportions of each type of isotope in a sample of hydrogen gas, so the average atomic mass of hydrogen is 1.008.

MATHS SKILLS

Standard form and making estimates

KEY WORDS

standard form
decimal point

Learning objectives:

- recognise the format of standard form
- convert decimals to standard form and vice versa
- make estimates without calculators so the answer in standard form seems reasonable.

Here, we're looking at using expressions in standard form and seeing how this helps us understand the size of very small particles such as atoms and ions.

When we talked about an atom earlier we used a model to describe it. We imagined it as a sphere with a radius of about 0.000 000 000 1 m. We also saw that the radius of the nucleus of the atom is about 0.000 000 000 000 01 m. It is very awkward to keep writing so many zeros, because it is easy to lose count and it is not so easy to see the comparison between one number and the other. Let's look at another way of writing these numbers using **standard form**.

KEY INFORMATION

Standard form is used to represent very large or very small numbers.
A number in standard form is written in the form $A \times 10^n$, where $1 \leq A < 10$ and n is an integer. For numbers less than 1, n is negative.

Positive powers of ten for very large numbers

We write 1, 10 and 100 knowing what we mean. We can also write them as 1, 1×10 and $1 \times 10 \times 10$. We also know that 10×10 is 10^2. So 100 is 10^2. We can write the numbers 1, 1×10 and 1×10^2.

Standard form	M	HTh	TTh	Th	H	T	O	.	t
							1	.	0
1×10						1	0	.	0
1×10^2					1	0	0	.	0
1×10^3				1	0	0	0	.	0
1×10^4			1	0	0	0	0	.	0
1×10^5		1	0	0	0	0	0	.	0
1×10^6	1	0	0	0	0	0	0	.	0

10^6 is NOT 10 multiplied by itself 6 times. It is 10 multiplied by itself 5 times.

What about writing bigger numbers in standard form?

The decimal point is fixed and the position, or place value, of the most significant digit, shows how big a number is.

1. **Write 1 000 000 000 in standard form.**
2. **Write out the number 1×10^8**

REMEMBER!

1000 can be written as $1 \times 10 \times 100$, which is the same as $1 \times 10 \times 10 \times 10$. How many tens? Three. So 1000 is written 1×10^3 (one times ten to the power of three). The number 3 tells you how many tens are in the multiplication.

To write 1 000 000 in standard form take the first digit on the left, which is 1. Looking at the table, how many places do we have to move the to the right to reach the decimal point? We have to move it six places. So in standard form 1 000 000 is 1×10^6. The number 6 tells you how many tens there are when you write the number as a multiplication of 10 ($10 \times 10 \times 10 \times 10 \times 10 \times 10$).

Negative powers of ten for very small numbers

It is also possible to write numbers smaller than 1 in this form. 1 is divided by 10 it is 0.1. The number 1 has moved one place to the right of the decimal point. This is written as 1×10^{-1} in standard form. What is 1 divided by 100? The number 1 moves two places to the right of the decimal point to be 0.01. In standard form this is 1×10^{-2}.

standard form	O	.	t	h	th	Tth	Hth	millionth
1×10^{-1}	0	.	1					
1×10^{-2}	0	.	0	1				
1×10^{-3}	0	.	0	0	1			
1×10^{-4}	0	.	0	0	0	1		
1×10^{-5}	0	.	0	0	0	0	1	
1×10^{-6}	0	.	0	0	0	0	0	1

Converting numbers to standard form

Standard form can also be used to represent numbers where the most significant digit is not one. For example, the ordinary number 6000 can be written as 6×1000, or 6×10^3 in standard form.

Remember that standard form always has exactly one digit bigger than or equal to one but less than 10. 0.3×10^4 is not in standard form. It is 3×10^3 in standard form.

Some big and small numbers that you have already met or will meet in Chemistry are:

Avogadro's number	Atomic radius (m)	Nuclear radius (m)	Mass of a gold atom (g)	Nanoparticle (m)
6.023×10^{23}	1×10^{-10}	1×10^{-14}	3.3×10^{-22}	1×10^{-7}

When you calculate with big and small numbers using a calculator it is essential that you first estimate what your answer should look like. Making an estimate of the result of a calculation can save you from making mistakes with your calculator. The best way to estimate the answer without a calculator is to round the numbers sensibly and then carry out the calculation in your head.`

3 **Write 0.000 000 000 000 000 001 in standard form.**

4 **Write out the number 1×10^{-9}**

KEY INFORMATION

To multiply two numbers in standard form you simply add the indices or powers of the tens. For example, $2 \times 10^{15} \times 3 \times 10^9$ is 2×3 with 10^{15+9}, which is 6×10^{24}. With smaller numbers $2 \times 10^{-15} \times 3 \times 10^{-9}$ is 6×10^{-24}

5 **Calculate:**

a $6 \times 10^9 \times 3 \times 10^3$

b $6 \times 10^9 \times 4 \times 10^{-2}$

c $\dfrac{6 \times 10^8}{2 \times 10^2}$

6 **If you were able to lay Avogadro's number of atoms in a straight line next to each other, how far would they stretch?**

7 **Calculate the mass of 3.0×10^{26} gold atoms using the mass of a single gold atom given in the data table.**

Check your progress

You should be able to:

use data to predict the states of substances	→	explain the changes of state	→	use state symbols in chemical equations
describe a change of state as a physical change	→	explain the difference between a physical change and a chemical change	→	explain how chemical changes involve rearrangement of bonds between particles
explain in terms of the particle model the differences between the states of matter	→	explain in terms of the particle model the distinction between physical changes and chemical changes	→	explain the limitations of the particle model when particles are represented by inelastic spheres
describe the work of Dalton and Thomson	→	explain how the work of Rutherford changed ideas about the atom	→	explain how the theoretical ideas of Bohr changed the idea of electron structure
explain that early models of the atom did not have shells with electrons	→	explain that early models of atoms developed as new evidence became available	→	explain why the scattering experiment led to a change in the atomic model
draw a diagram of a small nucleus containing protons and neutrons with orbiting electrons at a distance	→	describe that the nuclear radius is much smaller than that of the atom and that most of the mass in the nucleus	→	explain why the atom is mostly empty space
recall the relative sizes of everyday objects and compare these to the relative sizes of atoms	→	recall the typical size (order of magnitude) of atoms and small molecules	→	recall that typical atomic radii and bond length are in the order of 10^{-10} m
recall the relative charges and approximate relative masses of protons, neutrons and electrons	→	calculate the atomic masses of elements from the numbers of protons and neutrons	→	calculate numbers of protons, neutrons and electrons in atoms given the atomic number and mass number
describe the difference between isotopes of an element	→	calculate numbers of protons, neutrons and electrons in atoms given the atomic number and mass number of isotopes	→	complete data tables showing atomic numbers, mass numbers and numbers of subatomic particles from symbols
describe the difference between an atom and an ion	→	calculate numbers of protons, neutrons and electrons in ions given the atomic number and mass number of isotopes	→	represent numbers of subatomic particles of isotopes using standard symbols

Worked example

1 a Draw diagrams to show how substances change from solids to liquids.

The particles move faster when they are heated and break away from the solid structure to move more freely.

This answer shows both diagrams and has an added explanation.

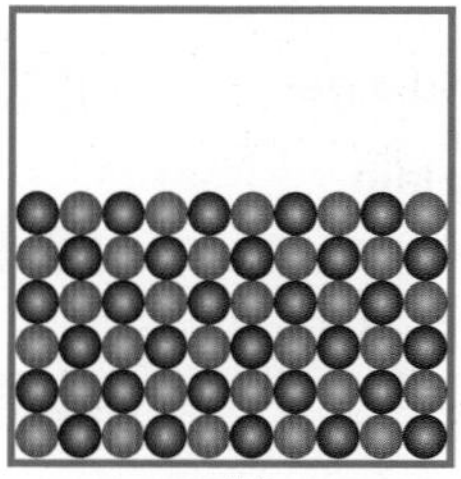

solid

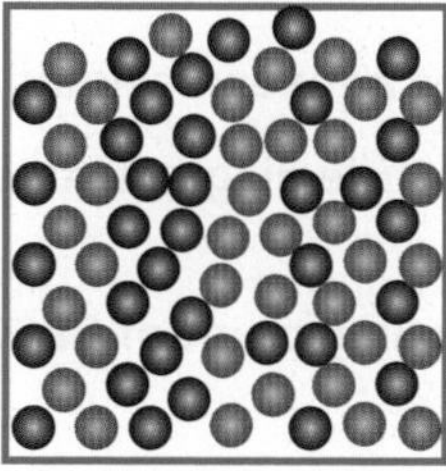

liquid

b Describe what kind of process is happening during this change and explain whether this is a physical change or a chemical change.

When a solid turns into a liquid it is called 'melting'. This is a physical change because the particles don't join together.

The process is correct and an explanation is given of why it is not a chemical reaction. The student should add that no new substance is made and that the substance will go back to the same solid if cooled.

2 Describe the structure of an atom that has atomic number 3 and mass number 7.

This atom has 3 protons and 3 electrons around them and 1 neutron.

The numbers of protons and electrons are correct. The number of protons is equal to the number of electrons and is the atomic number of an element.

The number of neutrons is incorrect and should be 7. The mass number is not equal to the sum of the numbers of protons, electrons and neutrons – it is equal to the sum of the numbers of just the protons and neutrons.

3 a Describe what is meant by 'isotopes' and draw an isotope of hydrogen that has 2 neutrons.

Isotopes are atoms that have the same number of protons but different numbers of neutrons. This means that their atomic number is the same but their mass numbers are different.

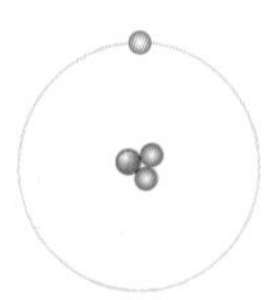

The description is correct and the diagram is correct.

b Describe the charges and masses of the subatomic particles you have drawn and explain how they relate to the atomic number and mass number.

The red dot is a proton that has a positive charge and a mass of 1. The blue dot is an electron that has no mass and a negative charge. The green dots are neutrons with no charge and a mass of 1. The atomic number is 1 and the mass number is 4.

The three descriptions of the charge and mass of the proton, neutrons and electron are correct. The atomic number is correct. However, even though the description of the mass of the electron is correctly given as 0, this has been included in the mass number (4). There are 2 neutrons and 1 proton so the mass number is 3.

End of chapter questions

Getting started

1. An atom has 3 protons, 4 neutrons and 3 electrons. What is its atomic mass?

   a 3 **b** 6 **c** 7 **d** 10

   1 Mark

2. Use two 'particle' diagrams to show the differences between a liquid and a gas. 2 Marks

3. A substance changes from a liquid to a gas. Use particle diagrams to explain why this is a physical change 2 Marks

4. A substance has a melting point of –33 °C and a boiling point of 52 °C. Explain which state it is in at 20 °C. 1 Mark

5. What was the difference between John Dalton's theory of atoms and J.J. Thomson's theory?

   a Dalton's theory included neutrons but Thomson's did not
   b Dalton's atoms were one particle but Thomson's included electrons
   c Dalton's included protons but Thomson's was only one particle.

   1 Mark

6. What is the approximate radius of an atom?

   a 10 m **b** 10^{10} m **c** 10^{-1} m **d** 10^{-10} m

   1 Mark

7. Fill in the missing data:

Particle	Mass	Charge
Proton		
Electron	0	
Neutron	1	

2 Marks

Going further

8. Element X has two isotopes with atomic masses of 7 and 8. The relative abundance of the two isotopes is 50% of each. Calculate the relative atomic mass of element X. 1 Mark

9. Chlorine has two naturally occurring isotopes. Each chlorine isotope has:

   a the same number of protons **b** the same number of neutrons
   c the same atomic mass **d** a different number of electrons

   1 Mark

10. Substances R and S are both gases. They combine together to make substance T. Draw two 'particle' diagrams to explain why this is a chemical change. 2 Marks

11. Explain how Rutherford's idea of atoms was very different to both Dalton's and Thomson's theories, and how Niels Bohr's theories developed Rutherford's theories. 4 Marks

12. Explain why the radius of a nucleus is much smaller than the radius of the atom. 2 Marks

More challenging

13. Identify the number of electrons in an atom of $^{31}_{15}P$. **1 Mark**

14. There are two atoms $^{28}_{14}Si$ and $^{30}_{14}Si$. Work out how many neutrons each atom contains. **1 Mark**

15. An atom, X, has 19 protons and 20 neutrons. Write the symbol for X using standard notation to show the atomic number and mass number. **2 Marks**

16. X has 19 protons and 20 neutrons and forms a +1 positive ion. State the number of electrons in:

    a the atom **b** the ion **2 Marks**

17. Explain, using ideas about subatomic particles, how atomic number, mass number and charge are related in a 2– ion. **4 Marks**

Most demanding

18. Explain why the element with an electron pattern of 2,8,6 is a non-metal.

    Explain why this element is less reactive than the one with an electron pattern of 2,6 or 2,7. **2 Marks**

19. Substance J is a solid and K is a liquid. They react together to make a solution L and a gas M. Draw particle diagrams to explain why this is a chemical change and explain the limitations of using this model. **4 Marks**

20. Using the information in topic 1.2 draw a timeline to show five major changes in theory of the structure of the atom. Explain why this development took place over an extended period of time. **4 Marks**

40 Marks

ELEMENTS, COMPOUNDS AND MIXTURES

IDEAS YOU HAVE MET BEFORE:

PURE SUBSTANCES AND MIXTURES

- Salt and sand can be separated by filtering.
- Iron and sulfur can be separated physically by a magnet.
- Crystals can grow when a warm saturated solution cools.

SEPARATING AND TESTING SUBSTANCES

- Chromatography separates inks and coloured dyes.
- Distillation separates alcohol and water.
- Filtering removes solid impurities.

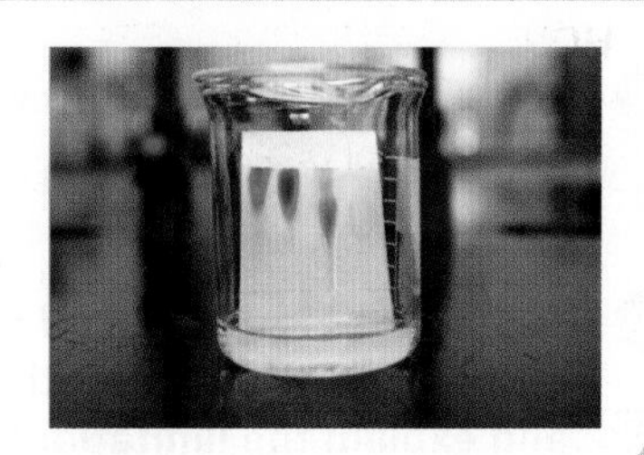

METALS AND NON-METALS

- Gold, iron, copper and lead are metals known for centuries.
- Oxygen and nitrogen are gases of the air.
- Sulfur is a yellow non-metal.

SOME ELEMENTS AND THEIR COMPOUNDS

- Helium is unreactive and is used in balloons.
- Sodium chloride is used to flavour and preserve food.
- Chlorine is used to kill bacteria in swimming pools.

STRUCTURE AND BONDING OF CARBON

- Carbon is an element that we know as soot and charcoal.
- Diamonds are hard, precious gems that are cut to sparkle.
- Graphite is used in pencil leads and as electrodes.

2

IN THIS CHAPTER YOU WILL FIND OUT ABOUT:

HOW CAN WE TELL IF A SUBSTANCE IS PURE?

- A pure substance has a specific melting point and boiling point.
- Some mixtures can be separated by distillation or filtration.
- Pure substances can be obtained by crystallisation.

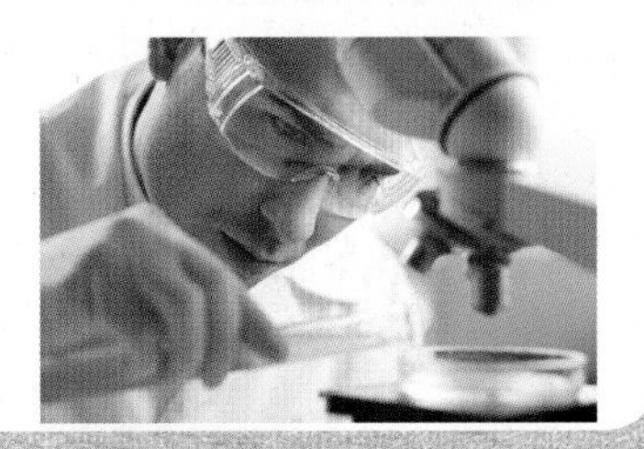

HOW CAN WE SEPARATE A SUBSTANCE TO ANALYSE IT?

- Dyes travel different distances along chromatography paper.
- Solvent-front distances can be measured.
- Ratios of spot position to solvent front can be calculated.

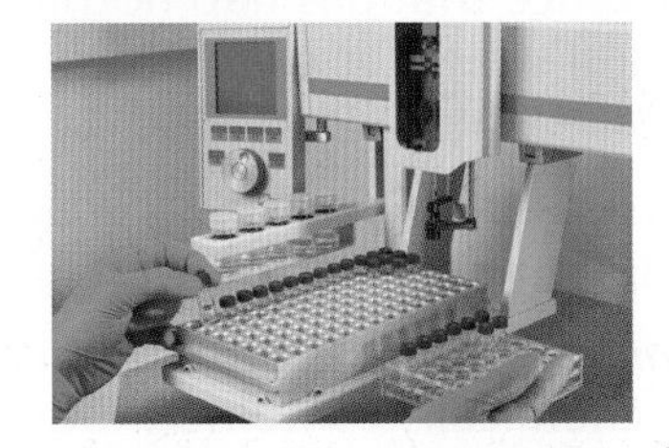

WHAT ARE THE CHARACTERISTICS OF METALS AND NON-METALS?

- Metals are shiny, sonorous and form basic oxides.
- Non-metals are gases or dull solids forming acidic oxides.
- Metals and non-metals have different atomic structures.

HOW IS THE PERIODIC TABLE CONSTRUCTED?

- Elements are arranged in order of their atomic number.
- Atoms with the same number of outer electrons form groups.
- Atoms with the same number of electron shells form periods.

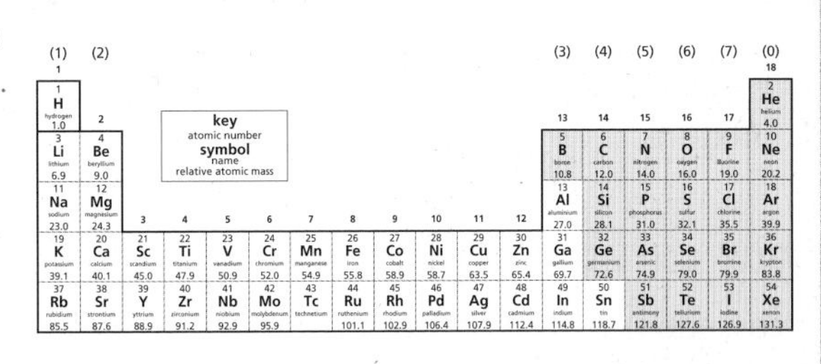

(1) 1	(2) 2	3	4	5	6	7	8	9	10	11	12	(3) 13	(4) 14	(5) 15	(6) 16	(7) 17	(0) 18
1 H hydrogen 1.0																	2 He helium 4.0
3 Li lithium 6.9	4 Be beryllium 9.0											5 B boron 10.8	6 C carbon 12.0	7 N nitrogen 14.0	8 O oxygen 16.0	9 F fluorine 19.0	10 Ne neon 20.2
11 Na sodium 23.0	12 Mg magnesium 24.3											13 Al aluminium 27.0	14 Si silicon 28.1	15 P phosphorus 31.0	16 S sulfur 32.1	17 Cl chlorine 35.5	18 Ar argon 39.9
19 K potassium 39.1	20 Ca calcium 40.1	21 Sc scandium 45.0	22 Ti titanium 47.9	23 V vanadium 50.9	24 Cr chromium 52.0	25 Mn manganese 54.9	26 Fe iron 55.8	27 Co cobalt 58.9	28 Ni nickel 58.7	29 Cu copper 63.5	30 Zn zinc 65.4	31 Ga gallium 69.7	32 Ge germanium 72.6	33 As arsenic 74.9	34 Se selenium 79.0	35 Br bromine 79.9	36 Kr krypton 83.8
37 Rb rubidium 85.5	38 Sr strontium 87.6	39 Y yttrium 88.9	40 Zr zirconium 91.2	41 Nb niobium 92.9	42 Mo molybdenum 95.9	43 Tc technetium	44 Ru ruthenium 101.1	45 Rh rhodium 102.9	46 Pd palladium 106.4	47 Ag silver 107.9	48 Cd cadmium 112.4	49 In indium 114.8	50 Sn tin 118.7	51 Sb antimony 121.8	52 Te tellurium 127.6	53 I iodine 126.9	54 Xe xenon 131.3

key: atomic number / symbol / name / relative atomic mass

WHY ARE DIAMONDS SO HARD AND GRAPHITE SO SOFT?

- Diamond is a giant structure with tetrahedral bonds.
- Graphite is made of rings of carbon atoms in layers.
- Weak forces hold graphite layers together which can slide.

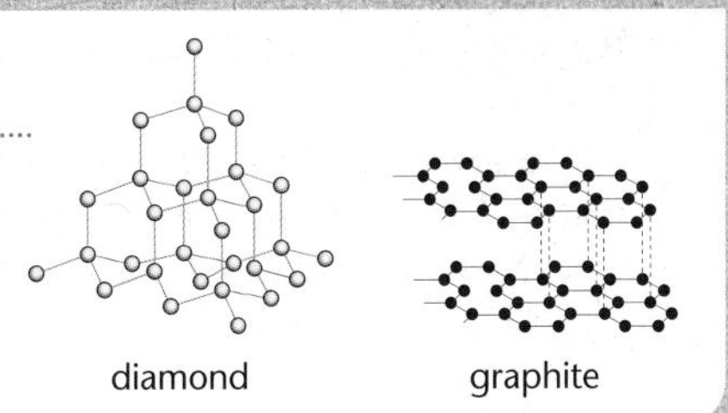

KEY CONCEPT

Pure substances

Learning objectives:

- describe, explain and exemplify processes of separation
- suggest separation and purification techniques for mixtures
- distinguish pure and impure substances using melting point (MP) and boiling point (BP) data.

KEY WORDS

pure
impure
melting point
boiling point

Physical separation

In everyday language, a **pure** substance can mean a substance that has had nothing added to it, so it is in its natural state, for example, pure milk. But this is not the chemical definition.

In chemistry, a *pure substance* is a single element or compound, not mixed with any other substance.

A *mixture* consists of two or more elements or compounds not chemically combined together. The chemical properties of each substance in the mixture are unchanged.

Mixtures can be separated by *physical processes*. These processes *do not* involve chemical reactions.

Separation processes include:

- filtration
- crystallisation
- distillation
- fractional distillation
- chromatography.

Figure 2.1 Separating 'pure' milk. This is not the chemistry definition of pure.

1. **Explain how you would separate salt and sand.**
2. **Explain how you would find out the colours in an ink mixture.**

What can we separate?

We can separate solids from solutions by *filtering*, for example we can filter out excess magnesium metal if we are making a solution of magnesium sulfate from magnesium and sulfuric acid. We can *crystallise* out samples of solids in order to assess their purity.

Two liquids with similar boiling points can be separated by *distillation*, for example, ethanol and water. *Fractional distillation* can be used to separate more than two liquids mixed together.

DID YOU KNOW?

Pure and impure milk can both be separated into their components by centrifuging. This means spinning round at high speeds so that layers of substances can be separated. Blood can also be separated in this way.

Chromatography can be used to separate dye colours, food colours, inks, different amino acids and to assess the purity of drugs and medicines.

3 **Explain why excess magnesium is added to sulfuric acid when they are reacted to form magnesium sulfate.**

4 **A liquid mixture of ethanol and water was found in an old bottle. Insoluble sediment and grit was present. Explain how you would separate the components of the mixture.**

KEY INFORMATION

You will need to know how to separate an odd collection of substances mixed together using the techniques you have learned.

Pure and impure substances

Pure elements and compounds melt and boil at specific temperatures. ***Melting point*** and ***boiling point*** data can be used to distinguish pure substances from mixtures.

If a substance is **impure** the melting point will be *lower* and the substance will melt over a *broad* range of temperature. If a substance is *pure* it will melt more *sharply* at a specific temperature.

If a substance is *impure* the boiling point will be *higher* than the pure substance.

5 **Sam and Alex each made a sample of the same substance. They measured their melting points. They then recrystallised their solid and measured the melting point again. They did this several times.**

Sam	46.5	46.6	46.7	46.7	46.7
Alex	42.8	43.7	44.5	44.9	45.1

Identify who made the purest sample and explain your reasoning.

6 **Akira and Ben tested their samples for their boiling points. The standard boiling point from data tables was 85.0°C. Their results were: Akira 86.2°C, Ben 87.1°C. Identify who had the purest sample and explain your reasoning.**

7 **A beaker of distilled water was thought to have been contaminated with an impurity. The boiling point of the water was 108°C. The contaminant was separated and found to consist of white crystals. The melting point of the crystals was 190–210°C.**

- **a** **Explain whether the distilled water was contaminated or not.**
- **b** **Suggest how the contaminant might have been separated from the water.**
- **c** **Explain whether the contaminant was pure or not.**

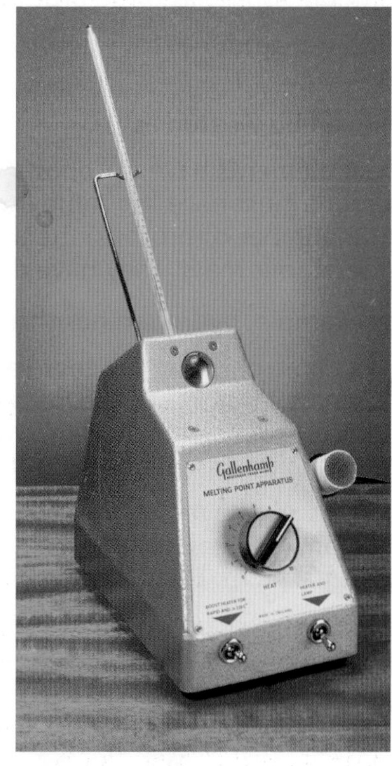

Figure 2.2 Using a melting point to establish purity

Relative formula mass

Learning objectives:

- identify the relative atomic mass of an element from the periodic table
- calculate relative formula masses from atomic masses
- verify the law of conservation of mass in a balanced equation.

KEY WORDS

atomic mass
conservation of mass
formula mass

When looking at the formula of a compound, the relative atomic masses of all the elements can be added to find the mass of the whole formula. The formula mass of the reactants in a chemical reaction must equal the formula mass of the products so that the conservation of mass is preserved.

Relative atomic mass

Atoms are made up of protons, neutrons and electrons. Protons and neutrons make up the nucleus with the electrons orbiting the nucleus in 'shells'. Most of the atom is empty space with almost all the mass in the nucleus.

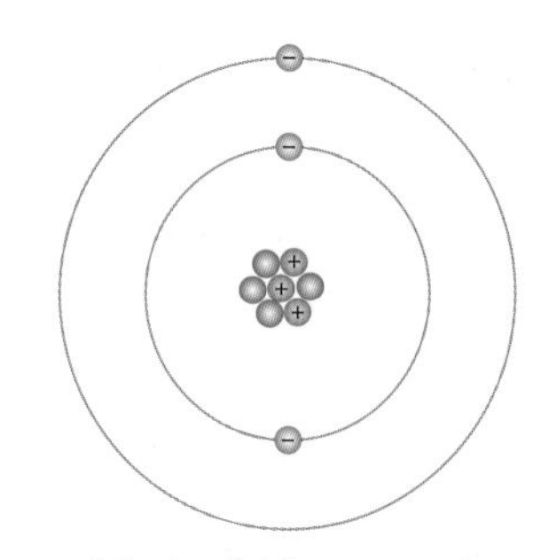

Figure 2.3 The lithium atom has three protons and four neutrons.

A lithium atom has three protons, so has an *atomic number* of 3.

A lithium atom has four neutrons, so has a *mass number* of 3 + 4 = 7

This can be written as $^{7}_{3}Li$.

Some elements have *isotopes*. Although all the atoms of an element must have the same atomic number (and so have the same number of protons) the atoms do not always have the same atomic mass. They have different numbers of neutrons.

For example, hydrogen has three common isotopes: $^{1}_{1}H$ $^{2}_{1}H$ $^{3}_{1}H$

In a sample of hydrogen, there are some atoms with mass 3, some with mass 2 but most have mass 1. Taken on *average*, the mass of a typical sample of hydrogen is 1.00794. This is not the mean of 1, 2 and 3 but takes into account the proportion of each isotope.

DID YOU KNOW?

This accurate value for hydrogen is usually rounded to 1.

Isotopes are the reason why, when you look on the periodic table, the atomic masses of some elements are given with numbers after a decimal point. Examples are Cl as 35.5 and Cu as 63.5.

1 **50% of bromine atoms have a mass of 79 and 50% a mass of 81. Work out the atomic mass of bromine.**

The carbon-12 standard

The mass of hydrogen used to be taken as the standard against which to measure the *relative* **atomic masses** (A_r) of other atoms. Later, it was found that better results were obtained if everything was compared the isotope of carbon that has an atomic mass of 12.

The relative atomic mass of an element is the average mass of an atom of the element compared to the mass of an atom of carbon-12. Values for each element are shown in the periodic table.

2. **Show the atomic structure of the three isotopes of carbon, $^{12}_{6}C$, $^{13}_{6}C$ and $^{14}_{6}C$, in terms of protons and neutrons.**
3. **Explain why relative atomic mass is defined in terms of average mass.**

Formula mass

The relative **formula mass** (M_r) of a compound is the sum of the relative atomic masses of the atoms in the numbers shown in the formula.

For example: M_r of MgO $= A_r$ of Mg $+ A_r$ of O

$= 24 + 16 = 40$

M_r Na_2O $= (23 \times 2) + 16 = 62$

M_r CO_2 $= 12 + (16 \times 2) = 44$

M_r $CuCO_3$ $= 63.5 + 12 + (16 \times 3) = 123.5$

M_r $Mg(OH)_2 = 24 + (16 \times 2) + (1 \times 2) = 58$

In a balanced chemical equation, the sum of the relative formula masses of the reactants equals the sum of the relative formula masses of the products.

	$ZnCO_3$	$\rightarrow$	$ZnO + CO_2$
M_r are:	$65 + 12 + (16 \times 3)$		$(65 + 16) + 12 + (16 \times 2)$
	$65 + 12 + 48$	$\rightarrow$	$81 + 12 + 32$
	125	$\rightarrow$	125

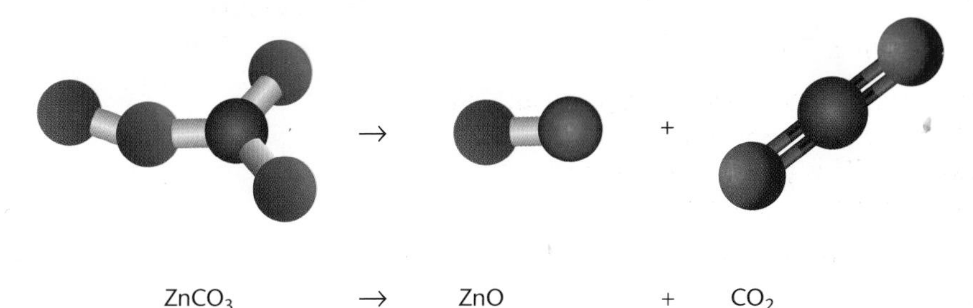

4. **Work out the formula mass of $MgSO_4$.**
5. **Calculate the formula mass of $Cu(NO_3)_2$.**
6. **Show that the relative formula masses of reactants and products are equal in this reaction:**

$$MgBr_2 + 2AgNO_3 \rightarrow Mg(NO_3)_2 + 2AgBr$$

7. **R is a molecule containing carbon and hydrogen. Work out the molecular mass and molecular formula of R.**

$$R + 5O_2 \rightarrow 3CO_2 + 4H_2O$$

KEY SKILLS

The formula mass of a formula with brackets can be worked out in two ways. Either M_r $Mg(OH)_2 = 24 + (16 \times 2) + (1 \times 2) = 58$; or $= 24 + (16 + 1) \times 2 = 58$

DID YOU KNOW?

This holds true for any quantity, so if the relative formula mass of the reactants is 94, then the relative formula mass of the products will be 94. This can be applied to any quantity, so 94 g of reactants will produce 94 g of products, or 94 tonnes of reactants will produce 94 tonnes of products.

KEY INFORMATION

The full-size '2' before a formula applies to all of the formula. The subscript '$_2$' after an element symbol applies to just the one element. Brackets are used if the '$_2$' applies to more than just one element symbol.

Mixtures

Learning objectives:

- recognise that all substances are chemicals
- understand that all substances are either mixtures, compounds or elements
- explain that mixtures can be separated.

KEY WORDS

chromatography
filtration
mixture
separation

You will have begun to use of a range of equipment to safely separate chemical mixtures and we need to extend this range of techniques. Filtering and distillation are probably familiar but fractional distillation can also be used to separate mixtures.

Mixtures

Many substances are made of mixtures. **Mixtures** can easily be **separated** because the chemicals in them are not joined together. Mixtures can be separated by filtration, crystallisation, simple distillation, fractional distillation and chromatography.

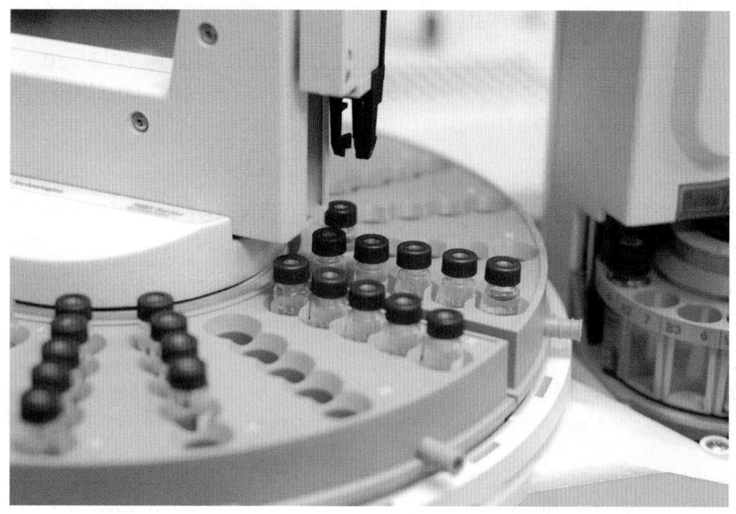

Figure 2.4 Separating mixtures

Let's take a mixture of salt and copper. To separate them we add water to the mixture. The salt dissolves but the copper does not. The salt solution can be filtered through filter paper, leaving the copper behind as a residue. The salt solution can be crystallised to make solid salt crystals. These physical processes do not involve chemical reactions and no new substances are made.

After separating, salt and copper can be mixed again.

1. **Draw a diagram of the equipment used to filter salt solution from copper.**
2. **Explain why a blend of copper and salt is not a compound.**

Separating mixtures

A mixture consists of two or more elements or compounds not chemically combined together. The chemical properties of each substance in the mixture are unchanged. Mixtures can be separated by physical processes. These processes do not involve chemical reactions.

These separation processes include:

- filtration
- crystallisation
- distillation
- chromatography

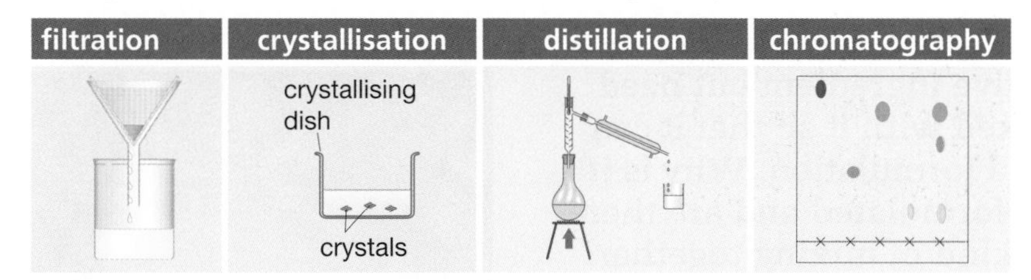

Figure 2.5 Techniques for separating mixtures.

3 **Which technique would be used to separate coloured inks in a mixture?**

4 **Which technique would be used to separate alcohol (boiling point 80 °C) and water?**

KEY INFORMATION

Filtration separates insoluble substances from soluble substances; distillation separates liquids that have different boiling points.

Fractional distillation

Some mixtures are very complex and have many different components. These mixtures can be separated either by using different techniques in sequence or by the same technique which includes multiple seperations.

An example of the first approach is filtration followed by crystallisation.

An example of the second approach is fractional distillation, where different liquids with different boiling points are separated at different points in the process.

Fractional distillation works by using a tall tower of gaps and surfaces, which are gradually cooler towards the top. The liquid mixture is heated at the bottom and the liquids boil together to make a mixture of gases. As each gas reaches a surface at the same temperature as its boiling point (or condensing point) that gas will condense and the condensed liquid will run off. The other gases continue up through the gaps until they reach the surface at their condensing temperature. Eventually nearly all the gases in the mixture will condense and be collected as separated liquids. The final gas is left at the top of the tower and is collected as a gas.

DID YOU KNOW?

Fractional distillation is used to separate the different substances in crude oil.

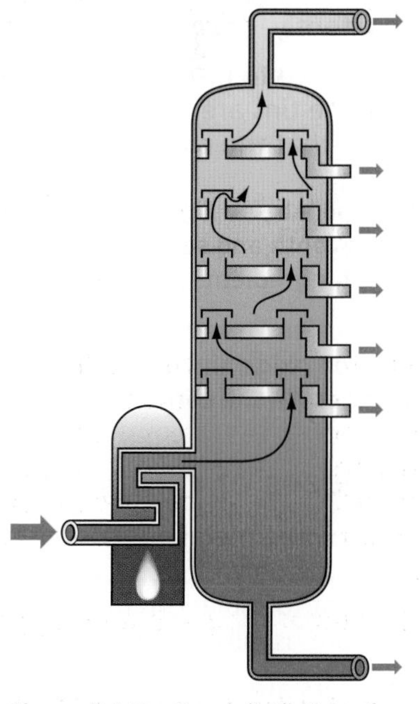

Figure 2.6 Fractional distillation of mixtures with different boiling points

5 **Suggest how you would collect a specimen of clean copper sulfate crystals from a mixture of solid copper sulfate, sand and alcohol.**

6 **Suggest the order of collection from a fractional distillation process of these liquids:**

a **(boiling point 85 °C)**
b **(boiling point 100 °C)**
c **(boiling point 35 °C)**
d **(boiling point 165 °C)**

Formulations

Learning objectives:

- identify formulations given appropriate information
- explain the particular purpose of each chemical in a mixture
- explain how quantities are carefully measured for formulation.

KEY WORDS

formulation
fertilisers
active ingredient
alloys

Medicines have to have an active ingredient but need to have other components mixed with it so that it can easily be taken. This is called a formulation. Why is it necessary to have a medicine formulated and are there other products that need this kind of mixing together?

Products using formulation

A **formulation** is a mixture that has been designed as a useful product. Formulations include food products. They are formulated so that they are consistent and the shopper will know that what they are buying will taste and feel the same every time they buy it. Who would want to buy tomato ketchup that tasted differently each time?

Can you think of other types of products that have to be consistent and are mixtures of specific components?

Paints, **alloys** and **fertilisers** are all substances that are formulated to specific recipes so that they can be used consistently, knowing that the recipe used is always the same.

Paints have to have the same ingredients, so that the contents of all the tins are the same. It would be useless to paint a room to find that the colour was different from each pot.

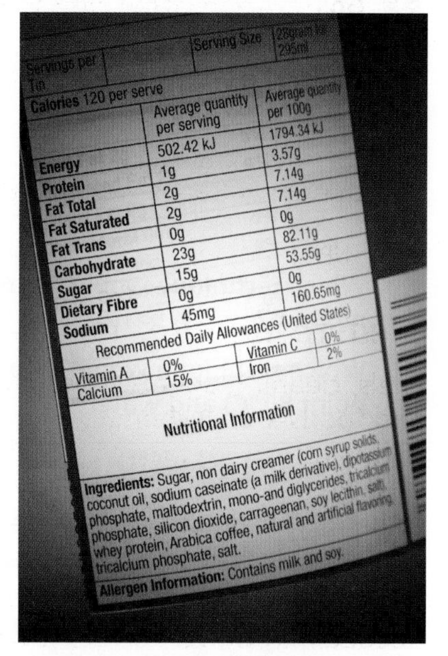

Figure 2.7 This food label shows the ingredients that have formulated this food.

KEY INFORMATION

The ingredients are listed in order by mass. There is a bigger quantity of the first ingredient.

1. **The label in Figure 2.7 has a list of 9 ingredients. Sugar is seventh and salt is last. What does this tell you about the mass of ingredients in the formula?**
2. **Portland cement contains calcium oxide, silica, aluminium oxide, iron oxide and gypsum. Explain why it is very important to mix these in the correct proportions.**

Purposes of component chemicals

Iron is a pure metal but it is fairly brittle and rusts easily. By making a formulation with other chemicals it can be made into an alloy. Different formulations make different alloys for different uses. The alloys are usually a type of steel which is iron mixed with carbon.

Fertilisers are formulated to add nutrients to the soil for different purposes. Some lawn fertilisers may have more nitrogen for shoot growth, some have phosphorus for strong root growth and some have potassium for withstanding drought and disease. Extra magnesium is sometimes in a fertiliser for use if leaves are turning yellow. Magnesium is an essential component of chlorophyll.

There are whole ranges of cleaning agents that are formulated to tackle different kinds of cleaning needs. Some cleaners that tackle grease need to be formulated differently if the grease is on a metal surface or if it is on a granite surface. Cleaners for sinks and bathrooms need a completely different formulation as they need to remove bacteria not grease.

Many products are, therefore, complex mixtures in which each chemical has a particular purpose.

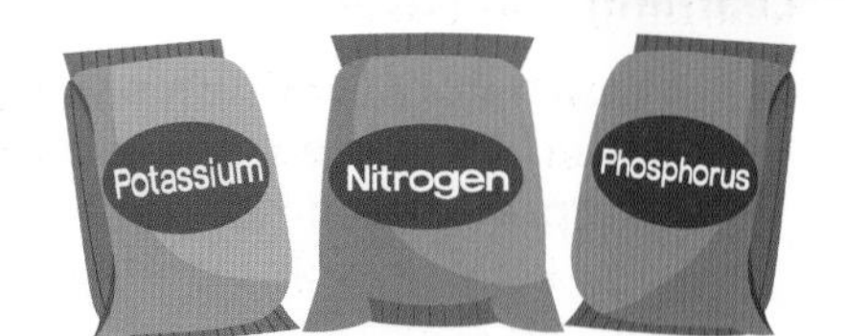

Figure 2.8 Fertilisers are formulated to have different amounts of nutrients such as nitrogen, phosphorus and potassium.

3. **Gold is often mixed with silver, copper and other metals. One of its uses is in electrical circuits. Explain why there are a variety of formulations and why they need to be precisely prepared.**

4. **NPK fertilisers are sold in a formulation by ratio of mass of these three main ingredients. Explain the difference between these two fertiliser formulations:**

NPK 4:1:3	NPK 4:2:2

Measuring quantities

Fuels, cleaning agents, paints, medicines, alloys, fertilisers and foods are all chemicals that need to be formulated. Formulations are made by mixing the components in carefully measured quantities to ensure that the product has the required properties.

The careful measurement is especially important in the formulation of medicines. The amount of **active ingredient** needed may be very small. This small amount may be difficult to take into the body and be difficult to sell. The active ingredient is added to a filler that bulks up the product to a sensible size and a lubricant is added, so that the tablet can then be easily taken. Pills, tablets, capsules and chewables are all made in this way. The filler materials are called excipients. For example, aspirin may contain the excipients starch, lactose and talc.

Because each tablet and capsule has to have exactly the same formula there are strict quality assurance processes that are put in place. These processes are laboratory tests that make sure that each tablet is the same and the quality of the ingredients is at the correct standard.

5. **Some medicines are quality-assured for correct formulation.**

Medicine	A	B	C	D	E	F
% active ingredient	6.25	6.25	6.25	6.25	6.27	6.25
% filler	69.27	69.27	69.37	69.27	69.25	69.27
% lubricant	24.48	24.48	24.38	24.48	24.48	24.48

Explain which medicines would be rejected.

6. **Most paints used to be formulated with a solvent and oil base. In the last two decades water-based paints were developed using new techniques. Suggest why the paints were developed with this different formulation.**

DID YOU KNOW?

In some countries there are different formulations of winter and summer petrol. They have different ratios of the ingredients so that the performance of the petrol is maximised for the particular season. Some volatile hydrocarbons are removed in summer.

REMEMBER!

You do not need to know the names of components in proprietary products.

Chromatography

Learning objectives:

- explain how to set up paper chromatography
- distinguish pure from impure substances
- interpret chromatograms and determine R_f values.

KEY WORDS

chromatography
R_f value
stationary phase
mobile phase

You may have already seen how chromatography with paper can be used to separate coloured dyes. It can also be used to identify colourless mixtures and to extract small amounts of pure substances such as medicines from plants. How can we use it to identify substances or solve mysteries?

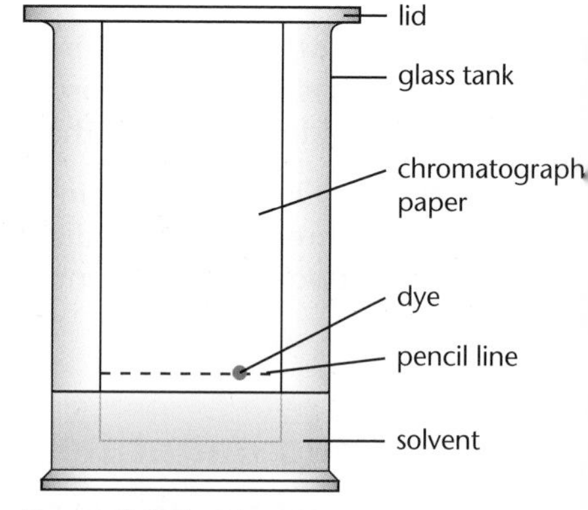

Figure 2.9 Carrying out chromatography.

Separating dyes

You may already have put spots of dye on **chromatography** paper and fixed the edge in water to watch the colour rise up through it.

This is a simple example where the chromatography paper is the **stationary phase** and the water is the **mobile phase**.

An example of how this technique can be used to solve questions is in testing food colourings. A spot of each standard food colour is dropped onto the start line. A spot of the unknown food colour mixture is dropped on the line at the end of the paper. The sheet is held in the solvent (water) and left for separation to take place. Figure 2.10 could be the resulting chromatogram.

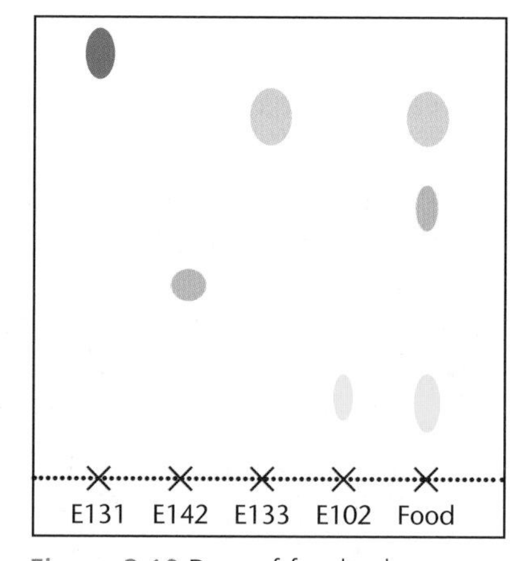

Figure 2.10 Dyes of food colours.

You can see that the four standard colours each have one spot that has travelled up the paper. This distance can be measured. The food colour mixture has three colours:

the yellow spot has moved the same distance as E102 yellow spot
the light blue spot is the same distance as the E133 light blue spot
there is also a green spot.

Has the green spot moved the same distance as the E142 green spot? The answer is no, so E142 is not in the mixture, it is another substance.

1. **Explain how you know there is no E131 in the test mixture.**
2. **Explain how you would find out which green colour is in the food.**

KEY SKILL

In this example the solvent shown in Figure 2.9 is water. Remember that the edge of the paper needs to just dip in the water. The spots must not be in the water.

Measuring R_f values

In the example above it is clear to see that the colours had moved different distances because they were all on the same chromatogram. A standard measure needs to be used so that separation can be compared in any situation. This is the R_f **value** where:

$$R_f = \frac{\text{distance moved by substance}}{\text{distance moved by solvent}}$$

This is the ratio of the distance moved by a compound (from the origin to the centre of the spot) to the distance moved by the solvent.

In this example (Figure 2.11) the substance has moved 45 mm and the solvent has moved 60 mm.

$$R_f = \frac{45}{60} = 0.75$$

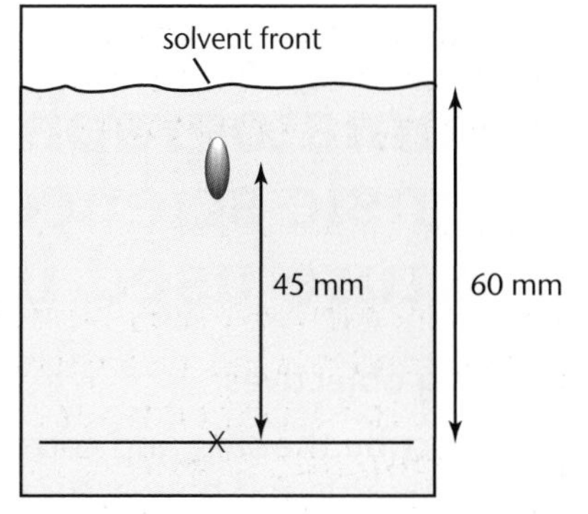

Figure 2.11 Measuring R_f values.

Different compounds have different R_f values in different solvents, which can be used to help identify the compounds.

In paper chromatography a solvent moves through the paper carrying different compounds different distances, depending on their attraction for the paper and/or the solvent.

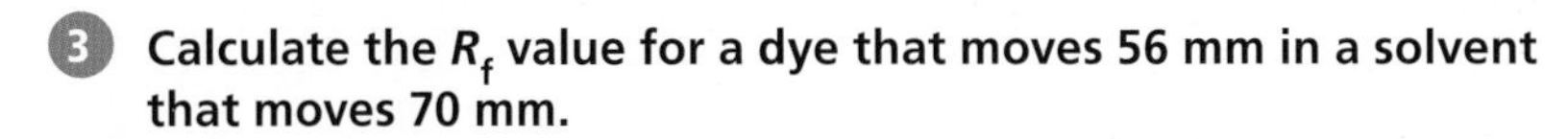

3 **Calculate the R_f value for a dye that moves 56 mm in a solvent that moves 70 mm.**

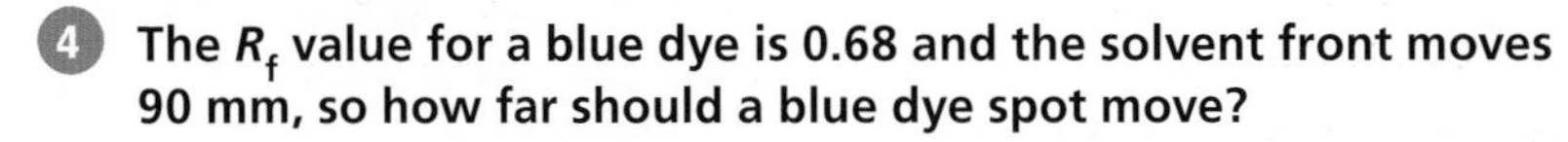

4 **The R_f value for a blue dye is 0.68 and the solvent front moves 90 mm, so how far should a blue dye spot move?**

HIGHER TIER ONLY

Assessing purity of drugs

In chromatography, separation depends on the distribution of substances between the stationary and mobile phases. Impure substances or mixtures will separate into their component parts. This happens with paper chromatography and with thin layer chromatography where a white solid thinly spread on a glass slide is used.

Thin layer chromatography is a method for finding the purity of a compound. The R_f value of a sample is compared against the R_f of a sample of substance known to be pure.

The compounds in a mixture may separate into different spots depending on the solvent but a pure compound will produce a single spot in all solvents. Because of this, chromatography can be used to assess the purity of compounds such as drugs and medicines. For example, a chromatogram such as Figure 2.12 can be developed.

Use Fig 2.12 to answer these questions:

5 **Explain which substances A, B and C contain.**

6 **Calculate the R_f values of each of the spots on the chromatogram.**

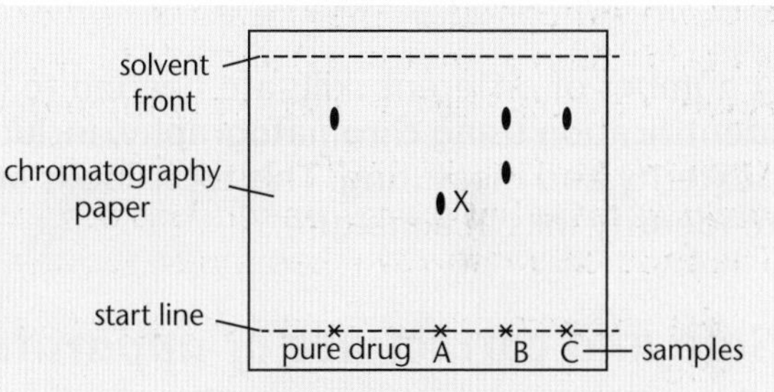

Figure 2.12 Chromatography to identify drug sample.

DID YOU KNOW?

The pure substance moves along the white solid of the stationary phase (usually silica gel or aluminium oxide) and can be scraped off at the end. It can then be extracted from the solid to be used as a pure substance. This is how we can extract small amounts of useful medicines from rare plants.

PRACTICAL

Investigate how paper chromatography can be used in forensic science to identify an ink mixture used in a forgery

KEY WORDS

chromatogram
solvent front
chromatography
elute

Learning objectives:

- describe the safe and correct manipulation of chromatography apparatus and how accurate measurements are achieved
- make and record measurements used in paper chromatography
- calculate R_f values.

Interpreting evidence is an important skill for forensic scientists. Chromatography can be used to separate mixtures, such as ink mixtures, so that patterns of separate inks appear as 'spots' on a chromatogram. Each separate ink travels a different distance when in the same solvent. Could you identify which colours make up the mixed ink dye used in a forgery?

These pages are designed to help you think about aspects of the investigation rather than to guide you through it step by step.

A number of different skills are needed to carry out an identification using **chromatography**, including manual dexterity and measuring. This topic looks at the skills needed to identify inks in a mixture.

DID YOU KNOW?

Paper chromatography is only one chromatographic technique that can be used, there is also thin layer chromatography and gas-liquid chromatography.

Safe and correct use of apparatus

A **chromatogram** is set up like this. Ink spots are placed along a start line and held in a small volume of water which acts as the solvent. The solvent rises up through the paper and 'elutes' the inks.

Think about these questions:

1. **The start line is drawn 1 cm from the bottom of the paper. Explain why the start line is drawn in pencil and not ink.**
2. **Name the piece of apparatus that makes the ink spots on the paper.**
3. **Explain why the level of the solvent must be below the ink spots.**
4. **Suggest when the developing chromatogram should be removed from the solvent.**

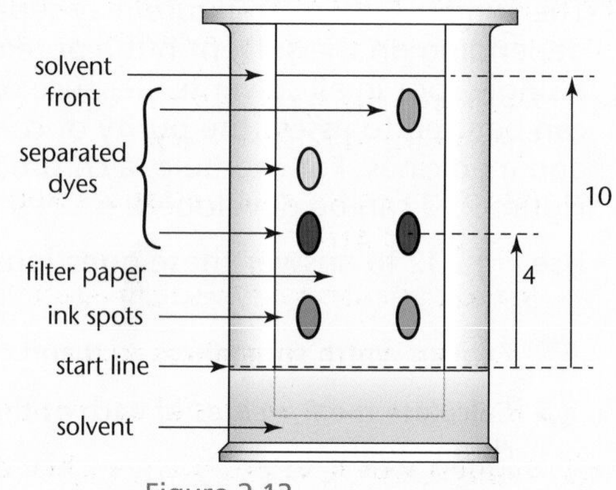

Figure 2.13

Once the solvent has risen up through the chromatography paper, the chromatogram is dried before measurement.

Sam and Alex were given an ink mixture that was used in a forgery and asked to identify which separate inks had made up the mixture.

They put spots of the separate ink along the start line, added a spot of ink mixture at the end of the paper and developed the chromatogram. They used water as the solvent to elute the spots.

Making and recording results

From the start line Sam and Alex needed to measure:

- the distance the solvent had travelled (the **solvent front**)
- the distance each spot of ink had travelled

They recorded these distances from the start line:

Chromatograms	Distance of red spot (cm)	Distance of blue spot (cm)	Distance of yellow spot (cm)	Distance of green spot (cm)	Distance of solvent (cm)	Distance of mixture spots (cm)
Sam's	4.2	5.4	6.8	10.3	12.4	4.3,5.3,10.4
Alex's	4.1	5.6	6.7	10.1	12.5	4.4,5.5,9.9
Jo's		4.5	7.8	8.1	10.0	3.2,4.4,8.2

5 Identify the colours that made up the forgery ink mixture.

6 Name the colour that was not part of the mixture.

Calculating R_f values

As the solvent front may not move the same distance each time, it important to calculate the ratio of distances travelled by solvent and ink. This is given as an index known as the R_f value.

$$R_f = \frac{\text{distance travelled by ink spot}}{\text{distance travelled by solvent front}}$$

7 Calculate the R_f value for the yellow spot on Alex's chromatogram.

Jo also made a chromatogram with the ink mixture and separate inks but did not leave the chromatogram in for as long. The solvent front did not travel as far as Sam's or Alex's.

8 Show that the R_f value for Jo's blue spot is consistent with the R_f value for Alex's blue spot.

9 Predict a value for Jo's red spot using an R_f value calculated from Sam's data.

10 Jo has one anomalous result (a result that does not fit the pattern).

a Identify the anomalous result.

b Explain what Jo should do to check the result.

11 Suggest why R_f values change when the solvent is changed.

KEY INFORMATION

To get a *reliable* result the data should be able to be obtained by another experimenter. Sam and Alex have similar results. Jo has similar results too, except for one anomalous result that does not fit the pattern started by Sam and Alex.

MATHS SKILLS

Use an appropriate number of significant figures

KEY WORDS

choromatogram
proportion
ratio
R_f value

Learning objectives:

- measure distances on chromatograms
- calculate R_f values
- record R_f values to an appropriate number of significant figures.

Chromatography can be used to separate mixtures using different solvents. As the solvent moves along a piece of chromatography paper, different parts of the mixture travel different distances with it.

Can you remember how to calculate the R_f value of a spot on a chromatogram?

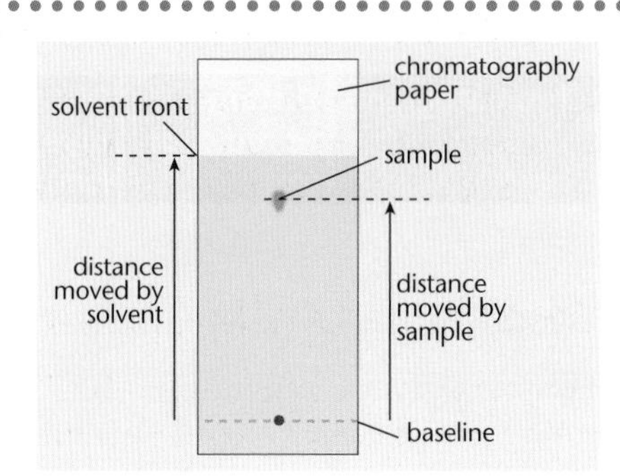

Figure 2.14

Measuring distances

There are two distances that need measuring on a developed chromatogram:

- the distance from the baseline to the solvent front
- the distance from the baseline to the spot from the mixture.

Both these can be measured with a ruler in cm or mm. Is your technique or your ruler accurate enough to measure these distances to 1/10th of a millimetre (0.1 mm)? Probably not, although you may be able to see to 0.5 mm. But is the apparatus set up well enough to achieve this level of accuracy? Probably not.

Look at Figure 2.14 and use it to answer these questions.

1. **Measure the distance from the baseline to the solvent front in mm.**
2. **Measure the distance from the baseline to the middle of the sample spot in mm.**

DID YOU KNOW?

'Accuracy' and, 'precision' have different meanings in science.

Precise measurements mean that you get the same reading each time (they are repeatably reproducible).

Accurate measurements mean that your measurements are very close to a standard or accepted value.

Calculating R_f values

To make sure that we can identify a particular chemical in a solvent by using a chromatogram we need to calculate the *ratio* of the two distances.

The reason we do this is because whatever distance the spot travels, it will always be in the same *proportion* to the distance travelled by the same solvent.

$$R_f = \frac{\text{distance travelled by the spot}}{\text{distance travelled by the solvent}}$$

For example, if the spot travels 8 cm and the solvent travels 10 cm then the ratio is 0.8.

This ratio will be the same if we measure in mm. Now the distances are 80 mm and 100 mm and again the ratio is 0.8.

If you had been short of time in your practical lesson and only allowed the solvent to travel 5 cm, then we would notice that the spot of the same sample chemical would travel 4 cm. Again the ratio would be 4/5 or 0.8.

Look back again at the chromatogram in Figure 2.14. Can you calculate the R_f value?

Now look at Figure 2.15 and use it to answer these questions.

3 **Measure the distance of the solvent front from the baseline. Also measure the distance travelled by each of the five spots.**

4 **Calculate the R_f values of each of the spots and suggest, with reasons, which of the samples (A, B or C) is likely to be the pure drug.**

KEY INFORMATION

R_f stands for 'retention factor' and is a ratio of the sample distance travelled divided by the solvent distance travelled. The *lower* the retention factor the more *strongly* the sample bonds to the chromatography paper.

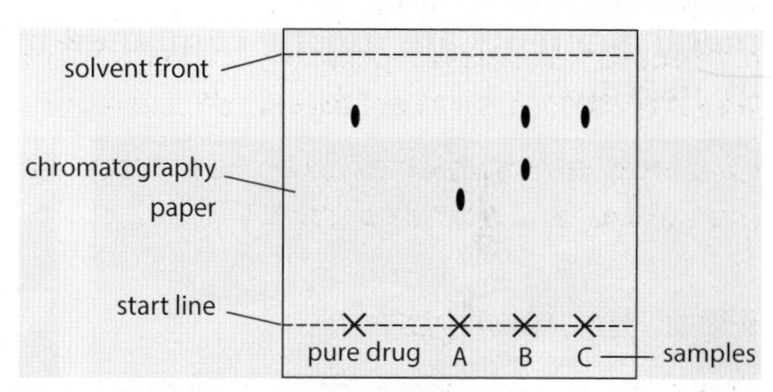

Figure 2.15 A paper chromatogram for measuring drug purity.

Applying significant figures

When we have calculated the R_f values, to how many significant figures should we record the result?

Here are examples of values measure on three chromatograms:

Experiment number	Distance travelled by sample spot (cm)	Distance travelled by solvent (cm)	R_f value shown on calculator	R_f value to sig fig
1	4	8	0.5	0.5
2	13.5	18.0	0.75	0.75
3	4.1	8.6	0.47674418	0.48

- In experiment 1, each value is measured to 1 significant figure (sig fig), so the R_f is calculated to 1 sig fig.
- In experiment 2, one measurement value is given to 3 sig figs while the other has 2 sig figs. The R_f value is written with 2 sig figs, which is the same as the value with the *lowest* number of sig figs.
- In experiment 3, both measurement values have 2 sig figs so the R_f value is written with 2 sig figs. Note that the result needs to be rounded up to 0.48.
- Now use the values you calculated in Q4 (from Figure 2.15) to answer the next questions.

0.476 has 3 sig figs and 0.47 has 2 sig figs, but the third place value (6) in 0.476 is greater than 5 so the second place value (7) is rounded up to 8.

5 **What is the R_f value of sample C, to the appropriate number of significant figures? Explain why you chose this number of significant figures.**

6 **Record the R_f values for samples A and B to the appropriate number of significant figures.**

Comparing metals and non-metals

Learning objectives:

- recall a number of physical properties of metals and non-metals
- describe some chemical properties of metals and non-metals
- explain the differences between metals and non-metals on the basis of their characteristic physical and chemical properties.

KEY WORDS

electrical conductor
lustrous
sonorous
tensile strength
thermal conductivity

Metals are useful materials because they have a wide range of properties. Gold jewellery does not corrode and it has an appealing colour and lustre. Copper has good thermal conductivity so is used for saucepans. Non-metal elements are mostly combined in compounds to be useful.

Figure 2.16 Many metals are instantly recognisable as metals because they are shiny and sonorous.

Physical properties

Most metals are instantly recognisable because they are shiny whereas non-metals are not. There are the other physical properties of metals and non-metals.

Metals	Non-metals
lustrous	dull
sonorous	not sonorous
hard	soft, brittle or gas
high density	low density
high **tensile strength**	low or no tensile strength or gas
high melting point and boiling point	low melting point and boiling point
good conductors of heat	poor or no **thermal conductivity**
good **electrical conductivity**	poor or non-conductors of electricity

Figure 2.17 Dating from 1779, this is the first bridge made from cast iron. Iron was used to make this bridge because it is very strong.

Figure 2.18 Sulfur, calcium and nitrogen are non-metals. They are not shiny and have low boiling points.

1 Write down two physical properties of silver that make it more useful than sulfur for making cutlery.

2 Use Figure 2.18 and the table to explain why sulfur is not a metal.

DID YOU KNOW?

Other physical properties of metals include being malleable and ductile.

Chemical properties

Chemical properties of metals are the result of reactions with oxygen or acids. Although copper is resistant to attack by oxygen and acid (which is a reason why it is used for saucepans), many metals react with oxygen to make an oxide (such as calcium oxide and iron oxide).

Metals react with acids to make salts (such as zinc with sulfuric acid to make zinc sulfate).

One chemical property of a non-metal is the result of the reaction with oxygen. Carbon reacts with oxygen to make carbon dioxide. Carbon dioxide dissolves in water to make a mildly acidic solution. This is what fizzy water contains.

Figure 2.19 Bottles of lemonade and cola with acidic carbon dioxide solution.

3 Suggest a test to show that a solution of carbon dioxide is mildly acidic.

4 Predict the product of reacting magnesium with nitric acid.

Distinguishing properties

Sulfur and phosphorus both react with oxygen to make oxides. Both sulfur dioxide and phosphorus oxide turn universal indicator red. They are acidic oxides.	Calcium and potassium both react with oxygen to make oxides. Both calcium oxide and potassium oxide turn universal indicator blue. They are basic oxides.

Metals form basic oxides. Non-metals form acidic (or neutral) oxides.

KEY INFORMATION

You will soon need to know how to explain the chemical properties of metals and non-metals using knowledge about how they form ions.

5 You have a sample of 'unknownium' oxide. Explain how you would use universal indicator to see if 'unknownium' was a metal or a non-metal element.

6 An element makes an oxide that turns universal indicator red. Is the element a metal or a non-metal?

7 Element X has a low melting point of 63 °C, has low density, is a good electrical conductor and is malleable. Explain whether the oxide of X is likely to be acidic or basic.

Electron structure

Learning objectives:

- explain how electrons occupy 'shells' in an order
- describe the pattern of the electrons in shells for the first 20 elements.

KEY WORDS

electronic structure
electron shells
energy levels

The electrons of an atom are arranged in patterns. The electrons fill up shells in order until that shell can take no more electrons. The next electron goes into the next shell. These patterns are the key to the behaviour of atoms.

The 'build-up' of electrons

Electrons occupy shells around the nucleus. The **electron shell** nearest to the nucleus takes up to two electrons. The second shell takes up to eight electrons. The next electrons occupy a third shell.

Oxygen has an atomic number of 8. It has eight protons and so it has eight electrons in the space around the nucleus. The first two go into the first shell. As the first shell is now full, the next 6 electrons go into the second shell. The electron pattern for oxygen is 2,6.

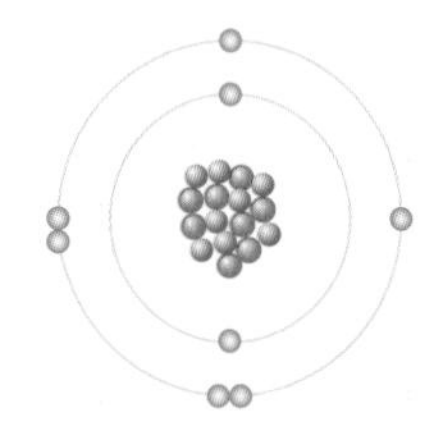

Figure 2.20 For oxygen, the eight electrons can written as 2,6 or be drawn like this.

1. **Draw the electron pattern for hydrogen and for lithium atoms.**
2. **Write down the electron pattern for nitrogen atoms.**

The shells are not fixed rings and are also known as **energy levels**.

Electron patterns and groups

The periodic table is arranged in order of 'proton number'.

There is a very important link between **electronic structure** and the periodic table. For example, consider an atom of an element that has the electronic structure 2,8,6.

- This element has three electron shells, so it is in the third row of the periodic table.
- It has six electrons in its outer shell, so it is in the sixth column.
- Using the periodic table, we can find that its atomic number is 16 and it is the element sulfur, S.

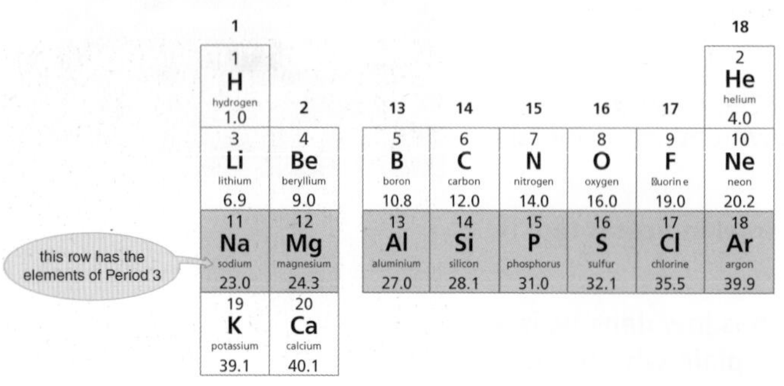

1	2	13	14	15	16	17	18
1 H hydrogen 1.0							2 He helium 4.0
3 Li lithium 6.9	4 Be beryllium 9.0	5 B boron 10.8	6 C carbon 12.0	7 N nitrogen 14.0	8 O oxygen 16.0	9 F fluorine 19.0	10 Ne neon 20.2
11 Na sodium 23.0	12 Mg magnesium 24.3	13 Al aluminium 27.0	14 Si silicon 28.1	15 P phosphorus 31.0	16 S sulfur 32.1	17 Cl chlorine 35.5	18 Ar argon 39.9
19 K potassium 39.1	20 Ca calcium 40.1						

Figure 2.21 Period 3 contains the elements from sodium to argon.

We can also work the other way. Find the element with the atomic number 12 in the periodic table. This is magnesium, Mg.

- It is in the third row, so it has three electron shells.
- It is in the second column, so it has two electrons in its outer shell.
- It has the electronic structure of 2,8,2.

The column of elements is known as a group. So column 2 is Group 2.

3 **An atom of an element has an atomic number 11.**
a Draw a diagram to show the pattern of electrons.
b Identify the element.
c Identify the group to which it belongs.

4 **Work out the electronic structure of the element that has an atomic number 9.**

5 **An element has a mass number of 40 and an electron arrangement of 2,8,8,2. Identify the element and work out the number of neutrons.**

Maximum numbers

In Figure 2.20 the electrons are drawn as dots on the rings.

In Figures 2.22 and 2.23 the electrons are drawn as crosses 'x' on the shells.

Both models are correct. When atoms combine, the electrons in the model need to be seen as electrons from different atoms so we use both 'dots' and 'crosses' to represent them.

The electronic structure of each of the first 20 elements can be worked out using:

- the atomic number of the element
- the maximum number of electrons in each shell.

The third shell takes up to eight electrons before the fourth shell starts to fill. Element 19, potassium, has the electronic structure 2,8,8,1.

Use the periodic table to help you to answer these questions.

6 **Work out the electronic structure of argon.**

7 **Use a blank periodic table sheet to draw the electronic structures of the first 20 elements, drawing each diagram in the correct box. What do you notice about the group number and the number of electrons in the outer shell?**

8 **The electronic configuration for the ion of an unknown isotope $^{26}X^{3+}$ is 2,8.**
a Work out the atomic number of element X.
b Determine the number of neutrons in X.
c Explain which group in the periodic table element X is in.

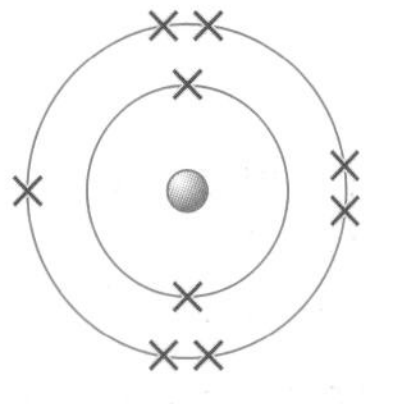

Figure 2.22 An atom of fluorine, 2,7.

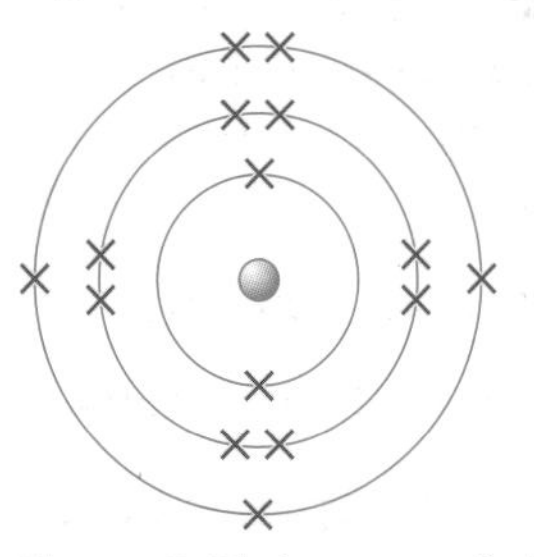

Figure 2.23 An atom of phosphorus, 2,8,5.

DID YOU KNOW?

It took many years for scientists to work out the theory of electrons occupying shells. They started with the behaviour of elements and then looked for patterns.

KEY INFORMATION

Do not try to use this method to work out the electronic structure of gold, because it has 79 electrons.

Metals and non-metals

KEY WORDS

ions
atomic structure
metalloids

Learning objectives:

- know that metals are found on the left of the periodic table and non-metals on the right
- explain the differences between metals and non-metals based on their physical and chemical properties
- explain that metals form positive ions and non-metals do not.

Whether an element is a metal or a non-metal depends on the electronic structure of its atoms. The element is classified as a non-metal or a metal depending on whether it needs to gain or lose electrons and whether its reactions are typical of a non-metal or a metal.

Metals and non-metals in the periodic table

3 Li lithium 6.9				7 N nitrogen 14.0	8 O oxygen 16.0		10 Ne neon 20.2
11 Na sodium 23.0	12 Mg magnesium 24.3			15 P phosphorus 31.0	16 S sulfur 32.1	17 Cl chlorine 35.5	
	20 Ca calcium 40.1						

Figure 2.24 Here are some metals and non-metals in the periodic table.

Looking at the periodic table you can see the metals lithium, sodium, magnesium and calcium on the left-hand side.

You can see the non-metals nitrogen, oxygen, sulfur, phosphorus, chlorine and neon on the right-hand side.

From your knowledge of chemistry so far try to draw a line that separates the metals from the non-metals.

DID YOU KNOW?

The line of elements separating the metals from the non-metals are called the metalloids.

key: atomic number / **symbol** / name / relative atomic mass

(1) 1	(2) 2	3	4	5	6	7	8	9	10	11	12	(3) 13	(4) 14	(5) 15	(6) 16	(7) 17	(0) 18
1 H hydrogen 1.0																	2 He helium 4.0
3 Li lithium 6.9	4 Be beryllium 9.0											5 B boron 10.8	6 C carbon 12.0	7 N nitrogen 14.0	8 O oxygen 16.0	9 F fluorine 19.0	10 Ne neon 20.2
11 Na sodium 23.0	12 Mg magnesium 24.3											13 Al aluminium 27.0	14 Si silicon 28.1	15 P phosphorus 31.0	16 S sulfur 32.1	17 Cl chlorine 35.5	18 Ar argon 39.9
19 K potassium 39.1	20 Ca calcium 40.1	21 Sc scandium 45.0	22 Ti titanium 47.9	23 V vanadium 50.9	24 Cr chromium 52.0	25 Mn manganese 54.9	26 Fe iron 55.8	27 Co cobalt 58.9	28 Ni nickel 58.7	29 Cu copper 63.5	30 Zn zinc 65.4	31 Ga gallium 69.7	32 Ge germanium 72.6	33 As arsenic 74.9	34 Se selenium 79.0	35 Br bromine 79.9	36 Kr krypton 83.8
37 Rb rubidium 85.5	38 Sr strontium 87.6	39 Y yttrium 88.9	40 Zr zirconium 91.2	41 Nb niobium 92.9	42 Mo molybdenum 95.9	43 Tc technetium	44 Ru ruthenium 101.1	45 Rh rhodium 102.9	46 Pd palladium 106.4	47 Ag silver 107.9	48 Cd cadmium 112.4	49 In indium 114.8	50 Sn tin 118.7	51 Sb antimony 121.8	52 Te tellurium 127.6	53 I iodine 126.9	54 Xe xenon 131.3

Figure 2.25 Top section of the periodic table

1. **Find element 53. Is this element a metal or a non-metal?**
2. **Is element 26 a metal or a non-metal?**

Positions of elements in the table

Magnesium is a metal because of its **atomic structure**. It has two electrons in the outer shell, which can be easily lost from the atom. The electrons join another atom that needs more electrons. Because electrons have been lost, a positive **ion** has been made.

Magnesium makes positive ions so it is a metal.

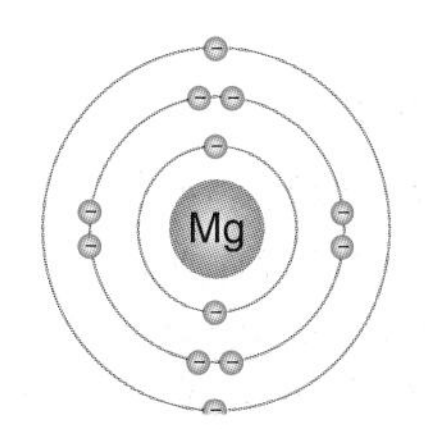

Figure 2.26 A magnesium atom with two electrons in the outer shell.

3. **Explain why aluminium is a metal, using knowledge about its atomic structure.**
4. **Explain why element number 8 is a non-metal. Use ideas about atomic structure.**

DID YOU KNOW?

Metal atoms are bonded together by their outer electrons. The atoms pack together and the outer electrons delocalise, which means that the outer electrons move through the metal as a 'sea' of electrons.

Electron transfer in metals and non-metals

When a metal, such as magnesium, reacts with oxygen, the metal loses electrons and oxygen gains electrons.

magnesium + oxygen → magnesium oxide

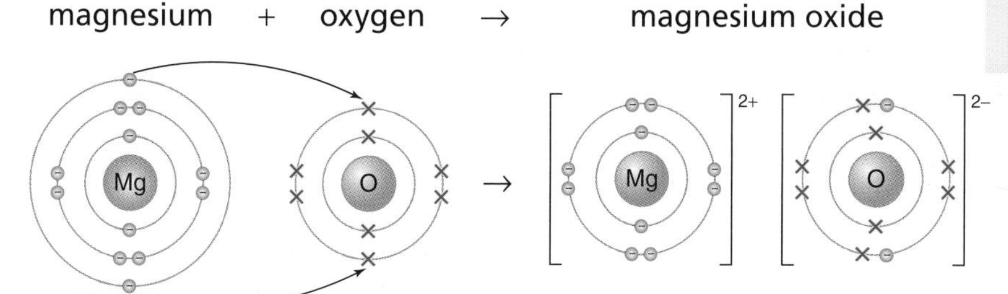

Figure 2.27

This is because metal atoms have just a few outer electrons which they 'lose' to form ions. Oxygen accepts the electrons.

When a non-metal such as chlorine reacts with a metal such as sodium, the non-metal atom gains an electron from the metal atom.

sodium + chlorine → sodium chloride

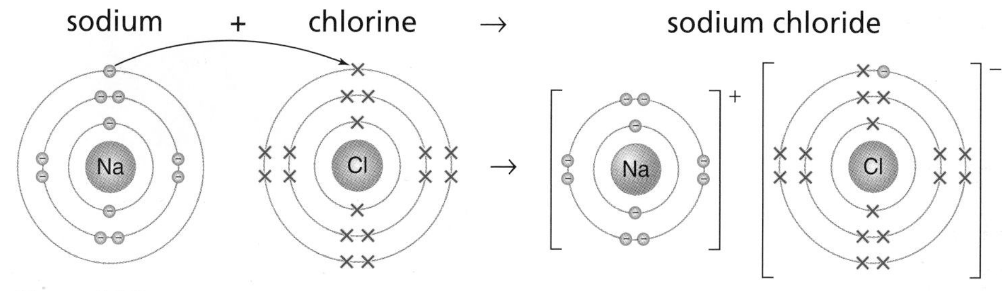

Figure 2.28

This is because non-metal atoms have empty spaces in their outer shell in which they 'gain' other electrons from metals to form negative ions. The non-metal does not form a positive ion.

5. **Explain why fluorine is a non-metal that can react with a metal like potassium to form potassium fluoride.**
6. **The elements with atomic numbers 3 and 9 can react together. Explain why and work out the formula of the product.**

KEY INFORMATION

Metals make positive ions and non-metals do not.

Chemical bonds

KEY WORDS

covalent
delocalised
ionic
metallic

Learning objectives:

- describe the three main types of bonding
- explain how electrons are used in the three types of bonding
- explain how bonding and properties are linked.

Why do sodium chloride, carbon dioxide and copper look so different and behave so differently? It is all to do with the number of electrons in each of the atoms that make up the substances and how these electrons are arranged. They may be in a localised arrangement or in a delocalised arrangement.

Types of bonds

There are three types of strong chemical bonds:

ionic **covalent** **metallic**

	Ionic	Covalent	Metallic
This bonding occurs in:	compounds from metals combined with non-metals	compounds of non-metals combined with non-metals and in most non-metallic elements	metallic elements and alloys
In this bonding the particles are:	oppositely charged ions	atoms that share pairs of electrons	atoms that share electrons that move about
		:F:F:	

1 **Compare ionic bonding and covalent bonding.**

How do atoms join together?

Atoms can join together by *transferring* electrons or by *sharing* electrons – and some atoms do both.

When metals and non-metals *transfer* electrons the atoms become charged. These charged atoms are called *ions*. The metal ions are positively charged. The non-metal ions are negatively charged. The positive ions and negative ions are attracted by *electrostatic forces*.

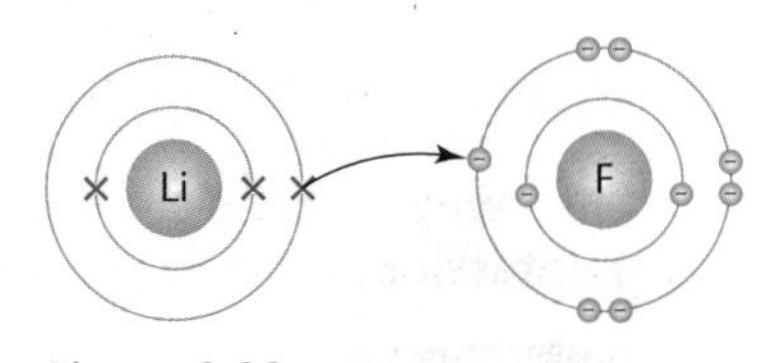

Figure 2.29

They are bonded together by electrostatic attraction. This is called *ionic bonding.*

Non-metals can also *share* electrons between atoms. This type of bonding is known as *covalent bonding*.

Metals are bonded together as positive ions by electrons that move around. This is called *metallic bonding*.

These three types of bonding produce different kinds of substances.

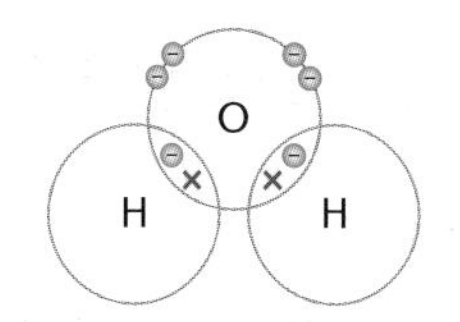

Figure 2.30

	Bonding		
	Ionic	**Covalent**	**Metallic**
Structures are:	large crystals made from ions attracted to each other by electrostatic attraction	molecules made from atoms bonded by covalent bonds	lumps or sheets of metal made from atoms packed together so that the delocalised electrons move through the fixed positive ions
Examples are:	sodium chloride and magnesium oxide	carbon dioxide and water	copper and gold
At room temperature they are:	solids	gases and liquids	solids
Melting points are:	high	low	high
Do they conduct electricity?	Not as a solid, only when melted.	Not at all or poorly.	Yes, good conductors.

2. **Substance R is a good electrical conductor as a solid and liquid and has a high melting point. Suggest the type of bonding.**
3. **Describe what a delocalised electron does in a metal.**
4. **Suggest why covalently bonded substances such as carbon dioxide cannot conduct electricity.**

DID YOU KNOW?

Metals conduct electricity because they contain delocalised electrons.

HIGHER TIER ONLY

Why do atoms form bonds?

Atoms with an outer shell of eight electrons have a stable electronic structure. These atoms are unreactive. Examples of such unreactive atoms are the elements neon, argon and krypton (all from Group 0).

However, most atoms usually have too many or too few electrons in their outer shell to be stable.

It is easiest for an atom that has one or two electrons in its outer shell to lose them. These are metal atoms that form positive ions.

It is easiest for an atom that has one or two spaces in its outer shell to fill to gain electrons. There are two ways this can happen: non-metal atoms can form negative ions and non-metal atoms can share electrons.

5. **Explain why it is the *electrons* that are involved in bonding.**

KEY INFORMATION

Delocalised electrons are often called a 'sea' of electrons in metals.

Ionic bonding

Learning objectives:

- represent an ionic bond with a diagram
- draw dot and cross diagrams for ionic compounds
- work out the charge on the ions of metals and non-metals from the group number of the element (1, 2, 6 & 7).

KEY WORDS

dot and cross
electron transfer
electrostatic attraction
square bracket

Electron transfer between atoms can be drawn in many ways. One way is by drawing electron patterns in shells and using arrows to show the transfer. A conventional way is to draw a 'dot and cross' diagram. Diagrams help us to visualise what is happening at a level that we cannot see even with the most powerful microscope.

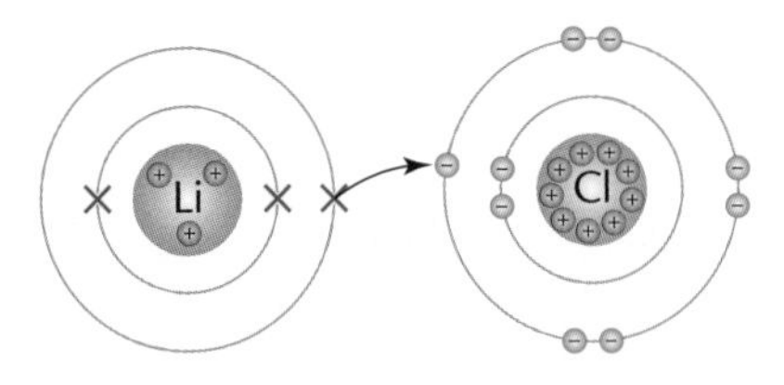

Figure 2.31 An electron is transferred *from* a metal atom *to* a non-metal atom.

Making ions

Noble gases have stable atoms. Except for helium, their outer shell contains eight electrons.

A metal atom has a small number of electrons in its outer shell.	A non-metal atom has spaces in its outer shell.
It needs to lose these electrons to get a stable electronic structure.	It needs to gain electrons to be stable.
The electrons transfer *out* from the metal atom to make a stable atom.	The electrons transfer *in* to the non-metal atom to make a stable atom.

The electrons transfer *out from* a metal atom *in to* a non-metal atom.

A metal atom becomes a positive ion	A non-metal atom becomes a negative ion

The positive ions and the negative ions are then attracted to one another. These attractions spread to a number of other ions to make a solid lattice.

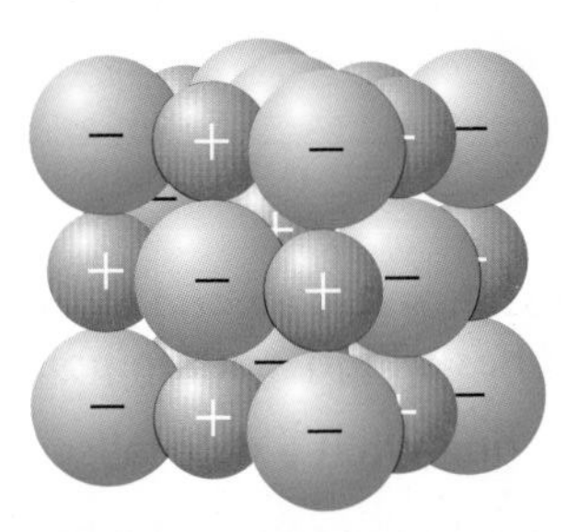

Figure 2.32 Ions of opposite charge are attracted together as a solid lattice.

1 **Explain how a metal atom becomes a positive ion.**

2 **Explain how solid ionic lattices are formed from elements using sodium chloride as an example.**

Figure 2.33 Ions held together forming crystals

Dot and cross diagrams

When a metal atom reacts with a non-metal atom, electrons in the outer shell of the metal atom are transferred.

Metal atoms lose electrons to become positively charged ions.	Non-metal atoms gain electrons to become negatively charged ions.
The ions produced by metals in Groups 1 and 2 have the electron structure of a noble gas (Group 0) and a positive charge.	The ions produced by non-metals in Groups 6 and 7 have the electron structure of a noble gas (Group 0) and a negative charge.

DID YOU KNOW?

The charge on the positive ion corresponds to the group number for Group 1 and Group 2 ions.

The **electron transfer** during the formation of an ionic compound can be represented by a **dot and cross** diagram. For example, for sodium chloride:

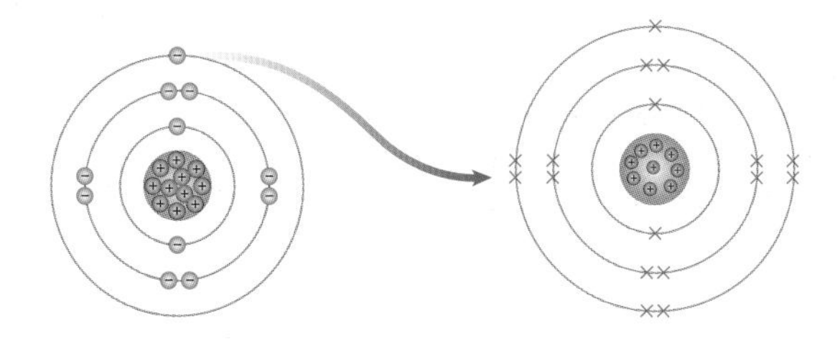

Figure 2.34

Sodium atoms have the electron pattern (2,8,1) and they have 1 outer electron. This electron is represented by 1 dot.	Chlorine atoms have the electron pattern (2,8,7) and they have 7 outer electrons. These electrons are represented by 7 crosses.

When the ions form, the 'dot' transfers out to the space in among the 'crosses'.

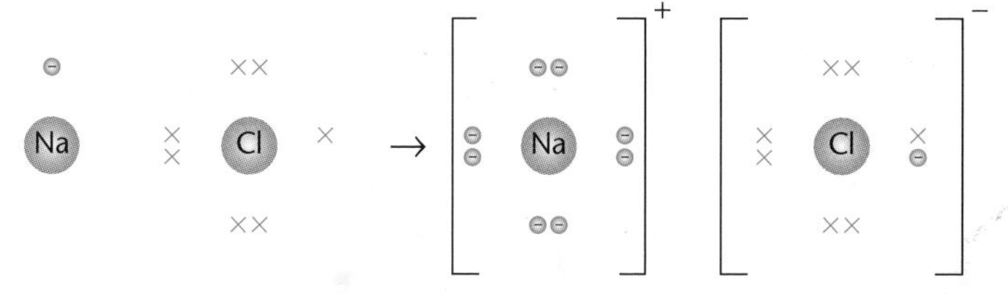

Figure 2.35

The sodium ion has an electron pattern (2,8) and is positively charged overall.	The chloride ion has an electron pattern (2,8,8) and is negatively charged overall.

- Both ions have a stable electron pattern.
- The ions are attracted together by the **electrostatic attraction** of opposite charges.
- They attract to make large packs of ions that form crystals.

3 **State the electron arrangement of a stable neon atom.**

4 **Predict how many electrons a fluorine atom needs to become an ion.**

5 **Explain why a chloride ion is negative. [Hint: use ideas about the nucleus of the atom].**

KEY INFORMATION

In dot and cross diagrams only the outer electrons are represented. No inner shells are shown. The ions are drawn with square brackets and the charge on the outside top right.

DID YOU KNOW?

When a Group 1 atom loses one electron a (positive) 1+ ion is formed, for example, $Na \rightarrow Na^+$.

When a Group 2 atom loses two electrons a (positive) 2+ ion is formed, for example, $Mg \rightarrow Mg^{2+}$

When a Group 7 atom gains one electron a (negative) 1– ion is formed, for example, $F \rightarrow F^-$. (Note: 7 + 1 = 8)

When a Group 6 atom gains two electrons a (negative) 2– ion is formed, for example, $O \rightarrow O^{2-}$. (Note: 6 + 2 = 8)

Ionic bonds

If an atom loses electrons a positive ion is formed. This is because there will be fewer negatively charged electrons than the number of positively charged protons in the nucleus.

If an atom gains electrons a negative ion is formed. This is because there will be more negatively charged electrons than the number of positively charged protons in the nucleus.

If magnesium makes a 2+ ion and chlorine makes a 1– ion, then *two* chloride ions are needed to bond to *one* magnesium ion to make $MgCl_2$.

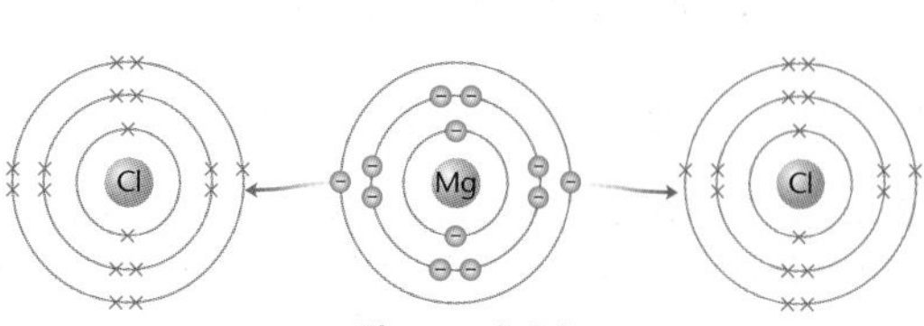

Figure 2.36

6 **Draw an electron transfer diagram and a dot and cross diagram to show the bonding of sodium oxide.**

Ionic compounds

Learning objectives:

- identify ionic compounds from structures
- explain the limitations of diagrams and models
- work out the empirical formula of an ionic compound.

KEY WORDS

3D representation
electrostatic
empirical formula
giant lattice

Positive ions and negative ions attract to form giant lattice structures with the electrostatic forces acting in all directions. Models and diagrams help us to understand this type of bonding and the structures that are formed.

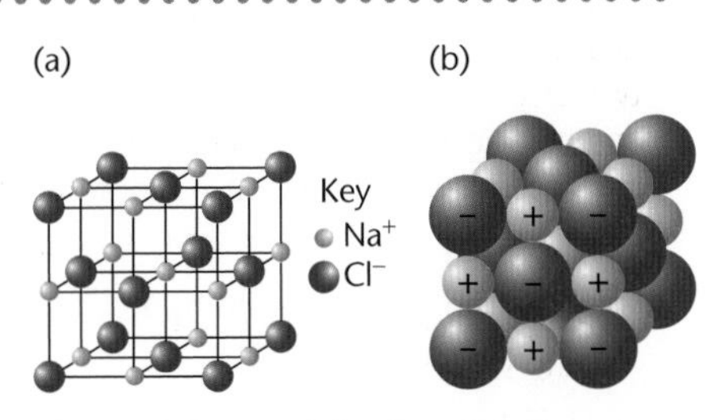

Figure 2.37 Models of NaCl (a) ball and stick (b) close packed.

Giant structures

Ionic bonding occurs when oppositely charged ions are held together by strong **electrostatic** forces of attraction.

An ionic compound is a giant structure of ions in a lattice. One example of an ionic compound is sodium chloride.

The structure of sodium chloride can be represented in the following ways.

	Type of diagram:	
	ball and stick (a)	**close-packed (b)**
The diagram is helpful to show:	the 3D structure the charges on the ions the 3D arrangement of ions the *type* of ions in all directions	the 3D structure the charges on the ions the 2D arrangement of ions (look at the front face of Figure 2.37a) the closeness of ions
The diagram is unhelpful:	the ions are actually close together gives a false image of bond direction when it is only electrostatic attraction	difficult to see the 3D arrangement of ions
This diagram is best to represent:	the number and type of ions in 3 dimensions	the way that ions are packed close together

Diagram for question 1

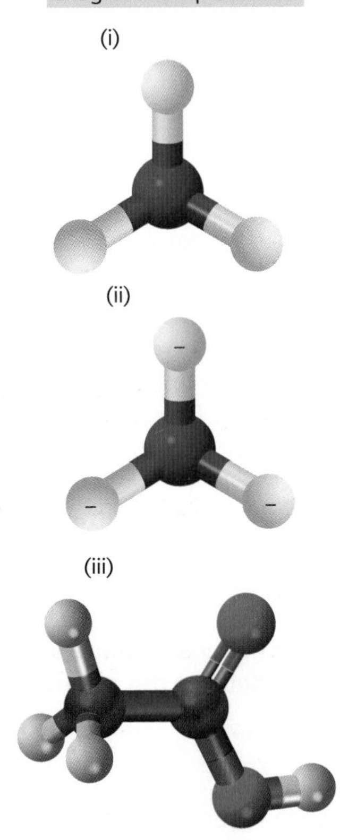

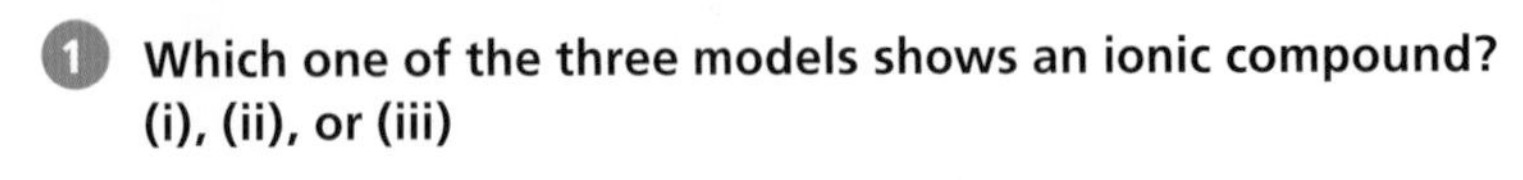

1 Which one of the three models shows an ionic compound? (i), (ii), or (iii)

Empirical formula

In magnesium oxide the ions are held together by electrostatic attraction to form a giant structure, in exactly the same way as in sodium chloride, except that the magnesium ion has a 2^+ charge and the oxide ion has a 2^- charge.

DID YOU KNOW?

In an ionic compound each atom has either gained or lost electrons to achieve a stable electronic structure.

Sodium chloride has the empirical formula NaCl which is one Na^+ ion to one Cl^- ion. The empirical formula is the simplest ratio of ions.

Magnesium oxide has the formula MgO which is one Mg^{2+} ion to one O^{2-} ion.

Other compounds that bond using ionic bonds are sodium oxide and magnesium chloride.

Sodium only has one electron to lose, but oxygen needs to gain two electrons. Two sodium atoms are needed to bond with one oxygen atom.	Magnesium needs to lose two electrons, but chlorine can only gain one electron. So one magnesium atom needs two chlorine atoms to make an ionic bond and make magnesium chloride.

(a) 2D electron transfer diagram of sodium oxide

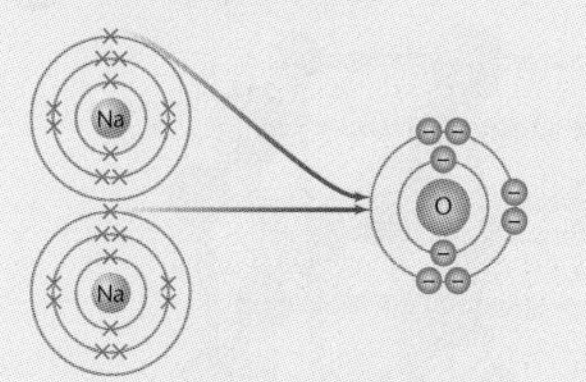

(b) Ball and stick diagram of sodium oxide

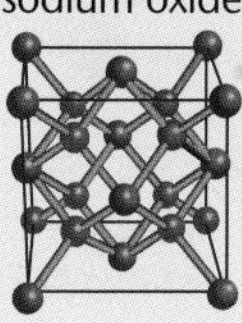

(a) 2D electron transfer diagram of magnesium chloride

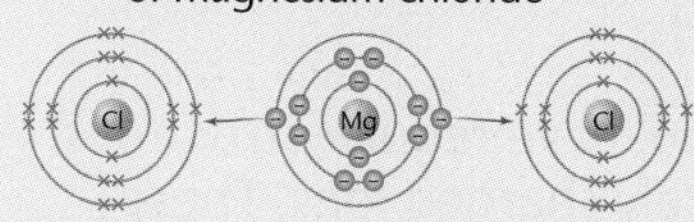

(b) Ball and stick diagram of magnesium chloride

(a) dot and cross diagram

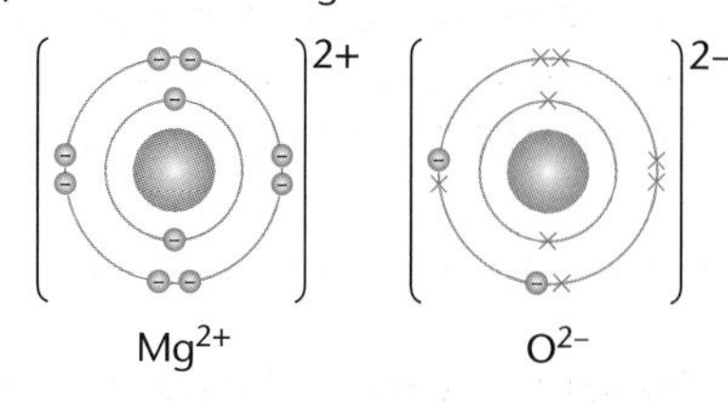

(b) 3D close packed diagram

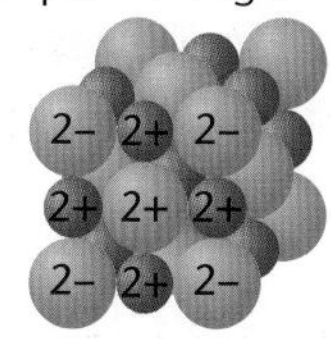

Figure 2.38 The bonding in magnesium oxide.

2 **Potassium fluoride is formed from K^+ and F^- ions. Show the ionic bonding in potassium fluoride using an electron transfer diagram and a dot and cross diagram.**

3 **Describe why, in terms of electrons, potassium sulfide (shown in the diagram to the right) has an empirical formula of K_2S.**

Diagram for question 3

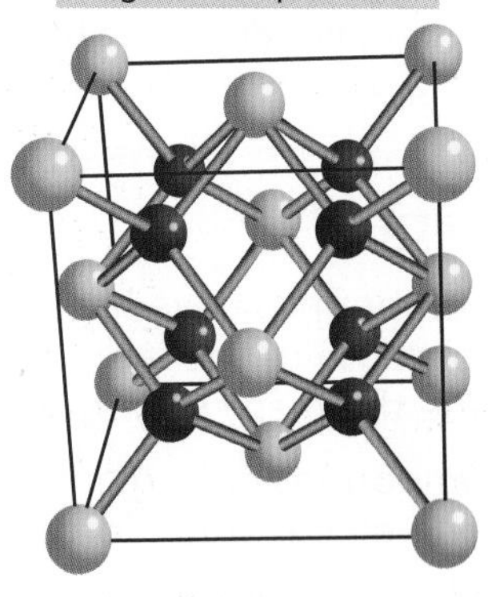

HIGHER TIER ONLY

Equations of ion formation

$Na - e^- \rightarrow Na^+$	$Cl + e^- \rightarrow Cl^-$
$Mg - 2e^- \rightarrow Mg^{2+}$	$O + 2e^- \rightarrow O^{2-}$

To make $MgCl_2$:

$Mg - 2e^- \rightarrow Mg^{2+}$

$2Cl + 2e^- \rightarrow 2Cl^-$ Empirical formula $MgCl_2$

4 **Draw a dot and cross diagram to show the formation of calcium chloride and write its empirical formula.**

5 **Determine the empirical formula of the following ionic compounds: a Na^+ and N^{3-}; b Al^{3+} and O^{2-}.**

REMEMBER!

Mg loses 2 electrons.
O gains 2 electrons.

Properties of ionic compounds

Learning objectives:

- describe the properties of ionic compounds
- relate their melting points to forces between ions
- explain when ionic compounds can conduct electricity.

KEY WORDS

conducting electricity
energy
high melting point
ions free to move

Ionic compounds have giant structures that require a lot of energy to break down. As solids they do not conduct electricity. However, they do conduct electricity under certain conditions, once their giant lattice is broken down.

Properties of ionic compounds

Ionic compounds, such as sodium chloride, have regular structures. They make giant ionic lattices of oppositely charged ions. These ions have strong electrostatic forces of attraction in all directions.

These compounds:

have **high melting points** and high boiling points	because	the forces of attraction holding the ions together are very strong
conduct electricity when melted or dissolved in water	because	the **ions are free to move** and so charge can flow

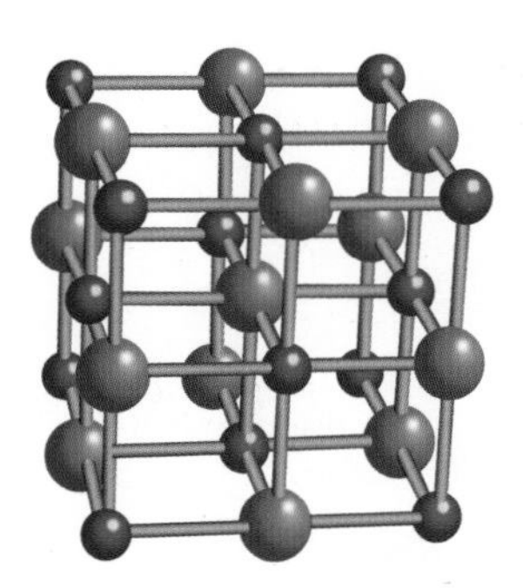

Figure 2.39 There is a strong attraction between the positive and negative ions of sodium chloride, forming a giant lattice in the shape of a cube.

The forces of attraction have to be overcome for sodium chloride to melt so a great deal of energy is required to overcome the lattice down to individual, moving ions.

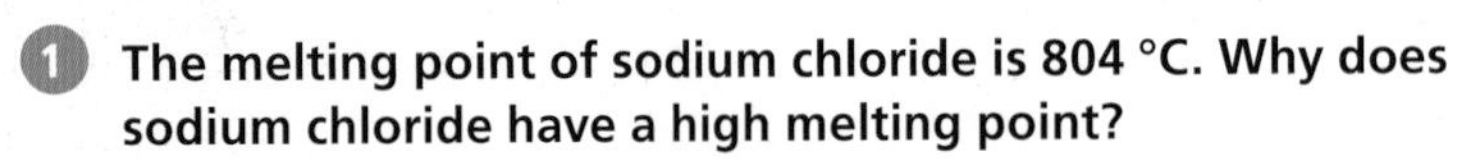

1 The melting point of sodium chloride is 804 °C. Why does sodium chloride have a high melting point?

Why do ionic compounds have these properties?

These compounds have high melting points and high boiling points because of the large amounts of energy needed to overcome the many strong bonds.

Ionic compounds have giant ionic lattices in which there are strong electrostatic forces of attraction in all directions between oppositely charged ions.

These forces have to be overcome for the particles to move freely and more randomly as a liquid or a gas.

When melted or dissolved in water, ionic compounds can conduct electricity.

This is because ionic compounds, such as sodium chloride, are made of charged ions that become free to move when melted or dissolved.

DID YOU KNOW?

Each magnesium atom donates two electrons to the oxygen atom, which makes a stronger ionic bond than the bonds made by sodium transferring one electron to chlorine atoms.

They conduct electricity when they are melted (molten) because the ions are free to move and so charge can flow between electrodes.

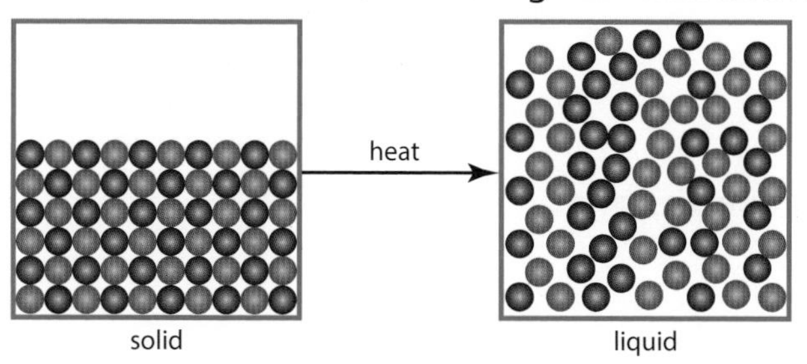

Figure 2.40 Energy is needed to overcome the strong attraction between the positive and negative ions of ionic compounds, for the solid to become a liquid when melting.

Ionic compounds such as sodium chloride can dissolve in water. When it is dissolved in water it makes a solution. Sodium chloride solution can conduct electricity because the ions move and split up from each other. Two electrodes are put into the solution to allow the electricity to pass through.

2 **Substance D has a high melting point and conducts electricity in solid and liquid state. Explain whether D is ionic or not.**

3 **Explain why sodium chloride does not conduct electricity when it is a solid.**

HIGHER TIER ONLY

Comparing NaCl to other ionic compounds

Sodium chloride NaCl is a typical ionic compound. It dissolves in water, does not conduct electricity when solid, but does when melted – it has a high melting point (804 °C).

Magnesium oxide MgO and potassium chloride KCl are other ionic solids. Look at the data table to see which solid will require the most energy to overcome the forces between their ions.

	Melting point °C	Conducts electricity when solid	Conducts electricity when molten
sodium chloride	804	No	Yes
magnesium oxide	2800	No	Yes
potassium chloride	770	No	Yes

These melting points are high because in all the compounds the ions form giant lattice structures that need a lot of energy to break down.

A lot more energy is needed to overcome the forces in MgO than in NaCl and KCl.

4 **Suggest why magnesium oxide has a higher melting point than sodium chloride or potassium chloride.**

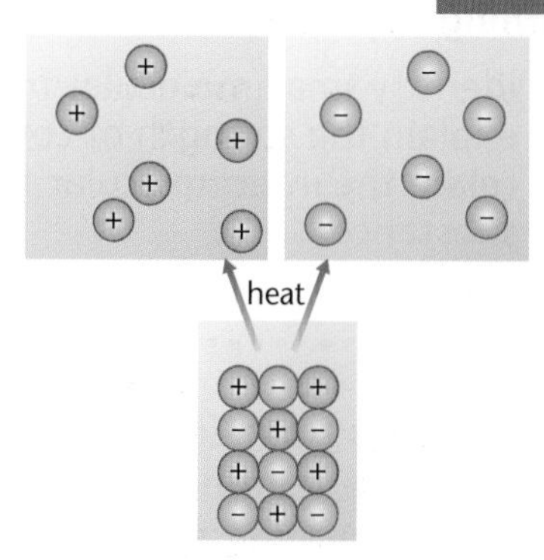

Figure 2.41 Ionic solid melting to release ions that are free to move.

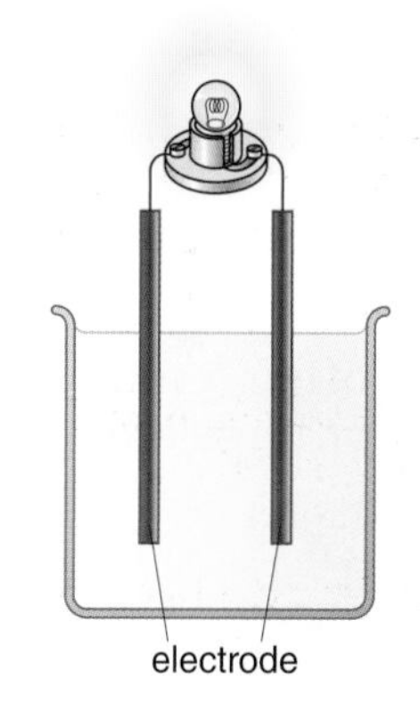

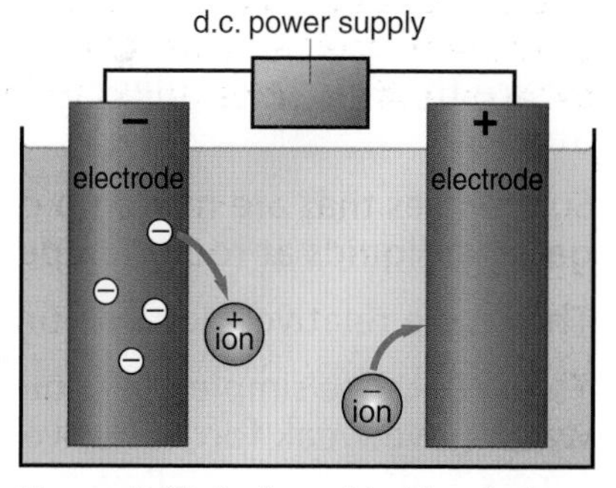

Figure 2.42 Sodium chloride solution can conduct electricity because the ions are free to move.

KEY INFORMATION

You only need to know the specific lattice structure of sodium chloride. In this example we are using evidence to make a reasoned suggestion. You do not need to know the details of magnesium oxide or potassium chloride.

Properties of small molecules

Learning objectives:

- identify small molecules from formulae
- explain the strength of covalent bonds
- relate the intermolecular forces to the bulk properties of a substance.

KEY WORDS

covalent bond
free electrons
intermolecular force
molecule

Researchers have found that the polar caps of Mars are probably made of both water ice and dry ice from carbon dioxide. However, at room temperature on Earth carbon dioxide is a gas and water is a liquid. These substances are made up of small molecules and their properties are related to the size of their molecules.

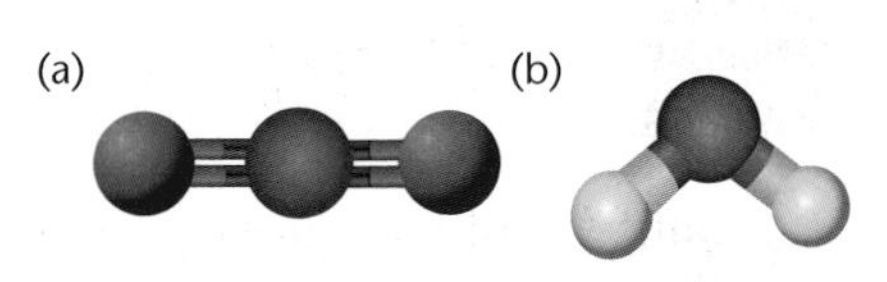

Figure 2.43 Models of small molecules (a) carbon dioxide (b) water.

KEY INFORMATION

The subscript number tells you how many atoms of the same type are bonded. For example H_2O means there are two hydrogen atoms bonded to one oxygen atom in one molecule of water.

Small molecules

Carbon dioxide and water are examples of simple, small **molecules**.

You can often recognise a small molecule from its formula. It has two or three atoms joined together, with no charge.

Atoms	O	Mg	Cl	Na	S
Ions	O^{2-}	Mg^{2+}	Cl^-	Na^+	SO_4^{2-}
Molecules	H_2	Cl_2	CO_2	Br_2	H_2O

Notice that the molecules have no charge and more than one atom.

Substances that are made up of small molecules are usually gases or liquids at room temperature.

They have relatively low melting points and boiling points.

This is because simple molecules such as carbon dioxide and water have weak forces *between* the molecules. These weak forces easily break to allow the molecules to move randomly around and away from each other.

These substances do not conduct electricity because the molecules do not have an overall electric charge.

1 Explain why carbon dioxide has a boiling point of –78.5 °C.

gas

Figure 2.44 CO_2 at low temperatures and CO_2 above boiling point

Intermolecular forces

Carbon dioxide and water have simple molecules that have strong **covalent bonds** within the molecule. The carbon atom does not break its bonds with the oxygen atoms when it changes state. These are strong covalent bonds formed by electrons sharing.

These substances have only weak forces *between* the molecules (**intermolecular forces**). It is these intermolecular forces that are overcome, not the covalent bonds, when the substance melts or boils.

Covalent bonds are about 100 times stronger than weak intermolecular forces.

The intermolecular forces increase with the size of the molecules, so larger molecules have higher melting and boiling points.

Substance molecules	CO	CO_2	SO_2	SO_3
Melting point °C	–205	–55.6	–72	17
Boiling point °C	–191.5	–78.5	–10	45

2 **Suggest why the boiling point of carbon dioxide, CO_2, is higher than the boiling point of carbon monoxide, CO.**

3 **Propane, C_3H_8, has a melting point of –188 °C and butane, C_4H_{10}, –140 °C. Suggest why.**

4 **Nitrogen has a very strong triple covalent bond but a very low boiling point of –195.8 °C. Explain why.**

Predicting the properties of small molecules

There are weak *intermolecular* forces *between* molecules, which affect the physical properties of small molecules.

- They have weak intermolecular forces between the molecules so they are easy to separate and the substances have low melting points.
- There are no charge carriers available (neither **free electrons** nor free ions) so they do not conduct electricity.

However, it is difficult to predict trends in boiling points and melting points as there are other factors involved. Boiling point and melting point not only depend on the *size* of the molecule but on other factors such as its shape. So CH_4 (tetrahedral shape, at –164 °C) has a much lower boiling point than CO_2 (linear, at –78.5 °C) and the molecules cannot get as near to one another.

5 **Explain why pure water does not conduct electricity very well, but sea water does.**

6 **Pentane is a straight chain molecule. 2,2-dimethylpropane has exactly the same molecular formula but is more like a sphere. Explain which is likely to have the higher boiling point.**

7 **Fluorine, F_2, has a boiling point of –188°C and chlorine, Cl_2, a boiling point of –34°C. Suggest why they have different boiling points.**

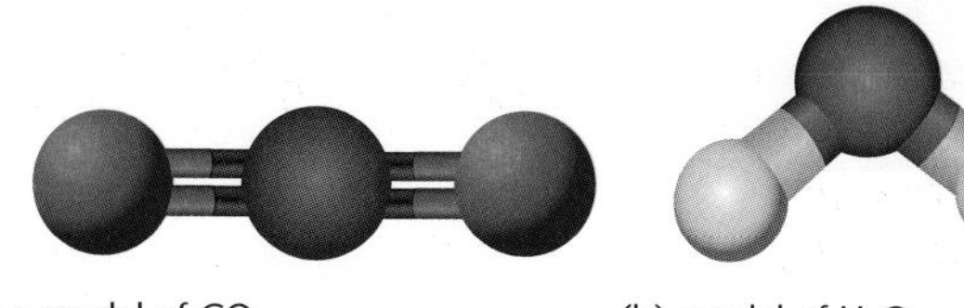

(a) a model of CO_2

(b) model of H_2O

Figure 2.46 The strong covalent bonds within molecules do not easily break.

DID YOU KNOW?

Small molecules with strong covalent bonds and weak intermolecular forces make up substances that boil on heating but do not undergo thermal decomposition.

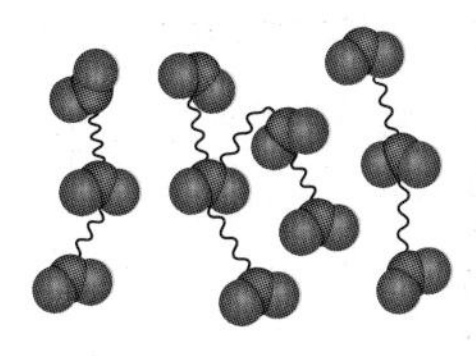

Figure 2.45 Forces *between* molecules have to be overcome for a substance to boil. These are *intermolecular* forces.

DID YOU KNOW?

Predicting boiling points is not as easy as it seems as there are other forces at work which depend on the *shape* of the molecule as well.

KEY INFORMATION

The atomic mass of C is 12 and of S is 32.

Covalent bonding

Learning objectives:

- recognise substances made of small molecules from their formula
- draw dot and cross diagrams for small molecules
- deduce molecular formula from models and diagrams.

> **KEY WORDS**
>
> covalent
> giant covalent structure
> polymer
> single bond

When two or more non-metal atoms bond together they *share* electrons. In this way they form small *individual* molecules. Such molecules can exist on their own, and don't have to be part of a large structure. They are freer to move around so the collection of molecules exists as liquids or gases.

Sharing electrons

When two or more non-metal atoms bond together they form molecules. Molecules are bonded by **covalent** bonds. These bonds are very strong. An example is a hydrogen molecule, H_2. Each atom has a single electron so the two atoms *share* their electrons as a pair.

This can be represented as:

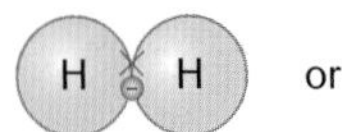

or H–H or H_2

Only one pair of electrons is shared so it is a **single bond**.

1 **Hydrogen and fluorine bond with a single bond. Draw a diagram to show this.**

Covalently bonded molecules

Other small molecules that are covalently bonded are hydrogen chloride, chlorine, water, ammonia and methane. These are bonded by single bonds when a pair of electrons is shared.

> **DID YOU KNOW?**
>
> Some covalently bonded substances may consist of small molecules, but some covalently bonded substances have very large molecules, such as **polymers**. (Topic 2.18). Some covalently bonded substances have **giant covalent** structures, such as diamond and silicon dioxide (Topic 2.17).

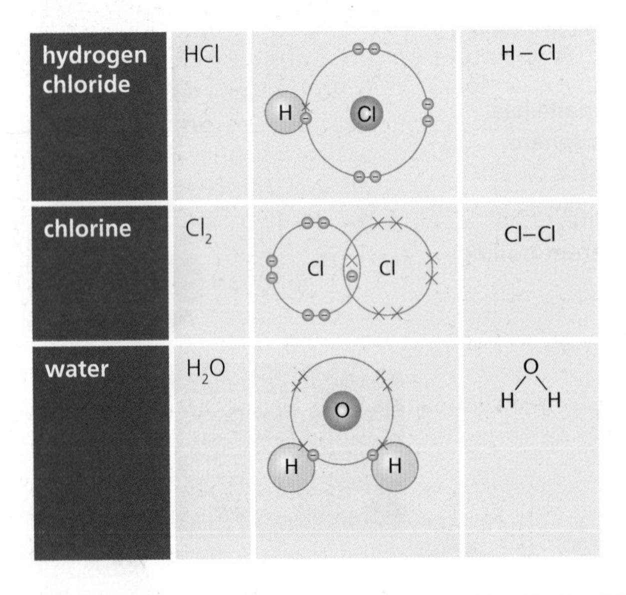

hydrogen chloride	HCl	[dot and cross diagram: H, Cl]	H – Cl
chlorine	Cl_2	[dot and cross diagram: Cl, Cl]	Cl–Cl
water	H_2O	[dot and cross diagram: O, H, H]	H–O–H (bent)

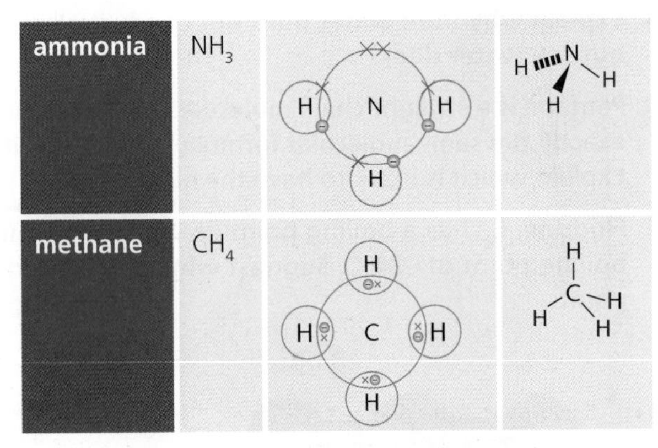

ammonia	NH_3	[dot and cross diagram: N, H, H, H]	H–N(–H)–H
methane	CH_4	[dot and cross diagram: C, H, H, H, H]	H–C(–H)(–H)–H

These 2D diagrams have their limitations as they do not show the 3D shape of the molecule.

Covalent bonds act in a particular direction. 3D ball and stick models or space-filling models can provide a better indication of the structures.

When atoms form covalent bonds they share pairs of electrons. This can be shown using a dot and cross diagram.

The outer electrons are written as pairs of electrons. The bond pair is written between the two symbols.

hydrogen	hydrogen chloride	chlorine
H H	H Cl	Cl Cl
water	**ammonia**	**methane**
O H H	H N H H	H H C H H

ball and stick HCl

space filling model HCl

ball and stick Cl_2

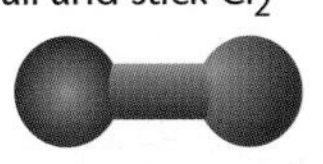

space filling model Cl_2

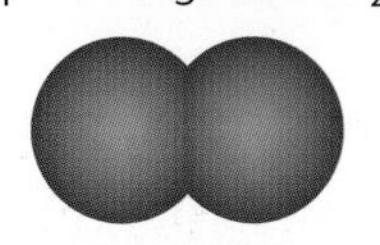

ball and stick H_2O

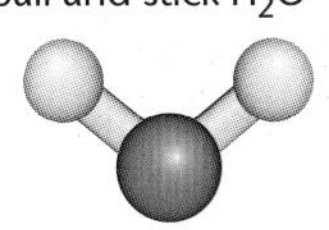

space filling model H_2O

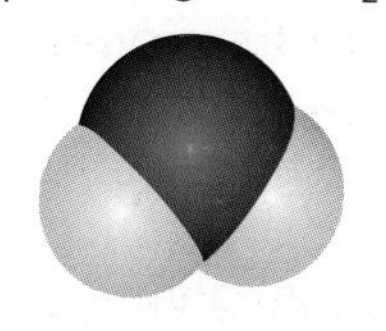

ball and stick NH_3

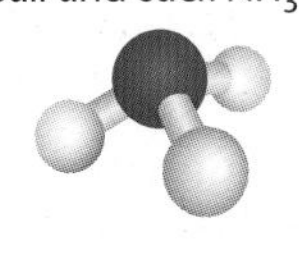

space filling model NH_3

ball and stick CH_4

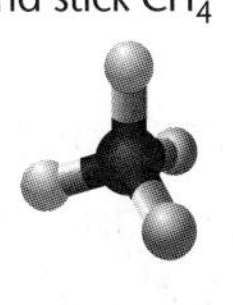

space filling model CH_4

2 **The displayed formula of hydrogen fluoride is H–F. Represent HF as a dot and cross diagram.**

3 **How many bond pairs are there in a molecule of ammonia?**

4 **Determine the molecular formula of methane.**

5 **Draw a dot and cross diagram for H_2S.**

Multiple bonds

One 'dot and cross' together represents a single bond, which is one bond pair of electrons. Two pairs of 'dots and crosses' side by side represent a double bond (as in O_2). Three electron bond pairs is a triple bond (as in N_2).

KEY INFORMATION

The inner shells of electrons are not shown. There is no charge needed after the square bracket because the molecule is neutral.

6 **Explain why there are two dots and two crosses between the oxygen atoms in diagram (e).**

7 **Use dot and cross diagram (e) to explain why a molecule of nitrogen N_2 has a triple bond.**

8 **Draw a dot and cross diagram to show the bonding in carbon dioxide CO_2 (Hint: it is a linear molecule with a C atom in the middle).**

(b) O═O

(d) space filling model O_2

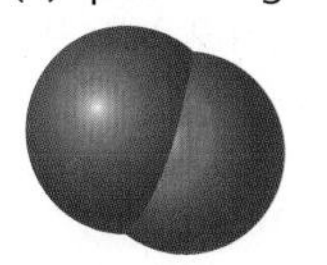

(e) dot and cross diagram O_2

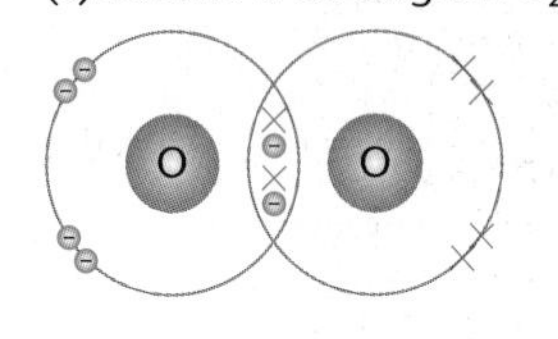

(b) N≡N

(d) space filling model N_2

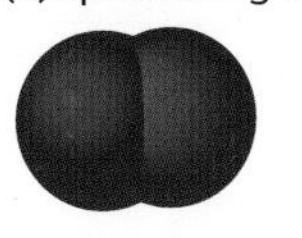

(e) dot and cross diagram N_2

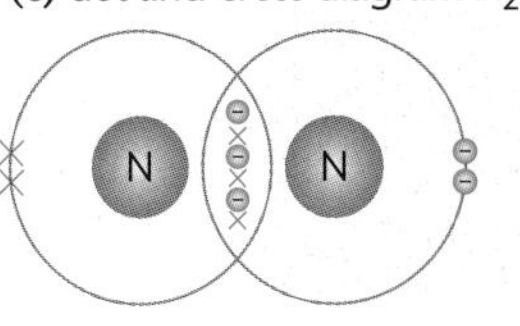

Giant covalent structures

Learning objectives:

- recognise giant covalent structures from bonding and structure diagrams
- explain the properties of giant covalent structures
- recognise the differences in different forms of carbon.

KEY WORDS

diamond
giant covalent structure
graphite
silicon dioxide

Amethyst, pink quartz and diamond are gems that have giant structures, with bonds in all directions, which make them hard.

Quartz is a giant covalent structure made of **silicon dioxide**. It is the second most abundant mineral in the Earth's crust. It is the biggest component of sand and can also be coloured with other substances, as in some gems. **Diamonds** are not abundant and are giant structures made of carbon.

Figure 2.47 Pure quartz is silicon dioxide which is colourless. It can be coloured naturally with iron or manganese to make amethyst.

Structure and bonding

Substances that consist of **giant covalent structures** are solids. All of the atoms in these structures are linked to other atoms by strong covalent bonds. These bonds must be overcome to melt or boil these substances. These substances with giant structures have very high melting points.

Examples of giant covalent structures are:

- diamond (a form of carbon)
- silicon dioxide (silica or quartz).

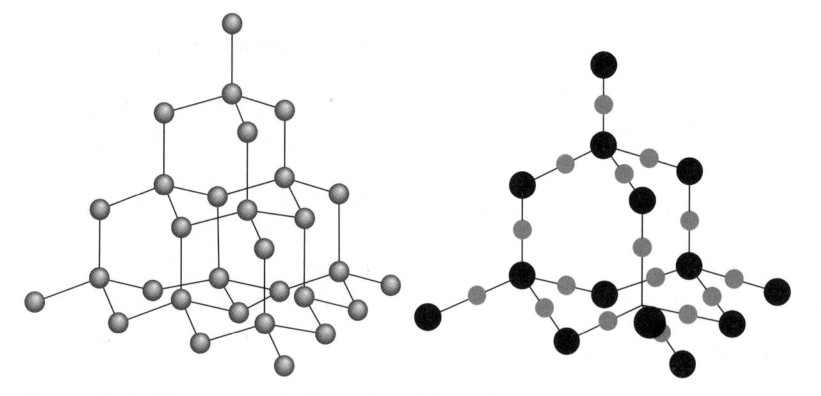

Figure 2.48 Diamond and silicon dioxide both have giant covalent structures. The ● represent C atoms, the ● represent Si atoms, the ● represent O atoms.

KEY INFORMATION

Remember you need to be able to recognise the different structures of carbon.

1. **Describe one difference and one similarity in the giant covalent structures of diamond and silicon dioxide.**
2. **Describe the difference between a molecule like water, H_2O, and a giant covalent structure like silicon dioxide.**

The table shows their physical properties.

Diamond	Silicon dioxide
lustrous (shiny), transparent and colourless	white crystalline solid
very hard	very hard
very high melting point	very high melting point
insoluble in water	insoluble in water
does not conduct electricity	does not conduct electricity

3. **Explain why the properties of silicon dioxide are quite similar to the properties of diamond.**
4. **Which two properties of diamond make it suitable for use in high temperature drilling tools?**

Three-dimensional structures

Diamond is not the only giant covalent structure that carbon forms. It also forms **graphite**. The structure of graphite is rings of carbon atoms in layers. The bonds do not act in all directions.

Covalent bonding usually means that all the available electrons are shared, so they do not conduct electricity. Graphite is an exception as graphite has delocalised electrons that can move through the layers of its structure.

Large structures such as quartz and diamond are hard because of a regular three-dimensional pattern in the bonding of their atoms.

The melting point depends on the energy required to break all the bonds joining the atoms or molecules. As the giant covalent structure has many bonds, in many directions, the energy needed will be high so the melting point will be high.

Covalent compounds are generally insoluble as the bonds are not ionic.

5. **Explain why graphite is not as hard as silicon dioxide.**
6. **Suggest why graphite bricks are suitable for use in furnaces.**

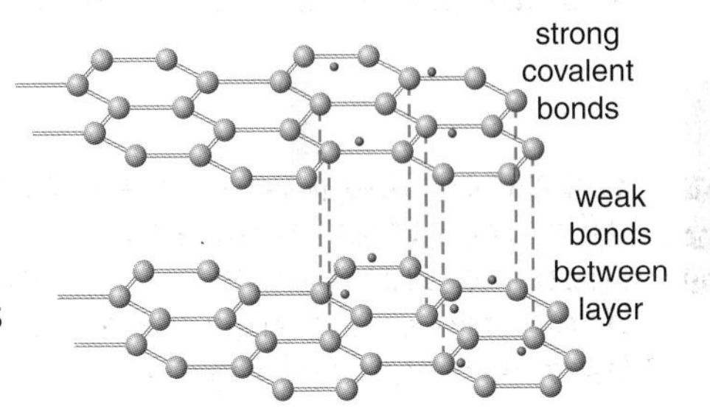

Figure 2.49 Graphite has a giant covalent structure, but it does conduct electricity.

DID YOU KNOW?

Diamond and graphite are both made of carbon atoms, but their structures are very different. These structures of the same element are called *allotropes*.

Diamond and graphite

Diamond and graphite are both made of carbon atoms, but their structures are very different.

7. **Suggest why graphite can be used in high temperature electrodes.**
8. **Suggest why diamond and silicon dioxide crystals do not conduct electricity.**

DID YOU KNOW?

Graphite has properties that work in one direction but not at right angles. Graphite is said to therefore be *anisotropic*.

Polymer structures

Learning objectives:

- identify polymers from diagrams showing their bonding and structure
- explain why some polymers can stretch
- explain why some plastics do not soften on heating.

KEY WORDS

intermolecular force
monomer
polymer
repeating unit

Polymers such as poly(ethene), PVC, PTFE and nylon are used in many areas of everyday life. These polymers are very large molecules. They are made in different ways and have different properties. Can we link their properties to their structure as chemicals?

Polymers

Polymers are very large molecules. The atoms in the polymer molecules are linked to other atoms by strong covalent bonds and so form a chain.

The forces *between* polymer molecules are weaker. These forces can be overcome so the polymer chains can move over each other so the polymer can be stretched, but are not so weak that the chains pull apart.

There are many forces between long polymer chains and so the substances are solids at room temperature.

1 Polythene is a solid at room temperature not a liquid. Suggest why.

strong covalent bonds

weaker forces between polymer chains

Figure 2.50 Strong bonds join atoms into a polymer chain. Weaker forces between the chains hold the chain together.

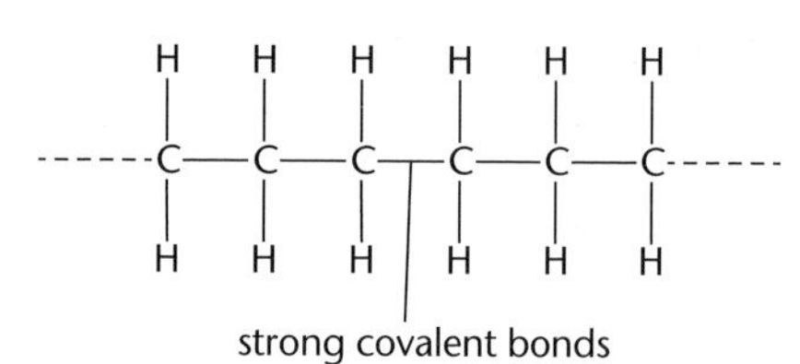

Figure 2.51 The bonds between the atoms in a polymer chain are strong covalent bonds.

Intermolecular forces

Chains of polymers are held together to make a bulk substance.

- The atoms of the **monomers** along the chains in a polymer are held together by strong covalent bonds. These are bonds within the polymer molecules and so are *intra*molecular bonds.
- The chains of the polymer are held together by weak forces of attraction. These are forces between polymer molecules and so are ***inter*molecular forces**.

Plastics are substances made from polymers such as poly(ethene), polyvinylchloride and nylon. They have weak intermolecular forces of attraction between the polymer molecules so they have low melting points and can be stretched easily as the polymer molecules can slide over one another.

Although the intermolecular forces between polymer molecules are weaker than the strong bonds along the chains, there are many of these forces so together they are relatively strong and so these substances are solids at room temperature. The weaker the force the lower the melting point.

KEY INFORMATION

Remember *intra*molecular bonds are bonds joining atoms *within* a molecule. *Inter*molecular forces are forces *between* chains of molecules.

KEY INFORMATION

Remember the two brackets have a bond through each one. This signifies that the bonds are continuing to join other repeat units.

Type of intermolecular force or bond	Property 1	Property 2
weak intermolecular forces of attraction	stretches easily	lower melting point
strong cross-links – chemical bonds between the chains	rigid	higher melting points

A polymer is a long chain and is represented by a **'repeating unit'**. In Figure 2.53 the repeating unit has two carbon atoms joined, with four hydrogen atoms, outside the bracket is an *'n'* – that means this unit is joined to itself *'n'* times into a very long chain.

2. **Explain how a polythene (poly(ethene)) bag stretches if carrying a heavy mass.**

3. **Suggest why the polymer melamine cannot stretch and is rigid.**

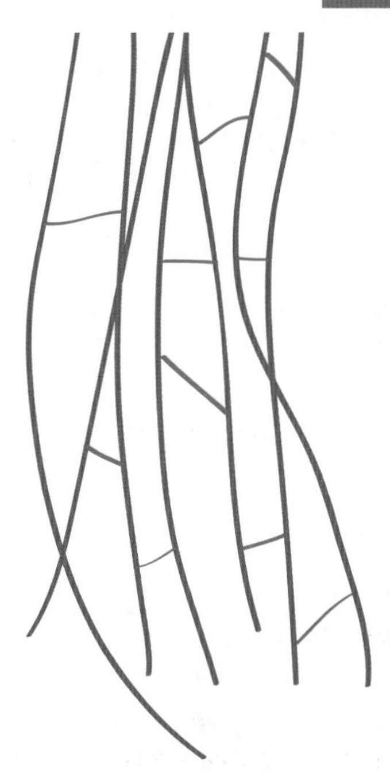

Figure 2.52 Polymer chains with weak intermolecular forces (as in Figure 2.49) can slide over one another, chains with cross-links, as in this figure, cannot slide.

Types of polymer

There are two main methods of making polymers. These are by processes of:

- addition
- condensation.

Addition polymers are long chains of units made from one molecule that has at least one double bond between carbon atoms. Examples of these polymers are poly(ethene), poly(propene), polyvinylchloride (PVC) and PTFE.

Their structure is represented by:

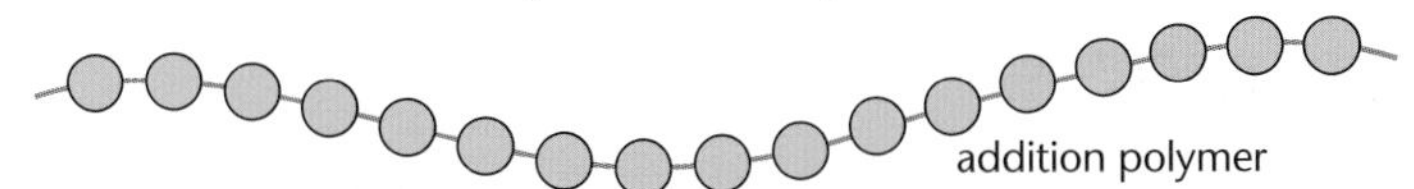

Figure 2.54 Model of polymers such as polyvinyl chloride

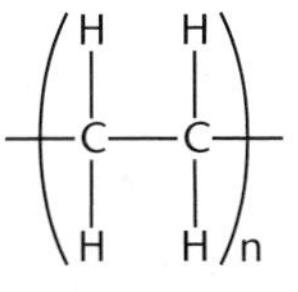

Figure 2.53 Show the repeating unit in brackets. *'n'* is a very large number.

4. **PVC is an addition polymer. CH_2CHCl is the monomer for poly(vinyl chloride), PVC. Draw the repeat unit for the polymer using this monomer. Use Figure 2.53 as a guide.**

HIGHER TIER ONLY

Condensation polymers are made of units using at least two different molecules (losing a water molecule in the process). Examples of these polymers are nylon and polyester.

Their structure is represented by:

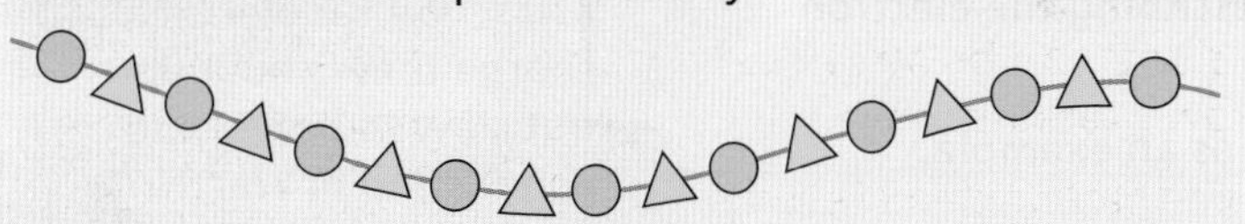

Figure 2.55 Model of polymers such as nylon

5. **Terylene is a polymer made from benzene–1,4–dicarboxylic acid and ethane–1,2–diol. It has a relatively high melting point.**
   - **a Explain whether it is a condensation or addition polymer.**
   - **b Suggest why it has a relatively high melting point.**

DID YOU KNOW?

Gore-Tex is a registered product made by expanding PTFE.

Figure 2.56 Gore-Tex is a popular polymer used by walkers.

Metallic bonding

Learning objectives:

- know that metals form giant structures
- explain how metal ions are held together
- explain the delocalisation of electrons.

KEY WORDS

delocalised electrons
metal ions
metallic bonds
'sea' of electrons

Think about copper, iron and gold. What is it about metals that makes them distinctive? Metals conduct electricity and can often be hammered into sheets, bent and twisted. This is because of the way that metal atoms are bonded together. Metallic bonds are strong. Metallic bonds involve the free movement of electrons through the giant structure.

Giant structures

Metals consist of giant structures of atoms arranged in a regular pattern.

The metals atoms have a small number electrons in their outer shell, which are free to move among the atoms.

The electrons move through the atoms making the bonds between the atoms strong.

These moving electrons are described as a **'sea' of electrons**. As they leave the atoms to move they leave behind a positive ion.

So the diagram of **metallic bonding** is drawn as:

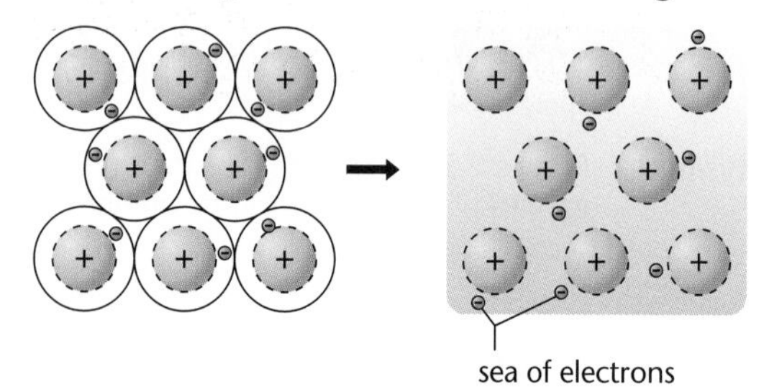

Figure 2.57 Metallic bonding

1. **Metals have a small number of electrons in their outer shell which are free to move. Show this by describing the electron pattern in magnesium.**
2. **Describe metallic bonding using lithium as an example.**

Holding a metal together

A metal is made of atoms that are close-packed and held together in a moving 'sea' of electrons.

The metallic bond is the strong electrostatic force of attraction between the positive **metal ions** and the 'sea' of delocalised electrons.

DID YOU KNOW?

We talk about *electrostatic attraction* for electrical charge that is built up but slow to move; and *conduction of electricity* for movement of charge. You know about static electricity from before when you rubbed balloons with woollen cloth and could stick them on a wall.

Figure 2.58 Close packed ions of a metal

The moving electrons are not localised to a particular atom but are *delocalised* and free to move through the whole structure.

The attraction between the **delocalised electrons** and the metal ions makes strong *metallic bonds*.

If the metals need to be melted then the metallic bonds need to be disrupted. This is done by heating the metal so giving the positive ions more energy so that they move with more energy. The electrons also move with more energy. In general the metals that only have one spare electron to contribute to the metallic bond melt at a lower temperature than those which donate two to the bond.

The sea of electrons is the key factor in one particular property of metals. The ability to conduct electricity depends on this sea of mobile, delocalised electrons. An electrical current is the movement of electrons through the lattice of ions.

All metals have this property, although some conduct electricity better than others.

Metal	lithium	calcium	aluminium	copper	silver
Relative electrical conductivity	1.08	2.98	3.50	5.96	6.30

3 **Explain the term delocalised electron.**

4 **Explain why silver is such a good conductor but solid silver oxide is not.**

HIGHER TIER ONLY

Consequences of electron delocalisation

A metal conducts electricity because delocalised electrons within its structure can move easily through it.

Metals often have high melting points and boiling points. This is because a lot of energy is needed to overcome the strong attraction between the delocalised electrons and the positive metal ions.

5 **Support the theory of the structure of metals by explaining how a metal expands when it is heated.**

6 **Suggest why aluminium is a better electrical conductor than sodium.**

KEY INFORMATION

This explanation is not the whole story though, there are other factors involved, such as lattice structures so be careful when making predictions.

KEY INFORMATION

Remember why metals form positive ions. Look at where metals are found in the periodic table and remember the electron pattern for metals.

Properties of metals and alloys

Learning objectives:

- identify metal elements and metal alloys
- describe the purpose of a lead–tin alloy
- explain why alloys are harder than pure metals.

KEY WORDS

alloy
distort
ductile
malleable

Making alloys has been carried out for centuries. That early human history is divided into the Stone age, the Bronze age and then the Iron age shows how important bronze was. Bronze was the first metal alloy – copper and tin were both extracted from their ores in rocks thousands of years ago and later the mixture of these metals was used to make bronze.

Figure 2.59 A bronze shield from the Bronze Age of the Greek Empire. Objects, tools and weapons were made from bronze as it is harder than either tin or copper.

Metals in alloys

Pure copper, gold, iron and aluminium are too soft for many uses. These metals are mixed with other metals to make **alloys**. Most metals we use every day are alloys.

An alloy is a mixture of a metal element with another element.

Alloy	bronze	brass	steel
Elements in alloy	copper and tin	copper and zinc	iron and carbon

Alloys have different properties to the metals that are in them.

Most metals have high melting and boiling points.

Sometimes a metal is needed that has a lower melting point that can then solidify to make a join. This metal alloy is called solder.

Solders are made of lead and tin. Solder has a melting point that is lower than either lead or tin.

Metal	lead	tin	solder
Melting point °C	327	232	183

1. **What is the difference between the composition of the alloys brass and bronze?**
2. **Suggest some advantages of using the alloy steel rather than iron.**
3. **'Alloys never contain non-metals.' Show whether this statement is true.**

Figure 2.60 Solder is an alloy with a lower melting point than tin or lead. If melted it can be used to make or mend metal objects or wiring.

Giant structures

In pure metals, atoms are arranged in layers. Pure metals are too soft for many uses and so are mixed with other metals to make alloys which are harder.

Metals have giant structures of atoms with strong metallic bonding. Overcoming this bonding takes a great deal of energy so this means that most metals have high melting and boiling points.

Through this giant structure of metal ions moves the 'sea of electrons' which gives the metal the ability to conduct electricity. An electrical current is the movement of delocalised electrons through the lattice of ions.

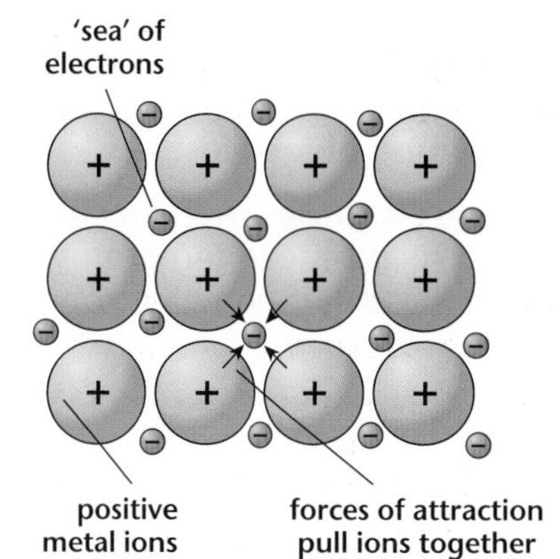

Figure 2.61

In the giant structures of metals, the layers of atoms are able to slide over each other. This means metals can be bent and shaped. They can be hammered flat (metals are **malleable**) and they can be pulled into a wire (metals are **ductile**) as the atoms can be moved over one another.

(a) 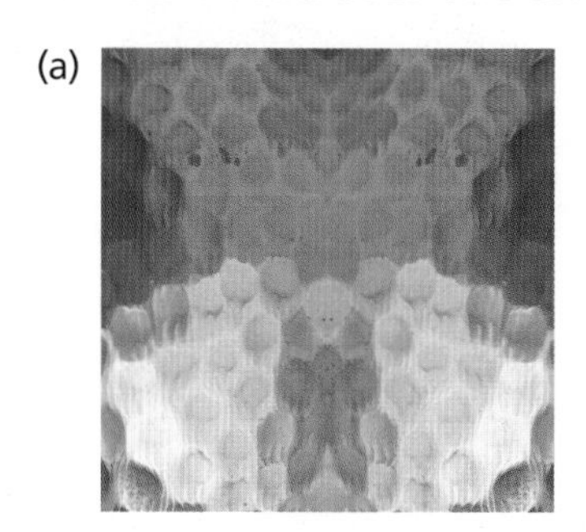

(b) 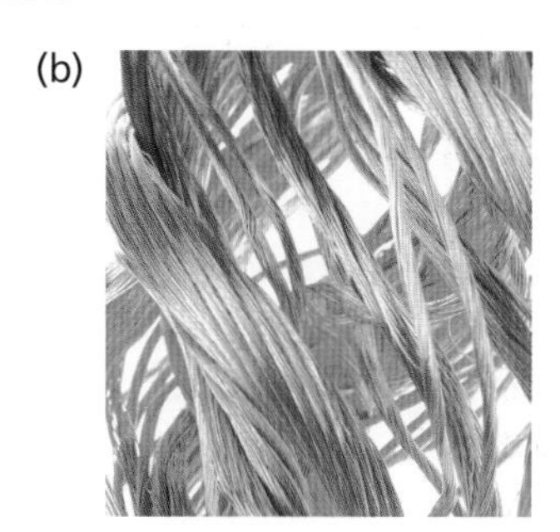

Figure 2.62 Copper is (a) malleable and (b) ductile.

The different sizes of atoms in an alloy **distort** the layers in the structure, making it more difficult for them to slide over each other, so alloys are harder than pure metals.

4 **Explain, in terms of the packed atom structure, why copper metal can be extruded into wires.**

5 **Explain why copper can conduct electricity.**

6 **Sterling silver contains 92.5% silver, the remainder being copper.**

a **Explain whether pure silver or sterling silver is more malleable.**

b **Calculate the mass of copper in 1.0 g of sterling silver.**

DID YOU KNOW?

Smart alloys are alloys that can return to their original shape if heated to a different temperature. This is called a 'shape memory'. Bendable glasses can be made from frames consisting of smart alloys. One smart alloy is 'nitinol'. It is made from nickel and titanium.

Aluminium

Aluminium is a very useful metal for use in aircraft as it has such a low density and (once it forms a layer of oxide) it is resistant to corrosion. However, it is too soft if it is not alloyed with other metals. The other metals used can be copper, magnesium, manganese and tin. For example aluminium and copper alloy is called duralumin.

Figure 2.63 Layers of alloy atoms, which are distorted layers due to the different sizes of the atoms.

7 **Explain, in terms of atoms, why alloys make the main metal harder and more useable.**

8 **Explain why adding copper to aluminium makes an alloy which is harder than both the constituent metals.**

9 **Copper is a very good electrical conductor. Suggest why alloying copper decreases its conductivity.**

REMEMBER!

Think about the different sizes of atoms

KEY CONCEPT

The outer electrons

Learning objectives:

- recognise when electrons transfer
- recognise when atoms share electrons
- predict when electrons are transferred most easily.

KEY WORDS

transfer
share
electrons
outer shell

Stable atoms

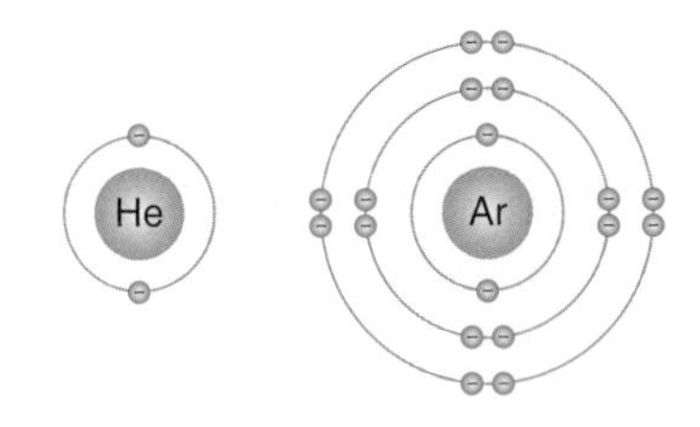

The noble gases are all very unreactive. Their atoms are very stable and do not react with other atoms. This is because their outer shells contain eight electrons, (except He which contains two electrons). This stable number of electrons in the outer electron shell means there is no tendency to transfer electrons.

Less stable atoms

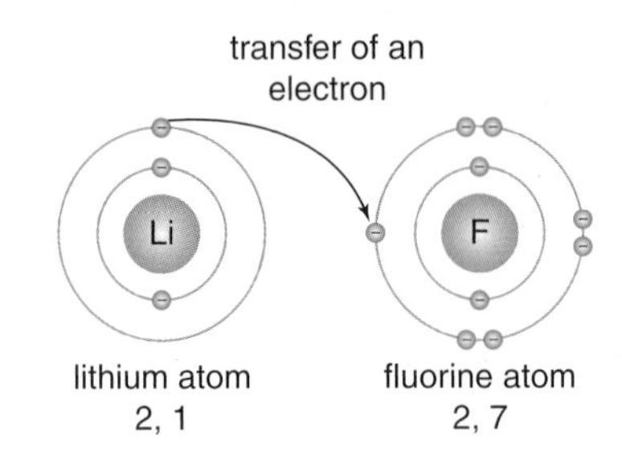

All other atoms are less stable. Their electrons move or share with other electrons to try to become as stable as the atoms with 8 electrons in their outer shell.

Chemical reactivity and chemical reactions depend on the number of electrons in the outer shell. (However, note that this is not always true in reactions of transition metals.)

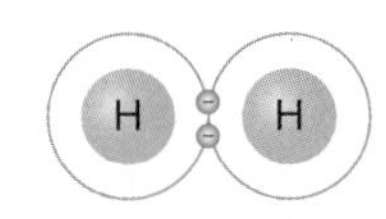

Three things can happen to the electrons in the outer shell:

- they can be transferred to the outer shell of another atom
- they can have other electrons added to their outer shell from another atom
- they can be shared with another atom.

1. **Name two noble gases that have eight electrons in their outer shell.**
2. **Predict the number of 'outer' electrons lithium has.**

Transferring or sharing?

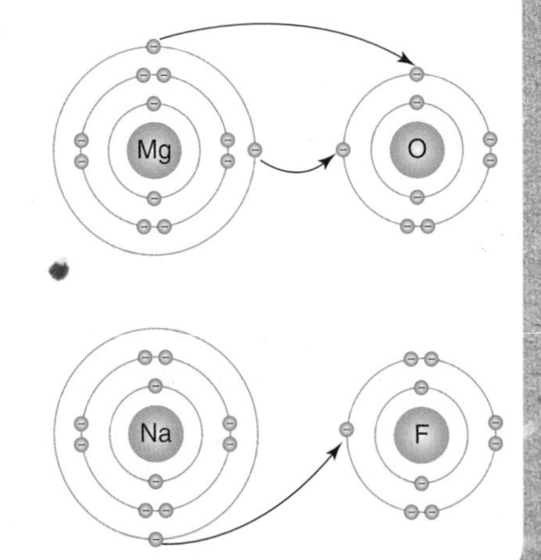

If an atom has one or two electrons in its outer shell these electrons will transfer out. They will transfer to the outer shell of another atom.

Adding in other electrons

If an atom has six or seven electrons in its outer shell, electrons from other atoms will add in to the 'spaces' to make up a stable outer shell. They will transfer in from the outer shell of another atom.

If an atom has an unstable number of electrons in its outer shell this atom will share its electrons with electrons from other atoms. Common examples of atoms that do this are carbon and hydrogen. They also share electrons with each other and with oxygen atoms.

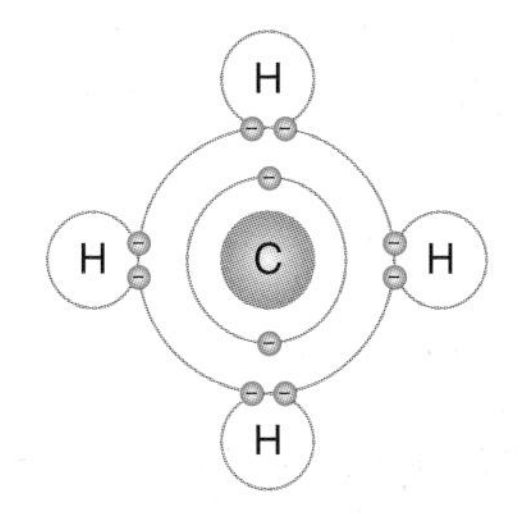

3 **Suggest how many 'spaces' in the outer shell an oxygen atom has.**

4 **Predict how many electrons are shared by two fluorine atoms.**

Transferring electrons

Some elements are more reactive than others. This can be because their outer electrons transfer *out* more easily than others or transfer *in* more easily than others.

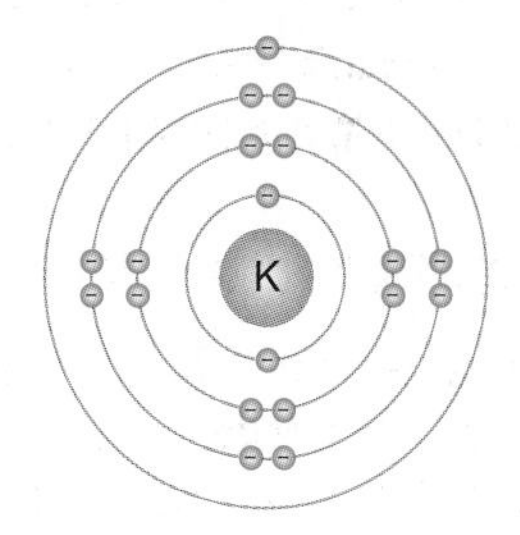

Potassium and lithium both need to lose one electron to become stable atoms.

Potassium reacts more quickly than lithium.

This is because the outer electron is further away from its nucleus in a potassium atom than in a lithium atom.

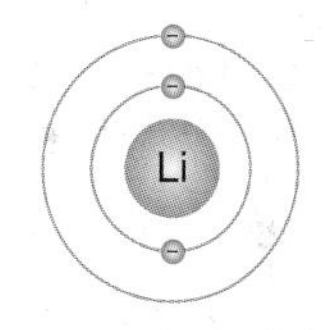

The 'pull' on the electron by the potassium nucleus is less than the 'pull' on the electron by the lithium nucleus. The electron of potassium is more easily lost (transferred out).

So potassium is more reactive.

Fluorine and bromine both need to gain one electron to become stable atoms.

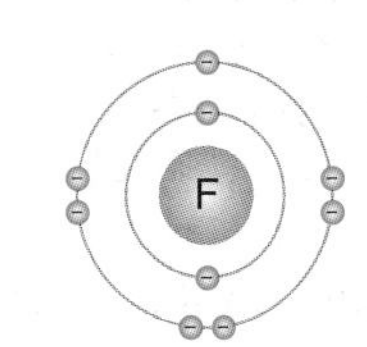

Fluorine reacts more vigorously than bromine.

This is because the outer electron shell is nearer to its nucleus in a fluorine atom than in a bromine atom.

The 'pull' on the electron coming in to fill the 'space' in the fluorine outer ring by the fluorine nucleus is more than the 'pull' on the electron by the bromine nucleus. The electron of fluorine is more easily gained (transferred in). So fluorine is more reactive.

5 **Explain, using ideas about 'the pull of the nucleus' why sodium is more reactive than lithium.**

6 **Explain the reactivity of chlorine within Group 7 in terms of the 'pull of the nucleus'.**

The periodic table

Learning objectives:

- explain how the electronic structure of atoms follows a pattern
- recognise that the number of electrons in an element's outer shell corresponds to the element's group number
- explain that the electronic structure of transition metals position the elements into the transition metal block.

KEY WORDS

electron shells
energy levels
group
period

We know that the periodic table is arranged in rows and columns and the elements are written in order of their atomic number. So why are all the elements in the last column unreactive gases? Why are all the elements in the first column highly reactive metals? The answer lies in the pattern of their electrons.

The order of elements and electron patterns

As we have seen, the elements in the periodic table are arranged in order of atomic number. Atomic number is the number of protons in an atom. As atoms are neutral, the atomic number also gives the number of electrons in an atom.

For example, the atomic number of hydrogen is 1, carbon is 6 and sodium is 11. This means that hydrogen is the first element in the table, carbon is the sixth and sodium is the eleventh.

We have also seen that electrons occupy energy levels (or shells). Each element has a pattern of electrons (known as its electronic structure) that is built up in a particular order.

The electronic structure of each of the first 20 elements can be worked out using:

- the atomic number of the element
- the maximum number of electrons allowed in each shell.

The third shell takes up to eight electrons before the fourth shell starts to fill. Element 20, therefore, has the electronic structure 2,8,8,2.

KEY INFORMATION

Remember:

- the first shell of electrons carries up to 2 electrons
- the second shell carries up to 8 electrons.

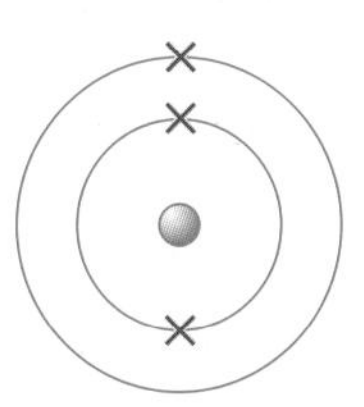

Figure 2.64 The electron pattern of a lithium atom.

1. **Which element has the atomic number 13?**
2. **Which element has an electronic structure of 2,8,7?**

Arrangement of groups and periods

The first row of the periodic table contains the elements hydrogen and helium. These two elements only have electrons in the first shell (energy level).

Lithium's third electron goes into the next electron shell. Lithium starts a new row in the periodic table. This second row is called the second **period**.

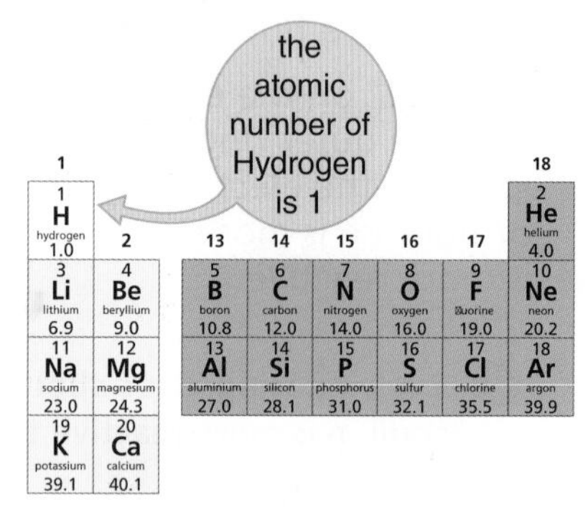

Figure 2.65 Lithium has an atomic number of 3. It has three protons and three electrons.

Let's consider an atom of the element that has the electronic structure of 2,8,2. We can work out that it is in the third row of the periodic table. This atom has three electron shells. It has two electrons in its outer shell. Its atomic number is 12. (You can work this out by adding up the number of electrons.) Looking at the periodic table the atom is of the element magnesium, 12.

So the element is therefore, in the third **period**.

Looking again, this atom only has two electrons in its outer shell, so it is in the second column. A column is known as a **group**. This second column is known as Group 2.

A column is known as a group. The group number refers to the number of electrons in the outside shell.

We have seen that the electron pattern for Li is 2,1. We can work out that the pattern for Na (11 electrons) is 2,8,1 and K (19 electrons) is 2,8,8,1. All of these elements have one electron in their outside shell. They are all in the first column. This column is known as Group 1.

3 **What is the pattern of electrons in the atom of the element with an atomic number 16? Identify the element, its group and its period.**

4 **To which group and period does the element chlorine belong?**

KEY INFORMATION

The periodic table is arranged in rows (called periods). Each period number is the same as the number of electron shells. Each period contains the elements whose outside shell of electrons is 'filling up'.

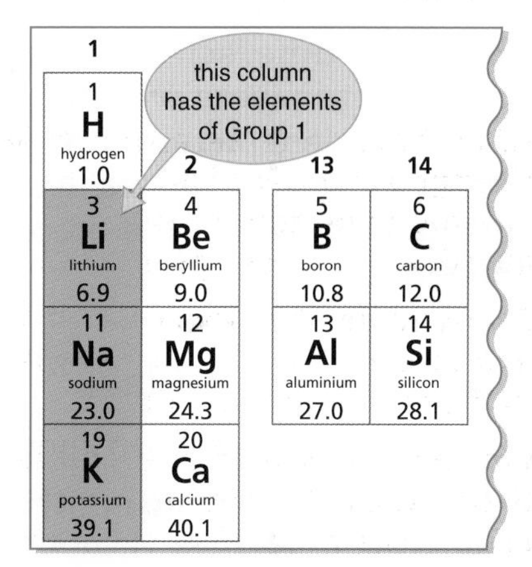

Figure 2.66 These elements are in the first column as they all have one electron in their outside shell.

Electronic structure and behaviour of elements

Element 20 has the electronic structure 2,8,8,2, and is in Group 2.

The fourth shell can take up to 18 electrons, but the next element (element number 21) is not in Group 3. Instead the next 10 elements (numbers 21 to 30) in period 4 are part of the first **transition element period**. Transition elements have characteristics that are similar to each other. For example, iron behaves more like copper than sodium.

Elements in groups often behave similarly to other elements in that group:

- the elements in Group 1 are all highly reactive metals
- the elements in Group 7 (the halogens) all react with Group 1 metals to make salts
- the elements in Group 0 are all unreactive (or noble) gases.

This pattern is because the *elements in each group have the same number of electrons in their outer shells.*

Use the periodic table to help you to answer these questions.

5 **What are the electronic structures of F and Cl? Why are they both found in Group 7?**

6 **Identify the elements in Group 6.**

Developing the periodic table

Learning objectives:

- describe the steps in the development of the periodic table
- explain how Mendeleev left spaces for undiscovered elements
- explain why the element order in the modern periodic table was changed
- explain how testing a prediction can support or refute a new scientific idea.

KEY WORDS

periodic
predictions
properties
patterns

Although some elements have been known about as parts of compounds since ancient times, many were only first isolated as elements in the last two centuries. For a long time, scientists trying to find patterns were working with an incomplete picture. It was like trying to do a 100-piece jigsaw puzzle with half of the pieces missing.

5	6	7	8	9	10	11	12
23 **V** vanadium 50.9	24 **Cr** chromium 52.0	?	26 **Fe** iron 55.8	27 **Co** cobalt 58.9	28 **Ni** nickel 58.7	?	30 **Zn** zinc 65.4
?	42 **Mo** molybdenum 95.9	43 **Tc** technetium	44 **Ru** ruthenium 101.1	?	46 **Pd** palladium 106.4	47 **Ag** silver 107.9	48 **Cd** cadmium 112.4
73 **Ta** tantalum 180.9	74 **W** tungsten 183.8	?	76 **Os** osmium 190.2	77 **Ir** iridium 192.2	?	79 **Au** gold 197.0	?

Figure 2.67 The element puzzle

The timeline of ideas

Some elements have been known since ancient times.

By 1829 Döbereiner had noticed that sometimes three elements had similar **properties**. He noticed **patterns** with:

lithium, sodium and potassium	calcium, strontium and barium	chlorine, bromine and iodine

These were called 'Döbereiner triads'.

Figure 2.68 Döbereiner

In 1860 a new list of more accurate atomic weights was published.

In 1865 John Newlands noticed that when he put the elements in atomic weight order (even though some then seemed to be in the wrong place) that there was often a pattern of similar properties every eight elements. He called his new *theory* the 'law of octaves'.

Döbereiner and Newlands noticed *patterns* in the properties of elements but did not make **predictions**.

REMEMBER!

You need to be able to explain that if germanium had not fitted into Mendeleev's pattern then the evidence would have *refuted* his predictions, but instead, it did fit his pattern so the new evidence *supported* his predictions.

1. **Describe the 'law of octaves'.**
2. **Suggest why there were so few elements for Döbereiner to test his theory on in 1829.**

Allowing for predictions

By 1869 the Russian scientist, Dmitri Mendeleev had also put the elements in order of their atomic weights. He also saw that some elements seemed to be in the wrong order. But crucially he:

- decided to swap some elements round so that the patterns of chemical behaviour fitted better
- was able to imagine that there were undiscovered elements
- brilliantly decided to leave gaps in his **periodic** table for later discoveries and used the patterns of chemical behaviour to decide where to leave these gaps.

He gave special names to these unknown elements in his gaps. He took the name of the element above the gap in that group and put the prefix 'eka' in front of the name. One unknown was eka-silicon (beyond silicon). This element was eventually discovered and was named germanium. Altogether three of these new elements were discovered within Mendeleev's lifetime.

Figure 2.69 Dmitri Mendeleev

3 **Mendeleev named one element eka-aluminium. It was later discovered. Identify the element.**

4 **Explain why some elements appeared to be in the wrong order.**

Discovering the unpredictable

Mendeleev died in 1907 and so did not live to see the discovery of sub-atomic particles, which gave the final evidence allowing the modern periodic table to be developed in the order of the *atomic number* of elements. His theory was supported, not refuted, by later evidence.

His theory was developed in 1869. Evidence from 1932 finally supported his theory, 63 years later. The evidence of isotopes finally explained why the order based on atomic weight was incorrect. The discovery of the neutron explained why the order of elements in the modern periodic table needs to be by the number of protons.

DID YOU KNOW?

Mendeleev did not win a Nobel Prize; however, he does have an element named after him. Look up which number this is.

5 **Mendeleev left gaps and made predictions for undiscovered elements in the his periodic table. Explain how the discovery of germanium supported this approach.**

6 **Search for an image of a new periodic table.**

a **Give the atomic number of the last element before the central block begins.**

b **Identify the missing elements in Figure 2.67.**

7 **Identify two elements in the periodic table that would be in the wrong position if they were ordered by atomic mass.**

8 **A student stated that 'The modern periodic table is complete'. Briefly explain whether you agree with the student.**

Diamond

Learning objectives:

- identify why diamonds are so hard
- explain how the properties relate to the bonding structure and in diamond
- explain why diamond differs from graphite.

KEY WORDS

diamond
directional bonds
electrical non-conductor
tetrahedral bonding

Diamonds **are made from carbon and are the hardest natural substance known. This makes them ideal for industrial cutting tools and grinding wheels. When diamonds form they have cleaving planes that allowing them to be shaped and cut to give reflective surfaces that make them one of the most beautiful jewels.**

Figure 2.70 Diamonds of beauty and diamonds of use

Structure of diamonds

Diamonds are made of carbon atoms. Each carbon atom forms four bonds.

The four bonds form as far from each other as possible. They form in the shape of a tetrahedron.

In diamond, each carbon atom forms four covalent bonds with other carbon atoms in all directions.

A giant structure is made.

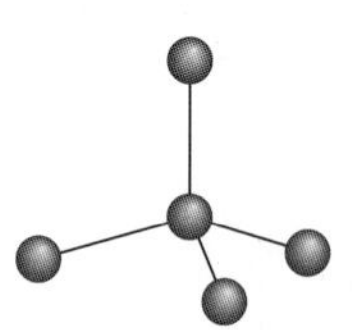

Figure 2.71 A carbon atom model showing four bonds in the shape of a tetrahedron.

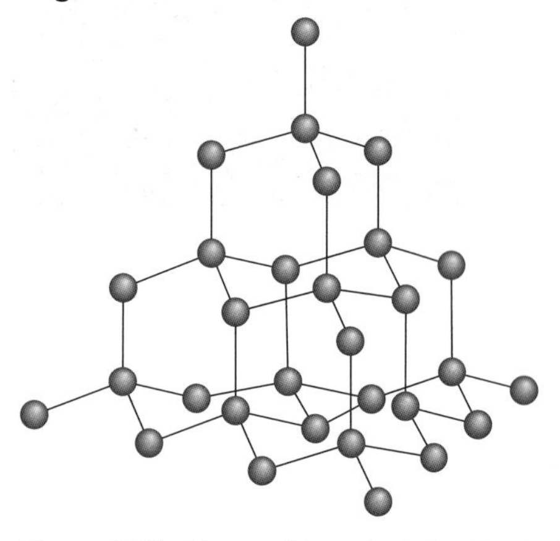

Figure 2.72 Diamond is a giant structure with bonds in all directions.

Diamond is very hard, has a very high melting point and does not conduct electricity. There are many strong bonds in diamond in different directions in a 3D structure, which makes diamond very hard.

1. **Describe how a carbon atom bonds.**
2. **Explain why diamonds are so hard.**
3. **Both silicon dioxide, SiO_2, and diamond are giant covalent structures. Suggest why silicon dioxide has a lower melting point than diamond.**

Covalent structures

Each carbon atom in diamond makes four covalent bonds with other carbon atoms in all directions in a giant covalent structure, so diamond is very hard. As all the bonds need to break for it to melt, this requires a great deal of energy. Diamond has a very high melting point. All the outer electrons of carbon are used so it does **not conduct electricity**.

4 Explain the difference in electrical properties between diamond and graphite.

5 Explain why each carbon atom in diamond can form four covalent bonds.

DID YOU KNOW?

Diamond does, however, conduct thermal energy. It does this by transmitting the energy down its covalent bonds as it is in a rigid lattice.

Atomic structure and covalent bonding

Carbon atoms have four electrons in the outer electron shell.

In diamond every carbon atom is joined to four others in a three-dimensional **tetrahedral** lattice. The atoms join together by strong covalent bonds that involve electron sharing. This arrangement gives strength in all directions, and requires a large amount of energy to break the bonds, giving a very high melting point of 3350 °C.

No free electrons exist in the structure, so diamond does not conduct electricity.

Diamonds are useful as cutting tools and are so rare and expensive that techniques were developed to create synthetic diamonds known as industrial diamonds. These are not as optically perfect as natural cut diamonds but have the same useful properties and are used in many industries as it is the hardest substance.

KEY INFORMATION

Carbon atoms in diamond make strong bonds in all directions which is why diamonds are hard.

Figure 2.73 The properties of industrial diamonds make them ideal as cutting tools if not as jewels.

6 Explain how the properties of diamond make it so useful as an abrasive for grinding other materials.

Graphite

Learning objectives:

- describe the structure and bonding of graphite
- explain the properties of graphite
- explain the similarity to metals.

KEY WORDS

delocalised
electrons
graphite
hexagonal rings
weak bonds

Graphite is a carbon structure that is black and slippery and sticks to paper. It makes good 'lead' for pencils.

The slipperiness also makes graphite a good lubricant, even though it is a solid. Powdered graphite is often used to lubricate door locks and is also used in furnaces and as brake linings. It is a good electrical conductor so it is used in batteries and as powder in carbon microphones.

DID YOU KNOW?

Graphite pencils were first made in 1565. All the graphite came from a mine near the Honister Pass in Cumbria. Clay was added to make different grades of pencil hardness. The graphite in pencils was called 'lead' because the Roman writing implement, the stylus (used for writing on wax tablets), was made of the metal lead.

Graphite layers

Like diamond, graphite is also made from carbon atoms. In graphite the carbon atoms are arranged differently. The carbon atoms only make three bonds with other carbon atoms.

The carbon atoms bond to make six-sided rings. The rings stack to make layers.

Graphite can be easily cut across the layers but not through the layers.

1. **State the number of bonds that carbon makes in a diamond b graphite.**
2. **Graphite can be used as a lubricant. State what this tells you about the forces between layers.**

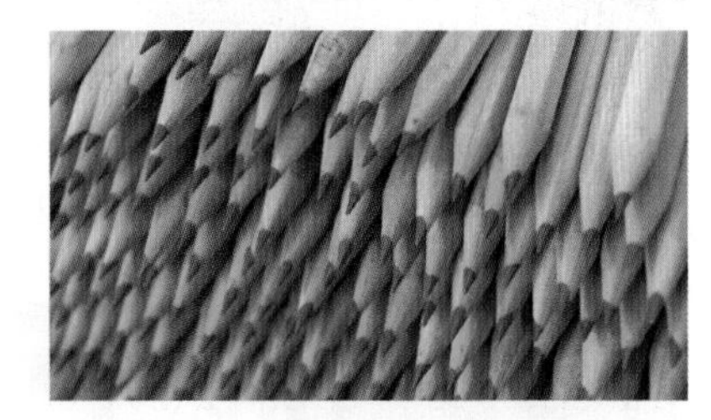

Figure 2.74 Graphite is slippery and can leave marks on paper.

Structure and properties

In graphite, each carbon atom forms three strong covalent bonds with three other carbon atoms, forming layers of **hexagonal rings** which have no covalent bonds between the layers.

As the bonds need a lot of energy to break, graphite has a high melting point.

Only **weak bonds** hold the layers together so the layers are free to slide over each other. There are no covalent bonds *between* the layers and so graphite is soft and slippery.

In graphite, only *three* **electrons** from each carbon atom form strong covalent bonds with electrons from other carbon atoms. The remaining one electron of each carbon atom is ***delocalised***. These delocalised electrons allow graphite to conduct electricity.

The delocalised electrons move easily along the layers. This is similar to the way that delocalised electrons move in metals. This is why both graphite and metals can conduct electricity. Diamond has no delocalised electrons so cannot conduct electricity.

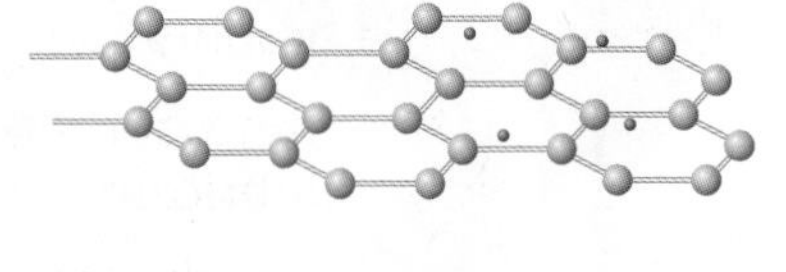

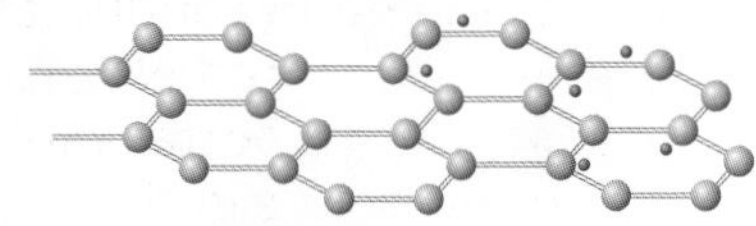

Figure 2.75 Graphite is made of hexagonal rings that stack as layers.

Graphite is also a good conductor of thermal energy, like metals, due to its delocalised electrons.

3. **Compare the organisation of electrons in diamond and graphite.**
4. **Suggest why graphite is a good thermal conductor.**

Uses related to structure

Graphite has a layered arrangement in which each carbon atom is covalently bonded to three others, forming single layers of regular hexagons.

This formation means each carbon atom has an unshared electron in its outer shell, free to move anywhere along the layer. So graphite is an electrical conductor and can be used as electrodes in electrolysis.

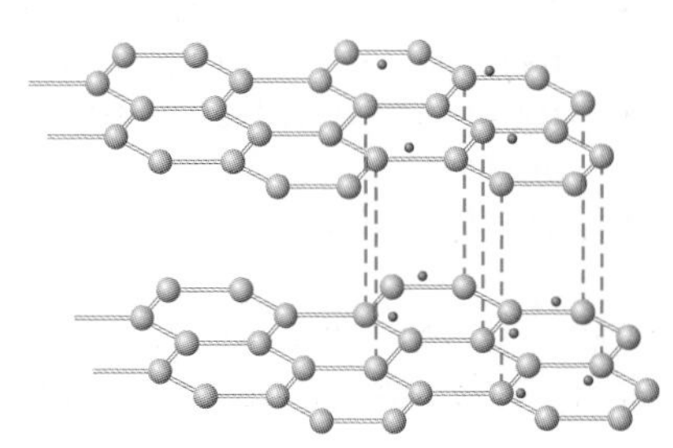

Figure 2.76 Delocalised electrons can move along layers so graphite is able to conduct electricity.

Graphite has a similar melting point to that of diamond so it is stable at high temperature. Due to these two properties graphite can also be used as electrodes in the electrolysis of *molten* electrolytes.

Figure 2.77 Graphite used as electrodes in a molten electrolyte

DID YOU KNOW?

Graphite splits cleanly between the layers, but is very difficult to cut across the layers.

The distance between the layers is greater than between bonded atoms and there are no bonding electrons, so the layers in graphite are only weakly attracted to each other. This means that when a force is applied, the layers can slide over each other, which means that graphite can act as a lubricant.

KEY INFORMATION

The weak bonds between the layers of hexagonal rings break and allow the layers to slip over one another.

Due to its high melting point it can also be used as a high-temperature lubricant.

Graphite is a good conductor of thermal energy, like metals, due to its delocalised electrons. Diamond has no delocalised electrons but is a better thermal conductor than graphite due to its strong bonds in the 3D lattice structure.

5. **Explain why (a) graphite is similar to a metal and (b) why graphite is classified as a non-metal.**
6. **Explain why graphite can be used as electrodes in molten electrolytes.**
7. **Explain why graphite has a similar melting point to diamond.**

Graphene and fullerenes

Learning objectives:

- explain the properties of graphene by its structure and bonding
- recognise graphene and fullerenes from their bonding and structure
- describe the uses of fullerenes, including carbon nanotubes.

KEY WORDS

cylindrical
fullerene
graphene
nanotube

Carbon does not only exist in the structures of diamond and graphite but also as ball-shaped or cylindrical-shaped structures called fullerenes. The structure of the first one to be discovered looked like a football and was called Buckminsterfullerene and other smaller ones are called 'buckyballs'. The tiny cylindrical tubes are fullerene nanotubes.

Graphene and fullerenes

Graphene is a single layer of graphite and so is one atom thick. It has properties that make it useful in electronics and composites.

It is made up of hexagonal rings of carbon atoms connected to one another by strong bonds.

Carbon atoms in rings can also form hollow 3D shapes. The first one found had 60 carbon atoms. It had rings of six carbon atoms and rings of five carbon atoms. It was named Buckminsterfullerene. Structures of this type are known as **fullerenes**.

Fullerenes are molecules of carbon atoms with hollow shapes. The structure of fullerenes is based on hexagonal rings of carbon atoms but they may also contain rings with five or seven carbon atoms.

Buckminsterfullerene (C_{60}) has a spherical shape, other fullerenes can be in the shapes of spheres or tubes.

There are already a number of medical uses for them but their potential use is still being researched.

Figure 2.78 Graphene – hexagonal rings of carbon one atom thick.

Figure 2.79 A traditional football can be used as a molecular model for C_{60}.

1. Compare the structures of graphene and fullerenes.

Carbon nanotubes

Fullerenes can be used for drug delivery into the body, as lubricants, and as catalysts. They can act as hollow cages to trap other molecules. This is how they can carry drug molecules around the body and deliver them to where they are needed and trap dangerous substances in the body and remove them.

Carbon can also be used to make very small structures called **nanotubes**. Carbon nanotubes are **cylindrical** fullerenes. They have very high length to diameter ratios. Their properties make them useful for nanotechnology, electronics and materials. Some of their special properties are:

- high tensile strength
- high electrical conductivity
- high thermal conductivity.

DID YOU KNOW?

C_{60} was named after the architect Buckminster Fuller as it looked like the geodesic dome that he created.

They are useful as:

- semi-conductors in electrical circuits
- catalysts
- for reinforcing materials, such as in tennis rackets.

Their potential has not yet been fully developed and new uses are being explored all the time.

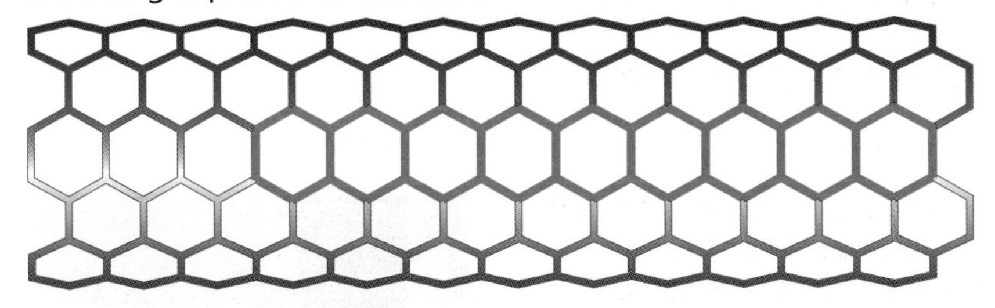

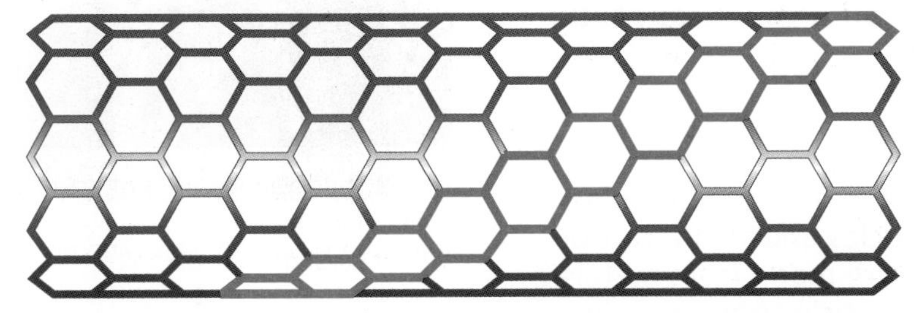

Figure 2.80 Carbon nanotubes. How many carbon atoms are in each ring?

Nanotubes can be used in catalyst systems because atoms of the catalyst can be attached to the nanotubes. The nanotube has a large surface area, so there is more chance that the reactants will collide with the catalyst.

2. **Compare the structures of diamond, graphite and a fullerene and explain their differences in bonding.**
3. **Tensile strength is a measure of how much pulling force it takes to break a material. Suggest why nanotubes have a very high tensile strength.**

Properties of graphene

Graphene is a two-dimensional compound as it is only one atom thick. Graphene was isolated in 2003 as a monolayer, and is thermally stable.

Graphene is an electrical conductor through a way known as 'ballistic transport'.

Graphene is the strongest material ever found, stronger than steel and Kevlar.

It is not only strong but elastic too and it can absorb white light. The potential uses of graphene are only just beginning to be researched and this is a major new area in materials science.

4. **Predict whether Buckminsterfullerene has a lower or higher melting point than diamond and use your knowledge of their structures to explain reasons for your prediction.**
5. **Compare graphene and graphite.**
6. **Suggest why graphene is so strong, unlike graphite.**

KEY INFORMATION

Nanoscience refers to structures that are 1–100 nm in size, of the order of a few hundred atoms. They have many applications in medicine in electronics, in cosmetics and sun creams, as deodorants, and as catalysts. New applications for nanoparticulate materials are an important area of research.

You will need to consider advantages and disadvantages of the applications of these nanoparticulate materials, evaluate the use of nanoparticles for a specified purpose and explain that there are possible risks associated with the use of nanoparticles.

Nanoparticles, their properties and uses

Learning objectives:

- relate the sizes of nanoparticles to atoms and molecules
- explain that there may be risks associated with nanoparticles
- evaluate the use of nanoparticles for specific purposes.

KEY WORDS

nanoparticle
surface area to volume ratio

Glassmakers have used nanoparticles for more than a thousand years. Nano-sized particles of gold make glass look red, green or orange – depending on their size. These nanoparticles were originally made by trial and error, but now we can design them for many different uses.

Figure 2.81 Nanoparticles of gold makes light look red as it shines through this stained glass window.

What are nanoparticles like?

Nanoparticles are between 1 and 100 nm in size. Molecules of buckminsterfullerene are nanoparticles and so are carbon nanotubes. Most other nanoparticles are made from metals, metal oxides and silicates. Each nanoparticle contains only a few hundred atoms.

When materials are present as nanoparticles, this can change the properties of the materials.

- Silver nanoparticles in clothes or deodorants stop bacteria growing in our sweat.
- Titanium dioxide nanoparticles in sunscreens protect us from ultraviolet light. Bigger particles could do the same, but they would leave a white coating on our skin.
- Many medical drugs are designed to work on one type of cell only. Nanoparticles can act like miniature envelopes to carry the drugs safely to the right cells.

DID YOU KNOW?

Gold nanoparticles embedded in plastic could be used to make sensitive artificial skin that would let amputees feel with their prosthetic limbs.

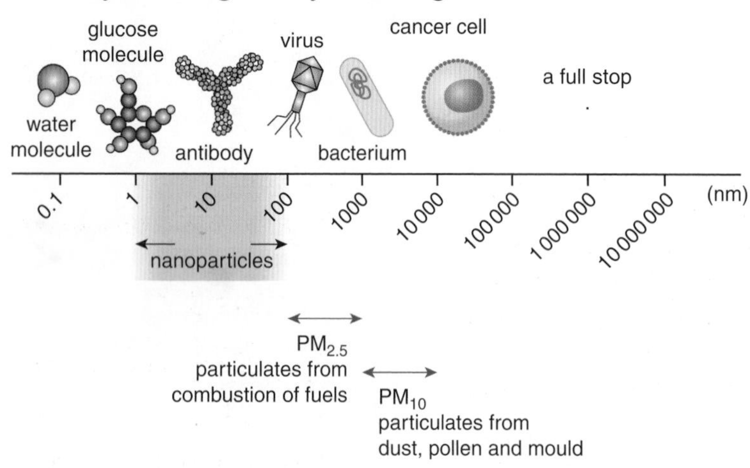

Figure 2.82 Nanoparticles are between 1 and 100 nm in size.

This diagram shows that although nanoparticles are incredibly small, they're between 10 and 1000 times longer than water molecules.

1 **How small must something be to be called a nanoparticle?**

2. **City air is full of particles called particulates. They form when fuels burn. Figure 2.82 compares their sizes with nanoparticles. Compare buckminsterfullerene (1 nm) with the smallest $PM_{2.5}$ particulates. How many times smaller is the buckminsterfullerene molecule?**

Using nanoparticles safely

$PM_{2.5}$ particulates from traffic fumes travel deep into our lungs and make lung problems such as asthma worse. Does this mean we should also be worried about nanoparticles?

There are many different nanoparticles and some could be hazardous. We should avoid breathing them in. Fortunately, the ones we use now are safely embedded in other materials, like stained glass and sunscreens. They don't get into the air and are not absorbed through skin. However, large quantities of free nanoparticles could be hazardous if they were released into the environment.

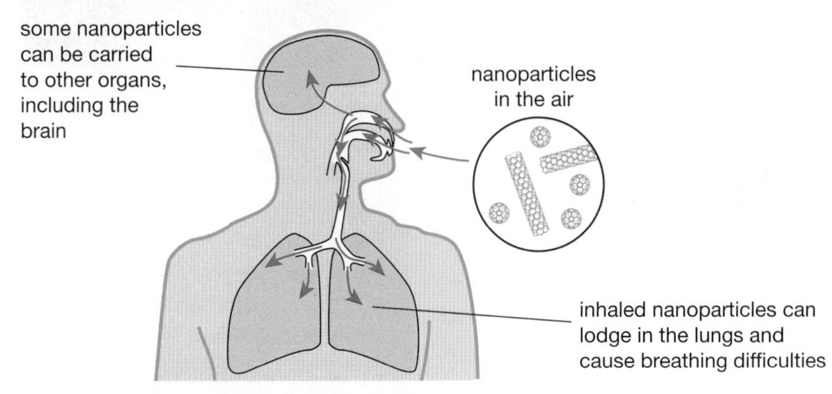

Figure 2.83 If nanoparticles are breathed in they can lodge in the lungs and damage tissues. Some can also be carried to other parts of the body.

3. **Many people worry that nanoparticles could be dangerous. Suggest one argument for this idea and one argument against.**

Changing properties

Imagine cutting up a block of gold. The more times you cut it the more surfaces you create and the more gold atoms you expose. As the blocks get smaller their surface area to volume ratio increases.

At first, the smaller cubes behave just like the original block. But when they become nanoparticles, new properties can appear. Gold displays different colours, while titanium dioxide particles become invisible and silver starts to destroy bacteria.

Other substances become more reactive as their surface area to volume ratio increases. This makes them better catalysts but could also make them more toxic if they got into living things.

4. **Look at the gold block in Figure 2.84. Imagine making one more cut to make eight equal cubes. Each side of each cube will be 0.5 cm long. What will the new surface area to volume ratio be?**

5. **Give two reasons why nanoparticles can be more useful than large pieces of the same material.**

KEY INFORMATION

New uses for nanoparticles are being found all the time. You should be able to evaluate whether their potential benefits outweigh any risks involved.

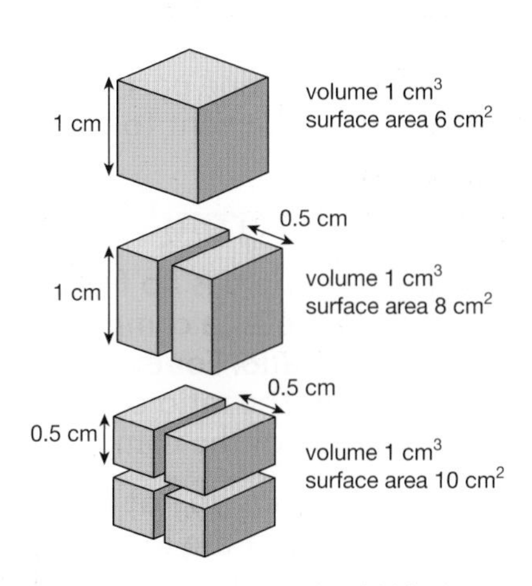

Figure 2.84 The original gold block has a surface area : volume ratio of 6 : 1. Each cut makes the surface area bigger.

MATHS SKILLS

Using ratios in mixture, empirical formulae and balanced equations

Learning objectives:

- use ratios, fractions and percentages to describe the composition of mixtures
- use ratios to determine the empirical formula of a compound
- explain how to balance equations in terms of numbers of atoms on both sides of the equation.

KEY WORDS

ratio
empirical formula

The ideal ratio of fuel to air for petrol engines that provides exactly enough air to completely burn the fuel is 14.7 : 1. This ratio is calculated from the ratio of molecules in the balanced equation for the combustion of octane (C_8H_{18}) in oxygen (O_2), together with the proportion of oxygen in air (20%).

Proportion of substances in a mixture

A Victoria sandwich cake recipe has 100g butter, 100g sugar and 100g flour. The **ratio** of the masses of butter, sugar and flour is 100 : 100 : 100. This ratio can be simplified by dividing by a common factor, 100. The simplest form of the ratio is 1 : 1 : 1. Another recipe has 150g butter, 150g sugar and 150g flour. The amount of each ingredient has increased, but the ratio of the masses of butter, sugar and flour is still 1 : 1 : 1.

Ratios are used in chemistry to describe the composition of mixtures. A gold alloy that is three parts gold and one part platinum is mixed in a 3 : 1 proportion of gold to platinum. A mixture with one part disinfectant to four parts water has a ratio of disinfectant to water of 1 : 4.

Empirical formulae

In a mixture there is no fixed proportion for the substances present. However a compound contains two or more elements in a specific ratio. For example the formula for water is H_2O, which means the ratio of hydrogen atoms to oxygen atoms in one molecule of the compound is 2 : 1.

Ionic compounds do not have molecules. Figure 2.85 shows a model of part of the ionic lattice of a sodium chloride crystal. In this model there are 13 sodium ions and 14 chlorine ions. However, the model only shows a small part of the lattice. Ionic compounds form a giant lattice – there could be billions of sodium ions and billions of chloride ions. If we counted all the ions we would find that for each sodium ion, there is one

MATHS

Ratios can also be expressed as fractions. A mixture with a ratio of 1 part liquid A to 4 parts liquid B has a ratio of A to B of 1 : 4. There are 5 parts in total. Liquid A is 1 part in 5 and liquid B is 4 parts in 5. As a fraction, liquid A is 1/5 of the mixture and liquid B is 4/5 of the mixture.

1 **a** Beth mixes 70 mL of concentrated orange squash with 490 mL of water. Write the composition of the mixture as a ratio in its simplest form.

b Caleb mixes 50 mL of the same orange squash with 250 mL of water. Write the composition of the mixture as a ratio in its simplest form.

c Whose squash is stronger?

2 Gold alloy A has 14 parts gold to 10 parts other metals. Gold alloy B has 18 parts gold to 6 parts other metals.

a Write the composition of each alloy as a ratio of gold to other metals.

b Which alloy has a higher proportion of gold?

chloride ion. This is because the sodium ion has a 1+ charge and the chloride ion has a 1− charge.

A ratio of 101 sodium ions to 101 chloride ions would still make the compound neutral. The simplest whole number ratio of the elements in a compound is called the **empirical formula**. The simplest (lowest possible) ratio of sodium to chloride ions is 1:1 so the formula of sodium chloride is NaCl.

The formula of other ionic compounds is determined in the same way, by looking at a diagram or model of the structure, or at the charges on the ions.

3 **Look at Figure 2.86. What is the empirical formula of potassium sulfide?**

4 **Complete the table to show the ratio of the different atoms in each compound.**

Chemical formula	sodium atoms		oxygen atoms		sulfur atoms
Na_2O		:		:	
Na_2SO_4		:		:	

5 **Determine the empirical formula of the following ionic compounds given the ions:**
a Mg^{2+} **and** SO_4^{2-}
b Mg^{2+} **and** Cl^-
c Fe^{2+} **and** Cl^-
d Al^{3+} **and** SO_4^{2-}.

Ratios of substances

During a chemical reaction, no atoms are lost or made, so symbol equations for chemical reactions are balanced in terms of the numbers of atoms of each element on both sides of the equation. Look at these two reactions:

$NaOH + HCl \rightarrow NaCl + H_2O$	$NaOH + H_2SO_4 \rightarrow Na_2SO_4 + H_2O$
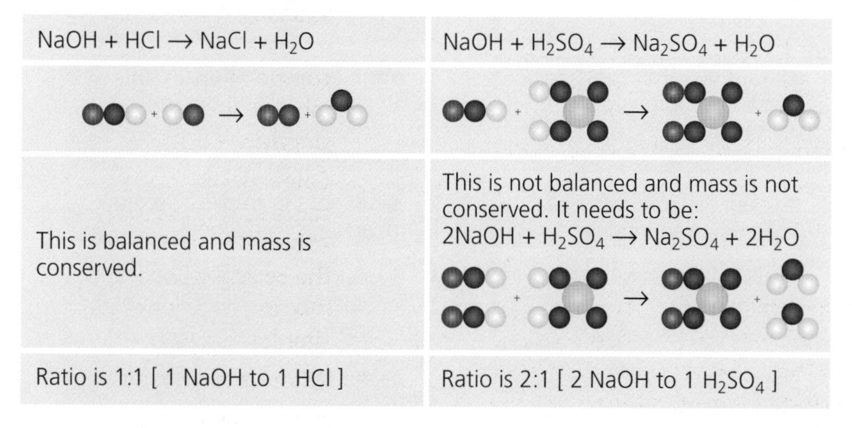	
This is balanced and mass is conserved.	This is not balanced and mass is not conserved. It needs to be: $2NaOH + H_2SO_4 \rightarrow Na_2SO_4 + 2H_2O$
Ratio is 1:1 [1 NaOH to 1 HCl]	Ratio is 2:1 [2 NaOH to 1 H_2SO_4]

Notice that balanced equations show the ratio by amounts of reactants and products. In order to balance equations, you can add a number in front of a formula in the equation.

6 **What is the ratio of reactants in this reaction?**
$Mg + 2HCl \rightarrow MgCl_2 + H_2$

7 **What is the ratio of reactants in this reaction?**
$3Fe + 2O_2 \rightarrow Fe_3O_4$

KEY INFORMATION

The small (subscript) number in a chemical formula tells you how many atoms there are of the element whose symbol comes just before the number. If there is no small number, it means the number is 1.

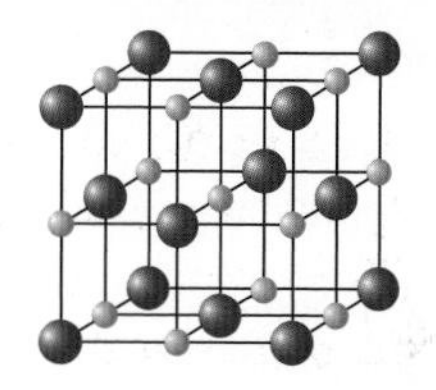

Figure 2.85 Model of sodium chloride

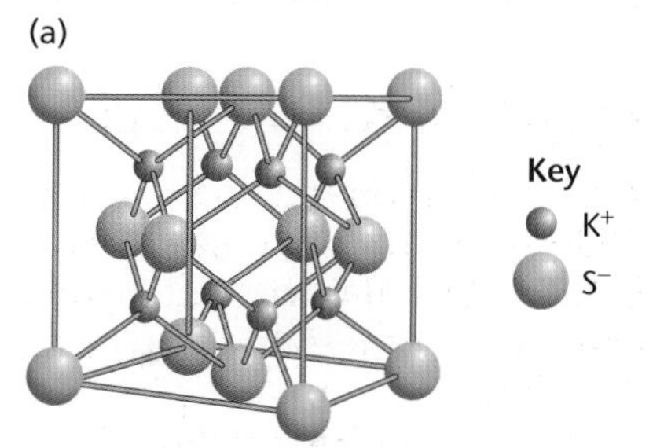

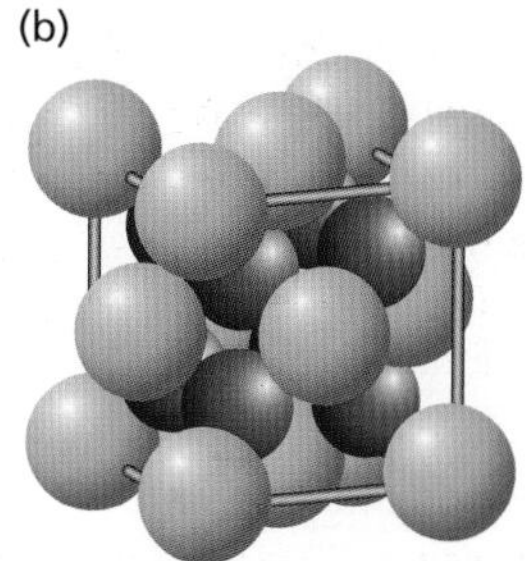

Figure 2.86 Models of potassium sulfide (a) ball and stick (b) close packed

8 **Determine the ratio of $H_2O : O_2$ in this reaction.**

$4NH_3 + 5O_2 \rightarrow 4NO + 6H_2O$

9 **N_2 reacts with O_2 in the ratio 1 : 2. A single product is formed. Give the balanced chemical equation for the reaction.**

Check your progress

You should be able to:

describe how to separate mixtures of elements and compounds →	use word equations to describe chemical reactions →	use balanced equations to describe reactions
be able to calculate a relative formula mass from the sum of the relative atomic masses →	calculate the sum of the relative formula masses of reactants and products →	show how the relative formula masses of reactants are equal to the relative formula masses of products
describe, explain and identify examples of processes of separation such as filtration, crystallisation and distillation →	suggest separation and purification techniques for mixtures →	distinguish pure and impure substances using melting point and boiling point data
describe how to set up paper chromatography →	distinguish pure from impure substances →	interpret chromatograms and determine R_f values
describe how Mendeleev was able to leave spaces for elements that had not yet been discovered →	explain why the modern periodic table has the elements in order of atomic number →	explain how Mendeleev was able to make predictions of as yet undiscovered elements such as eka-silicon
describe the pattern of the electrons in shells for the first 20 elements' →	explain how the electronic arrangement of atoms follows a pattern up to the atomic number 20 →	explain how the electronic arrangement of transition metal atoms put them into a period
describe a number of physical properties of metals and non-metals →	explain that atoms of metals have 1, 2 or 3 electrons in their outer shell →	explain that non-metals need to gain or share electrons during reactions and that metals need to lose electrons during reactions.
describe three main types of bonding →	explain how electrons are used in the three types of bonding →	explain how bonding and properties are linked
represent an ionic bond with a diagram →	draw a dot and cross diagrams for ionic compounds →	work out the charge on the ions of metal and non-metals from the group number of the element
identify ionic compounds from structures →	explain the limitations of diagrams and models →	work out the empirical formula of an ionic compound
describe that metals form giant structures →	explain how metal ions are held together →	explain how metallic bonding is enabled by the delocalisation of electrons
identify small molecules from formulae →	identify polymers from their unit formula →	relate the intermolecular forces to the bulk properties of a substance
explain how the properties relate to the bonding in diamond →	explain why diamond differs from graphite →	explain the similarity of graphite to metals
describe the structure of graphene →	explain the structures and uses of fullerenes →	compare 'nano' dimensions to dimensions of atoms and molecules

Worked example

1 **Suggest which piece of apparatus you would use to measure out exactly 25 cm^3 of acid or alkali solution each time.**

a pipette

The answer is correct.

2 **Suggest the difference between the melting points of a pure and an impure substance.**

An impure substance has a lower melting point than a pure one.

The answer is correct.

3 **Draw the spots that will appear on the chromatogram if the mixed dye contains colours Q, R and T.**

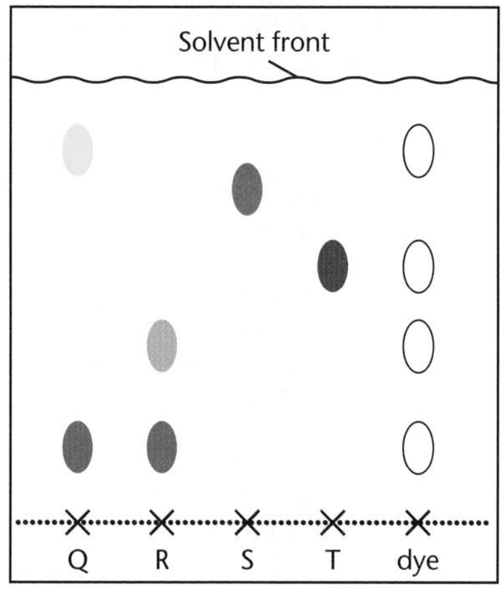

This answer is correct as the distances of all the spots are the same as the original and the red spot is not there.

4 **Calculate the R_f value for the red spot S.**

R_f value = solvent distance/spot distance
= 14/10.5 = 1.3 mm

This answer is incorrect as the R_f value is spot distance / solvent distance. Also there are no units to an R_f value. The answer should have been R_f = 10/14.5 = 0.69

5 **An element from Group 1, X, bonds with an element from Group 7, Y.**

a **Identify the type of bonding.**

a) metallic b) ionic (circled) c) covalent d) giant

The answer is correct.

b **Draw the dot and cross diagram for the resulting compound.**

Use X to represent the metal and Y to represent the non-metal. Show the outer shell only.

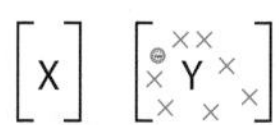

X has lost an electron so the 1+ charge is missing. **Y** has gained an electron so the 1– charge is missing. Both charges should be at the top right outside the brackets. Electrons should be kept in pairs in a circle.

6 **Explain why both diamond and silicon dioxide are hard with high melting points.**

They have covalent bonds that act in all directions.

This answer explains what the bonding is like but needs to be linked to the energy required to break the bonds for melting to happen.

7 **Fill in the missing data in the table.**

Substance	Metal	Small molecule	Giant covalent	Ionic
Melting point	*High*	*Low*	High	*High*
Conducts electricity	Yes	*No*	*Yes*	*Yes when melted, no when solid*

The metal, small molecule and ionic columns are all correct. The student has correctly stated the difference in conductivity between an ionic solid and liquid. Giant covalent structures do not normally conduct electricity. Graphite is an exception and this should be explained.

End of chapter questions

Getting started

1. Identify the type of bonding in potassium bromide, KBr.

   a metallic **b** ionic **c** covalent **d** giant — 1 Mark

2. Which substance is made of small molecules?

   a carbon dioxide **b** silicon dioxide **c** magnesium oxide **d** sodium sulfide — 1 Mark

3. What elements in fertilisers help plants to grow? — 2 Marks

4. Determine the electron arrangement in an atom with 11 electrons. — 1 Mark

5. Explain why an atom with an electron pattern of 2,8,1 is in Group 1 and Period 3 of the periodic table. — 2 Marks

6. Sand has been added to a solution of black ink. Describe how you would separate the sand from the solution and find which colours were in the ink. — 2 Marks

7. Match the symbol to the type of particle.

K^+	small molecule
NO_2	polymer
$-(XZ)-_n$	ion

2 Marks

Going further

8. What method is used to separate ethanol (Bp 80 °C) and water (Bp 100 °C)? — 1 Mark

9. **a** A tablet is made of 0.050 g of aspirin added to 0.79 g of starch and 0.11 g of talc. Calculate the percentage of aspirin in the tablet.

   b Aspirin is a mixture. Explain why the substances in aspirin have to be mixed in exactly the correct proportions. — 2 Marks

10. Substance R conducts electricity when solid and is malleable. What is the structure of R?

    a simple molecular b giant ionic c giant covalent d giant metallic — 1 Mark

11. Determine the structure of A and B. Justify your answer.

Substance	A	B
Mp and Bp / °C	–205 and –192	420 and 907
Conducts electricity	No	Yes when solid

4 Marks

12. Suggest one structural feature that graphene and fullerenes have in common. — 2 Marks

More challenging

13 Explain why silver is a good conductor of electricity. 1 Mark

14 Describe how the electrons are arranged in the bond in chlorine, Cl_2. 1 Mark

15 Draw a dot and cross diagram for PF_3. Only outer shell electrons need be shown. 2 Marks

16 Some physical properties for three substances are given in the table below.

Substance	A	B	C
Mp and Bp / °C	1085 and 2562	605 and 1360	3550 and 4830
Conducts electricity?	Yes when solid	Yes when liquid	No
Dissolves in water?	No	Yes	No

For substances A, B and C.

a Identify the structures, justifying your answer.

b Explain the electrical conductivity. 2 Marks

Most demanding

17 Methane, CH_4, has a low melting and boiling point. Explain why. 2 Marks

18 The table shows the data and properties for an element.

Deduce which type of element it is, the group it belongs to and why it reacts so violently with water.

Atomic number	Electron pattern	Reaction with water
37	2,8,8,18,1	Violently, giving off hydrogen and forming an alkali

2 Marks

19 Explain why the atom with an electron pattern of 2,8,6 is a non-metal. Explain why this atom is less reactive than the atom with an electron pattern of 2,6 or 2,7. 4 Marks

20 Some data on three substances is given in the table.

	Hardness / Mohs' scale	Thermal conductivity / $Wm^{-1}K^{-1}$
Sodium chloride, NaCl	2.5	6.5
Sulfur, S_8	1.5–2.5	0.205
Iron, Fe	4–5	80.4
Silicon dioxide, SiO_2	7	1.4

a Calculate the relative thermal conductivity of the giant metallic substance compared to the giant ionic substance. Explain the difference.

b Explain the difference in hardness between the giant covalent substance and simple molecular substance. 4 Marks

40 Marks

CHEMICAL REACTIONS

IDEAS YOU HAVE MET BEFORE:

CHEMICALS CHANGE DURING REACTION AND ARE NOT DESTROYED

- Iron and sulfur heated together make iron sulfide.
- Magnesium burns in air to make magnesium oxide.
- Wood burns to make ash, smoke and carbon dioxide.

GASES HAVE MASS

- Air in a balloon records a mass on a balance.
- Helium in a balloon has less mass than air in the same balloon.
- Chlorine gas 'rolls' along the ground.

EXPLAINING ENERGY CHANGES

- A temperature rise means that energy is being transferred out.
- Energy is transferred when something is moving.
- Energy can be stored.

NEUTRALISATION AND SALT PRODUCTION

- Acids turn universal indicator red.
- Alkalis turn universal indicator blue.
- Copper sulfate solution can be left to make blue crystals.

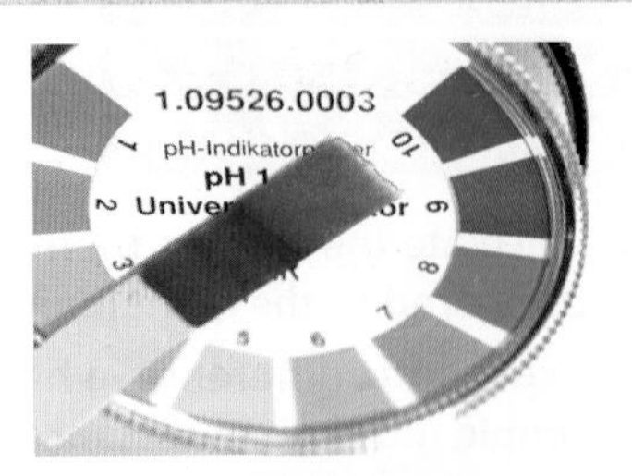

ELECTROLYSIS

- Copper and graphite conduct electricity.
- Copper put into silver nitrate makes a coat of silver.
- Zinc in copper sulfate gets a coat of copper.

3

IN THIS CHAPTER YOU WILL FIND OUT ABOUT:

HOW IS MASS CONSERVED IN CHEMICAL REACTIONS?

- The mass of the reactants is the same as the mass of products.
- The same number of atoms are on both sides of an equation.
- An equation needs to be balanced by multiples.

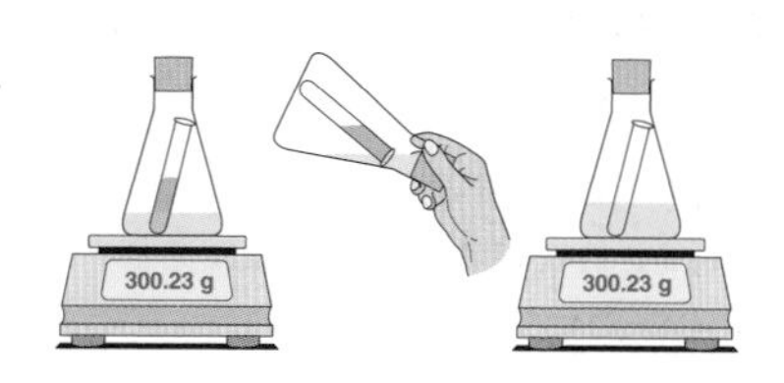

WHAT HAPPENS TO MASS CHANGES WHEN A GAS IS GIVEN OFF?

- A gas is often given off during a thermal decomposition reaction.
- Mass seems to reduce in a thermal decomposition reaction.
- A substance reacting with oxygen can seem to increase in mass.

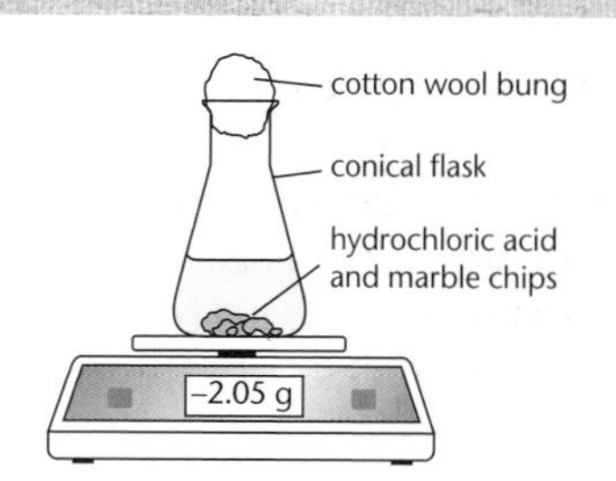

HOW CAN WE EXPLAIN ENERGY CHANGES?

- Collision theory explains that particles are moving and colliding.
- Reactions take place if successful collisions take place.
- Energy changes can be calculated using bond energies.

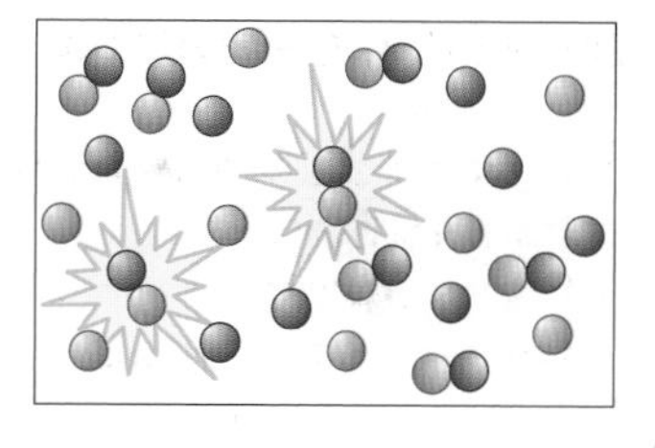

HOW DO ACID AND BASES PRODUCE NEUTRAL SALTS?

- Acids react with bases or alkalis to produce salts and water.
- Acids react with carbonates to make salts, water and carbon dioxide.
- pH is the concentration of hydrogen ions and measures acidity.

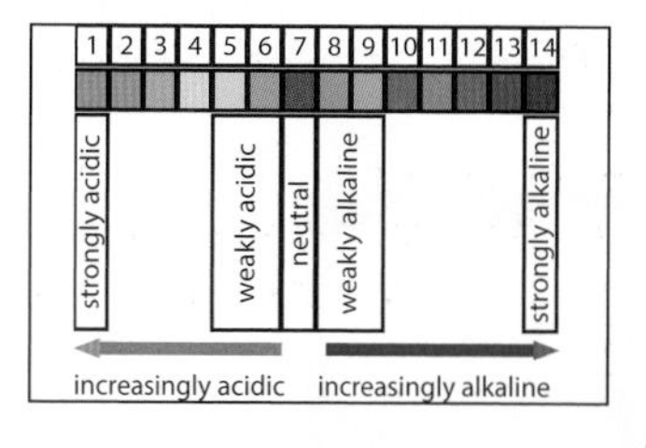

WHAT HAPPENS DURING THE PROCESS OF ELECTROLYSIS?

- Metals form ions that move towards the cathode.
- Non-metals form ions that move towards the anode.
- Electrolysis of solutions often produces hydrogen and oxygen.

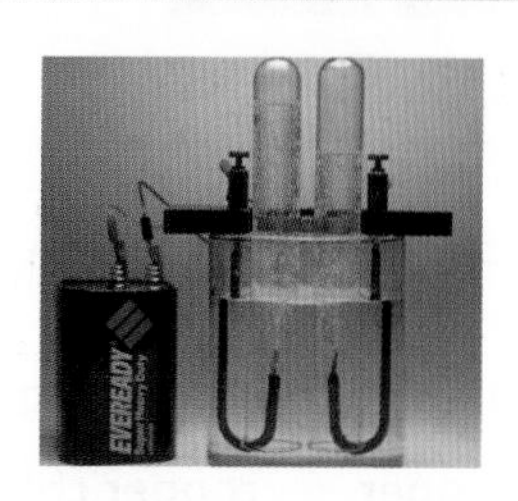

Elements and compounds

Learning objectives:

- identify symbols of elements from the periodic table
- recognise compounds from their formula
- identify the elements in a compound.

KEY WORDS

balanced
compound
element
equation
symbol

All the elements are listed in the periodic table. The elements in the formulae of any compound, no matter how large, can be identified by using the periodic table.

Elements and compounds

An **element** is a substance that cannot be broken down chemically.

A **compound** is a substance that contains at least two different elements, chemically combined in fixed proportions.

Periodic table elements 1–20

1	2	13	14	15	16	17	18
1 H hydrogen 1.0							2 He helium 4.0
3 Li lithium 6.9	4 Be beryllium 9.0	5 B boron 10.8	6 C carbon 12.0	7 N nitrogen 14.0	8 O oxygen 16.0	9 F fluorine 19.0	10 Ne neon 20.2
11 Na sodium 23.0	12 Mg magnesium 24.3	13 Al aluminium 27.0	14 Si silicon 28.1	15 P phosphorus 31.0	16 S sulfur 32.1	17 Cl chlorine 35.5	18 Ar argon 39.9
19 K potassium 39.1	20 Ca calcium 40.1						

Figure 3.1 Two sections of the periodic table. Can you find magnesium and oxygen?

Figure 3.2 Magnesium (metal) reacts with oxygen (gas) to make magnesium oxide (white powder)

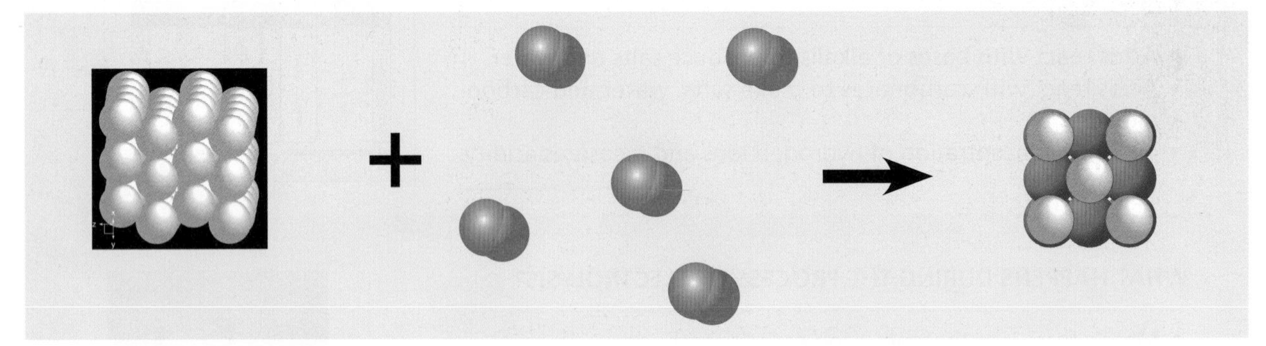

Figure 3.3 The reaction of the elements in Figure 3.2 can be represented by models of their atoms

1. **Identify the following substances as elements or compounds.**

   copper copper chloride copper sulfate

2. **Name the elements in beryllium chloride.**

Compounds and elements

Copper is an **element**. It cannot be broken down into any other substances, but salt can.

The chemical name for common salt is sodium chloride. Sodium chloride is a **compound**. Sodium chloride can be broken down to make sodium and chlorine, but this is not easy to do because the sodium and chlorine are chemically combined. We need to use electricity to make sodium and chlorine from sodium chloride.

Sodium and chlorine cannot be broken down any further. Sodium and chlorine are elements.

There are only about 100 elements but these can join together chemically to make an enormous number of compounds. They need chemical reactions to do this.

KEY INFORMATION

When chlor<u>ine</u> reacts to make a compound it chemically combines and becomes a chlor<u>ide</u>.

Similarly, bromine reacts to become a bromide and oxygen reacts to become an oxide.

3. **Identify the elements in potassium bromide.**
4. **Predict the products when lead iodide is split by electricity.**

Making the element copper into a compound

A chemical reaction is needed to make copper (an element) into a compound. If copper is burned in oxygen it forms copper oxide. Chemical reactions always involve the formation of one or more new substances, and often involve a detectable energy change.

Figure 3.4 Copper (an element) burning in oxygen (an element) to make copper oxide (a compound). This is normally done by heating the copper in crucibles.

DID YOU KNOW?

Compounds can only be separated into elements by chemical reactions.
To get the element copper back, the oxygen needs to be chemically removed. This is done using hydrogen. The oxygen combines with the hydrogen to make water, another compound.
copper oxide + hydrogen → copper + water

5. **What is the name of the compound made from sodium and oxygen.**
6. **Oxygen can be removed from iron(III) oxide by carbon monoxide. Identify the element and compound produced.**
7. **Substance D reacted with hydrogen to form zinc and water. Explain whether substance D is an element or compound.**

Atoms, formulae and equations

Learning objectives:

- explain that an element consists of the same type of atoms
- explain that atoms join together to make molecules
- explain how formulae represent elements and compounds.

KEY WORDS

compound
element
molecule

All substances are chemicals. Many people say "I don't want chemicals in my food" not realising that foods *are* chemicals. Our food is made of compounds and mixtures. Compounds are elements joined together in many different ways. In fact, we are all made of chemical compounds.

Figure 3.5 Vinegar is used in salad dressing. Its chemical name is ethanoic acid. Ethanoic acid is a compound made from carbon, hydrogen and oxygen.

Atoms and molecules

Elements are made up of atoms that are all the same.

Compounds are made from atoms (or charged ions) of different elements.

CH_3COOH is a compound made from three elements. These are carbon, C, hydrogen, H, and oxygen, O. You will know it as vinegar. Its chemical name is ethanoic acid. The atoms join by sharing their electrons. The compound is made of molecules.

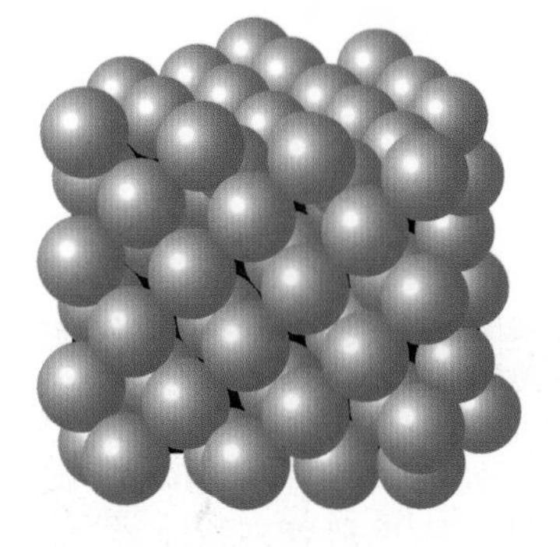

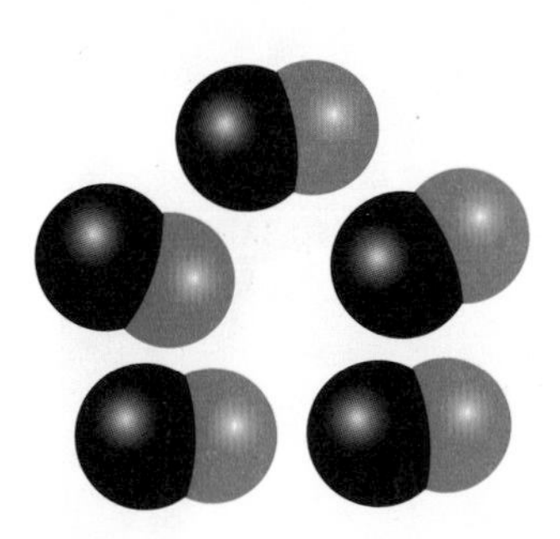

Figure 3.6 Gold is an element. Carbon monoxide is a compound. How could you tell by looking at these diagrams?

Figure 3.7 The molecule in vinegar CH_3COOH

If two or more atoms join together by sharing their electrons the atoms form a **molecule**.

Two examples of molecules are oxygen and water.

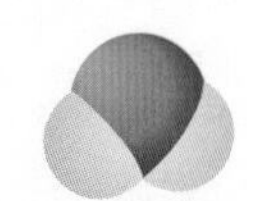

Figure 3.8 Which of these is a molecule of an element? How can you tell?

1 **Explain whether the following substances are elements or compounds.**

C (carbon), CO_2 (carbon dioxide), Cl_2 (chlorine) and SO_3 (sulfur trioxide)

2 **For the substances below, write down:**

a **the names of the elements**

b **how many different types of atoms they contain**

c **how many atoms are in the molecule overall.**

CO_2 S_8 Cl_2 SO_3 C_{60} C_4H_{10}

KEY INFORMATION

In a chemical formula the number that is a subscript on the bottom right is the number of atoms in the molecule, for example C_2H_6 has two carbon atoms joined to six hydrogen atoms.

Formulae

You can see which elements are in a compound by looking at its **formula**. For example, the compound magnesium oxide (MgO) contains Mg (magnesium) and O (oxygen).

Look back at Figure 3.1.

Elements from Group 1 and elements from Group 7 combine to make compounds in a fixed ratio of 1 : 1.

- The **formula** of lithium fluoride is LiF.

Elements from Group 2 and elements from Group 6 also combine to make compounds in a fixed ratio of 1 : 1.

- The formula of calcium oxide is CaO.

Elements from Group 2 and elements from Group 7 combine to make compounds in a fixed ratio of 1 : 2. The element that you need two of has a suffix '2' after the symbol.

- The formula of calcium fluoride is CaF_2.

Elements from Group 1 and elements from Group 6 combine to make compounds in a fixed ratio of 2 : 1. Again, the element that you need two of has a suffix '2' after the symbol.

- The formula of sodium sulfide is Na_2S.

3 Give the names of the elements in $MgSO_4$ and in CH_4.

4 Determine the formulae of lithium chloride, magnesium chloride and potassium oxide.

DID YOU KNOW?

Oxygen exists as a pair of atoms not as a single atom. Its formula is O_2. In a symbol equation, O_2 must be written not just O.

Equations and balancing

When magnesium reacts with oxygen it makes magnesium oxide.

The word **equation** is:

magnesium + oxygen → magnesium oxide

The **symbol** equation is:

Mg	+	O_2	→	MgO
1		2		1 1 ✗

This is not a **balanced** equation. We need to add '2' to the front of the formulae of magnesium and magnesium oxide. This gives:

2Mg	+	O_2	→	2MgO
2		2		2 2 ✓

KEY INFORMATION

In a balanced equation 'O_2' means a pair of atoms joined together in a molecule. '2Mg' means two separate atoms, where the '2' is added to balance the equation.

5 Write a balanced equation for the formation of sodium chloride from sodium, Na, and chlorine, Cl_2.

6 Write a balanced equation for the formation of aluminium oxide Al_2O_3.

7 Complete and balance the following equation by suggesting values for D, E and F:

$$\mathbf{D}\ N_2 + \mathbf{E}\ O_2 \rightarrow \mathbf{F}\ NO_2$$

Moles

Learning objectives:

- describe the measurement of amounts of substances in moles
- calculate the number of moles in a given mass
- calculate the mass of a given number of moles.

KEY WORDS

mole
chemical amount
relative formula mass
Avogadro number

Particles of a substance can be packed close together as in copper metal, diamond or ice or they can be moving rapidly and randomly as in water vapour or oxygen gas. What is the one measurement which allows us to compare the amount of substance in each chemical?

HIGHER TIER ONLY

Measurement of amounts

Chemical amounts are measured in **moles**. The symbol for the unit mole is 'mol'.

The measurement of amounts in moles can apply to:

atoms	ions	formulae
molecules	electrons	equations

For example:

In one mole of carbon (C) the *number* of atoms is the same as the number of molecules in one mole of carbon dioxide (CO_2).

The mass of one mole of a substance in grams is numerically equal to its **relative formula mass**.

Number of atoms in 1 mole C	= number of molecules in 1 mole CO_2
Carbon has a formula of C	Carbon dioxide has a formula of CO_2
C has a relative atomic mass of 12	CO_2 has a relative molecular mass of $12 + (2 \times 16) = 44$
Mass of 1 mole = 12 g	Mass of 1 mole = 44 g

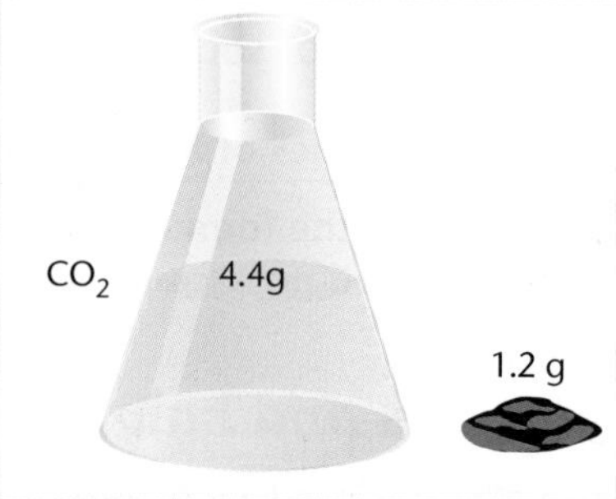

CO_2 4.4g

1.2 g

Figure 3.9 This lump of carbon (C) and jar of carbon dioxide (CO_2) may not look the same but they have the same amount of substance. There is the same number of atoms in the carbon lump as the number of molecules of carbon dioxide in the jar.

1. **Work out the mass of one mole of H_2O.**

   [Relative atomic mass of H is 1, relative atomic mass of O is 16]

2. **Work out the mass of three moles of KBr. [Use the periodic table to help you]**

Avogadro's constant

Single atoms cannot be measured for mass in an experiment – they are too small. Instead, they are measured in standard amounts.

The mole is a unit for a standard **amount** of a substance. One mole of any substance contains the same number of particles, atoms, molecules or ions as one mole of any other substance.

The number of atoms, molecules or ions in a mole of a given substance is the Avogadro constant.

The value of the **Avogadro constant** is 6.02×10^{23} per mole (or mol^{-1})

DID YOU KNOW?

Avogadro's hypothesis was developed in 1810, over 200 years ago.

3 **Calculate the number of particles in two moles of helium, He.**

Calculating molar mass

The name for the mass of one mole is 'molar mass' and its unit is g/mol or g mol^{-1}.

A_r is the abbreviation used for relative atomic mass. The mass of one mole of an element that consists of atoms is numerically equal to the relative atomic mass, A_r on the periodic table. For example:

- Ne has a relative atomic mass, A_r, of 20
- 1 mole of neon, Ne has a molar mass of 20 g/mol

The mass of one mole of molecules is found by adding up all the relative atomic masses in the formula.

Oxygen gas has the formula O_2.

- The relative atomic mass, A_r for oxygen is 16.
- In O_2, there are two oxygen atoms.
- The molar mass for O_2 is 16 + 16 = 32 g/mol.

The mass of one mole of a compound is found by adding up all the relative atomic masses in the formula.

- Calcium carbonate has the formula $CaCO_3$.
- Relative atomic masses A_r are Ca = 40, C = 12, O = 16.
- The molar mass of $CaCO_3$ is 40 + 12 + (3 × 16) = 100 g/mol.

If the formula contains brackets, these must be considered in the calculation:

- Magnesium nitrate has the formula $Mg(NO_3)_2$.
- Relative atomic masses, A_r of Mg = 24, N = 14, O = 16.
- Relative formula mass $Mg(NO_3)_2$ = 24 + (2 × 14) + 2 (3 × 16) = 148
- The molar mass is 148 g/mol.

4 **Work out the molar mass of:**

a **nitrogen gas, N_2**

b **zinc oxide, ZnO**

c **magnesium carbonate, $MgCO_3$**

d **ammonium sulfate, $(NH_4)_2SO_4$**

5 **Work out the number of moles of water in 72 g.**

6 **$2H_2 + O_2 \rightarrow 2H_2O$**

a **Work out the number of moles of O_2 needed to make four moles of water.**

b **Calculate the number of grams of H_2 needed to make four moles of water.**

KEY INFORMATION

If the formula contains brackets, the mass can be calculated in two ways, e.g. $Mg(NO_3)_2$ = 24 + (2 × 14) +2 (3 × 16) = 148 or 24 + 2 [14 + (3 × 16)] = 148

KEY CONCEPT

Conservation of mass and balanced equations

KEY WORDS

balanced equation
conservation of mass
products
reactants

Learning objectives:

- explain the law of conservation of mass
- explain why a multiplier appears as a subscript in a formula
- explain why a multiplier appears in equations before a formula.

When chemicals react, the atoms are just rearranged, they are not made or destroyed. As a result, the total mass is always conserved during a chemical reaction. Because of this, chemical reactions can be represented by balanced symbol equations.

The law of conservation of mass

The law of **conservation of mass** states that 'no atoms are lost or made during a chemical reaction'.

This means that the mass of the **products** equals the mass of the **reactants**.

In a chemical reaction, all the atoms in the reactants are *rearranged* into products. For example, for one type of reaction:

$$AB + CD \rightarrow AD + CB$$

the mass of atoms at the start equals the mass of atoms at the finish. This is consistent with the 'conservation of mass'.

Figure 3.10 Masses of reactants and products are balanced

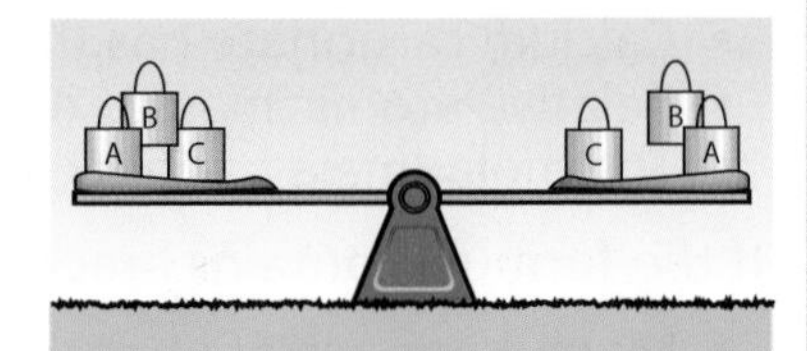

Figure 3.11 Numbers of atoms of the formulae are balanced

1 Identify the missing product to ensure conservation of mass.

$$XY + ZRT \rightarrow XR + ? + T$$

Balanced equations and formulae

Many compounds contain more than one atom of an element in a formula. The number of atoms of the element is written as a subscript after the element symbol.

For example, these formulae have more than one atom of some of their elements. If there is no number after the element symbol it means 1 atom.

```
        O
        ||
HO— S — OH
        ||
        O
```

Figure 3.12 The molecule has 2 H atoms, 1 S atom and 4 O atom

KEY SKILLS

Always work out or look up the formula first, then put it into an equation.

carbon dioxide	CO_2	O=C=O	magnesium chloride	$MgCl_2$	
water	H_2O	H–O–H	sodium sulfate	Na_2SO_4	
ethane	C_2H_6	$H_3C–CH_3$	calcium nitrate	$Ca(NO_3)_2$	

2 **Determine the number of hydrogen atoms in ethanoic acid CH_3COOH.**

3 **How many atoms of each element are in sodium thiosulfate $Na_2S_2O_3$?**

4 **How many atoms of each element are in aluminium sulfate $Al_2(SO_4)_3$?**

Balancing equations

This equation is balanced.

$$NaCl + AgNO_3 \rightarrow NaNO_3 + AgCl$$

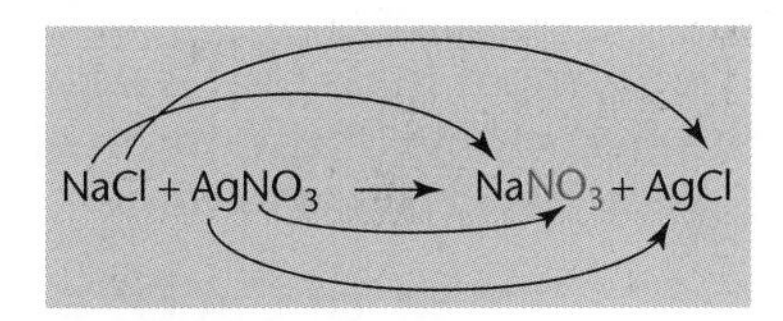

Here is another example:

$$Mg + O_2 \rightarrow MgO$$

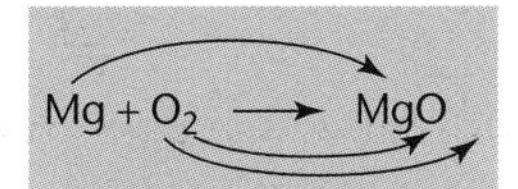

This equation is not balanced.

The two atoms of oxygen will each join with a magnesium atom to make two formula units of MgO.

$$Mg + O_2 \rightarrow 2\,MgO$$

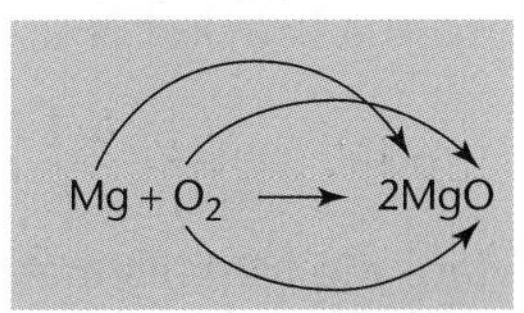

This equation is still not balanced.

If we make 2 formula units of MgO, there must have been 2 atoms of Mg as reactants to 'conserve mass'.

$$2Mg + O_2 \rightarrow 2\,MgO$$

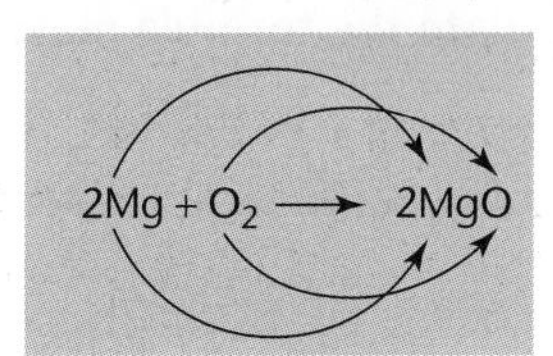

The equation is now balanced. There are the same number of atoms on the two sides and 'mass is conserved'.

DID YOU KNOW?

You must write the equation of the reaction that actually happened even if it is difficult to balance. You cannot alter the formulae of compounds even if this makes balancing easier.

5 **Balance the following equations by determining the numerical values for D, E, F and G.**

a $DNaOH + EH_2SO_4 \rightarrow FNa_2SO_4 + GH_2O$

b $DC_3H_8 + EO_2 \rightarrow FCO_2 + GH_2O$

KEY CONCEPT

Amounts in chemistry

Learning objectives:

- use atomic masses to calculate formula mass
- explain how formula mass relates to number of moles
- explain how number of moles relate to other quantities.

KEY WORDS

mole
molar mass
Avogadro number
molar volume

We have learned that there are different ways of measuring chemical quantities. The amount of chemical substance is known as the *mole*. We have learned how each quantity relates to the mole. Let's draw these altogether here in one place.

Group 1	…	…	Group 7
3 **Li** lithium 6.9			9 **F** fluorine 19.0
11 **Na** sodium 23.0			17 **Cl** chlorine 35.5
19 **K** potassium 39.1			35 **Br** bromine 79.9

The periodic table shows the atomic mass of each element.

Formula mass

The atomic mass of an element is written on the periodic table.

A formula mass is the sum of the atomic masses of elements in a substance.

The subscript $_3$ means three atoms in the molecule, for example:

$$CaCO_3$$

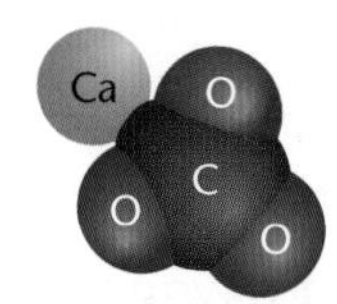

The formula mass (M_r) is equal to the atomic masses (A_r) of the five atoms of the three elements, one Ca atom, one C atom and three O atoms.

$$\begin{aligned} M_r \text{ of } CaCO_3 &= 40 + 12 + (3 \times 16) \\ &= 40 + 12 + 48 \\ &= 100 \end{aligned}$$

If there is more than one atom in a group attached to another atom a bracket is used, for example:

$$Ca(OH)_2$$

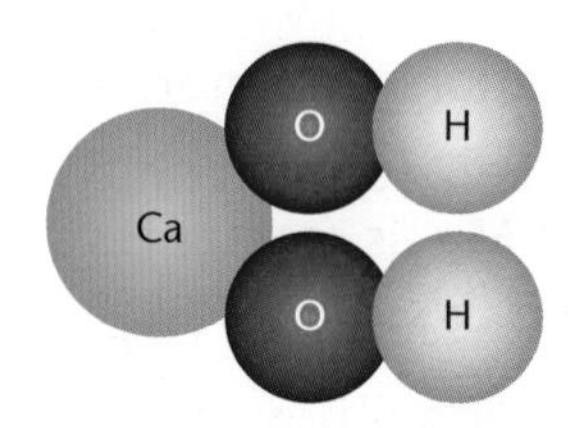

1. **Calculate the formula mass of magnesium sulfate $MgSO_4$.**
2. **Calculate the formula mass of a) $Ca(NO_3)_2$ and b) $Al_2(SO_4)_3$**

HIGHER TIER ONLY

Numbers of moles

The amount of substance of a chemical is the **mole**.

This relates to other quantities such as the **molar mass**, concentration and volumes of gas.

In summary, the mole is an amount of substance that:

- contains the molar mass in grams
- has the **Avogadro number** of particles (6.02×10^{23})
- relates to the molar mass in grams dissolved in 1 dm^3 of solution (1000 cm^3)
- occupies, if as a gas, the same volume as all other gases (the **molar volume** is 24 dm^3 at room temperature and pressure, rtp)
- reacts with other substances in whole number ratios that can be written as an equation, for example,

2 moles NaOH : 1 mole H_2SO_4

$2\,NaOH + H_2SO_4 \rightarrow Na_2SO_4 + 2H_2O$

2 +  → 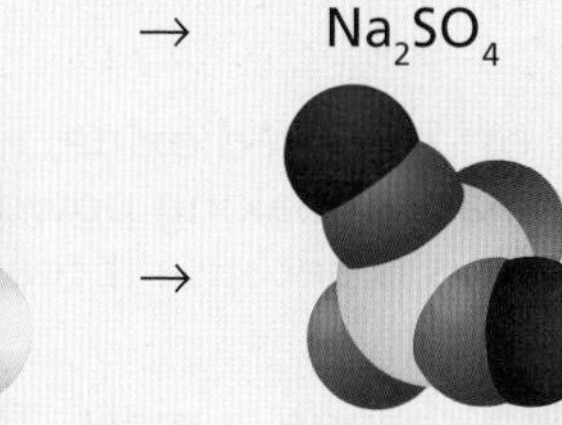 + 2 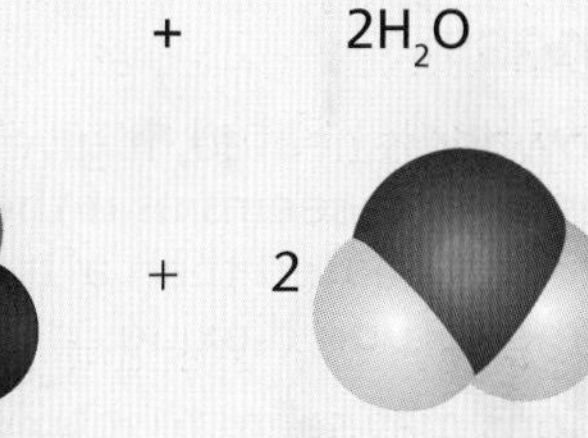

DID YOU KNOW?

We have seen the detail of these other quantity measurements in other pages in this chapter.

3 **Work out the molar mass of NaOH.**

4 **Calculate the number of particles that two moles of neon contains.**

5 **Calculate the mass in grams of NaOH needed to make a solution that has a concentration of 0.5 mol/dm^3.**

6 **Calculate the volume of gas (at rtp) occupied by 4 g of H_2**

Working with moles

The key to working with moles is the chemical equation. The balanced symbol equation indicates the number of moles that are involved in any reaction. For example,

$$ZnCO_3 \rightarrow ZnO + CO_2$$

1 mole of $ZnCO_3$ produces 1 mole of ZnO + 1 mole of CO_2

This is related to the molar masses and molar volumes:

M_r of $ZnCO_3$ = 125 M_r of ZnO = 81 M_r of CO_2 = 44

So 125 g $ZnCO_3$ will produce 81 g ZnO and 44 g CO_2

Or 125 g $ZnCO_3$ will produce 81 g ZnO and 24 dm^3 CO_2

KEY INFORMATION

You are not required to remember how to work with neutralisation calculations that are in 2 : 1 ratio, only 1 : 1 ratio.

7 **How many grams of ZnO would be produced by 6.25 g $ZnCO_3$?**

8 **Calculate the mass of CO_2 produced when 660 g of propane, C_3H_8, is combusted.**

$$C_3H_8 + 5O_2 \rightarrow 3CO_2 + 4H_2O$$

Mass changes when gases are in reactions

Learning objectives:

- explain any observed changes in mass in a chemical reaction
- identify the mass changes using a balanced symbol equation
- explain these changes in terms of the particle model.

KEY WORDS

gas
mass
particles
thermal decomposition

We have learned that mass must be conserved in a reaction. The sum of the masses of the reactants must equal the sum of the masses of the products. What is happening when some chemicals seem to lose mass when they are heated?

Losing mass

If baking powder is heated, it gives off carbon dioxide, which makes cakes 'rise'. It seems as if the mass of baking powder is more before it is heated than after. This is because it rearranges and 'loses' the carbon dioxide.

Some reactions may seem to involve a change in **mass**. This can usually be explained because a reactant or product is a **gas**. The mass of the gas is often not measured.

A gas can be driven off as a product or taken in as a reactant.

For example: when copper carbonate is heated, it reacts to make copper oxide and carbon dioxide. Where does the carbon dioxide gas go?

Sammy heats 5 g of copper carbonate.

After heating, Sammy measured the mass of the copper oxide.

It had a mass of 3.2 g.

How much gas was made?

The answer is 1.8 g.

How did Sammy work this out?

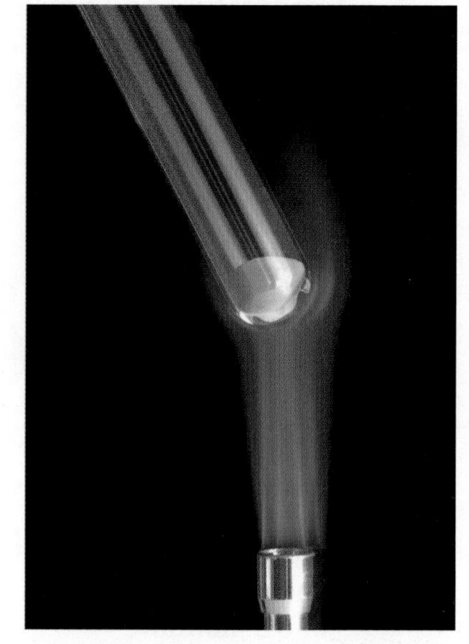

Figure 3.13 Heating copper carbonate

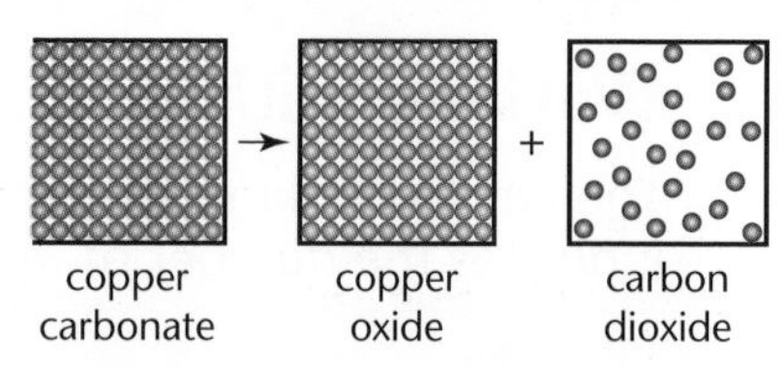

Figure 3.14 Solid copper carbonate when heated turns to solid copper oxide and the gas carbon dioxide

KEY INFORMATION

The conservation of mass means that the mass of the products must equal the mass of the reactants.

1. **Jane heats 2 g of zinc carbonate.**

   zinc carbonate → zinc oxide + carbon dioxide

   2 g of zinc carbonate makes 1.3 g of zinc oxide. Explain why the mass of zinc oxide is less than the mass of zinc carbonate.

2. **In the reaction in question 1, what mass of carbon dioxide is produced?**

3. **Copper sulfate crystals change into copper sulfate powder when heated. Water is given off. If 2.5 g of copper sulfate crystals make 1.6 g of powder, how much water is produced?**

Gaining mass

Sometimes one of the reactants is a gas.

For example: when a metal reacts with oxygen the mass of the oxide produced is greater than the mass of the metal.

Alesha heats 4 g of copper. When it cools the solid product has a mass of 5 g. Why is there an increase?

The answer is that the copper reacted with 1 g of oxygen to form copper oxide.

Figure 3.15 Blue, hydrated copper sulfate becomes white when heated and the water is driven off

4. **Draw three boxes. Draw the particle arrangement for the reactants and products in the reaction of copper + oxygen → copper oxide**
5. **When magnesium is heated in air it gains mass. If 2.4 g of magnesium makes 4 g of magnesium oxide, how much oxygen was added from the air?**

HIGHER TIER ONLY

Other reactions and limiting reactants

If magnesium carbonate is put into acid in a flask and the flask is on top of a digital balance, what is observed? The mass of the flask and its contents decrease.

This is because carbon dioxide is being given off.

$$MgCO_3 + 2HCl \rightarrow MgCl_2 + H_2O + CO_2$$

If the mass decrease is measured every two minutes a graph such as Figure 3.17 could be drawn to use in an investigation.

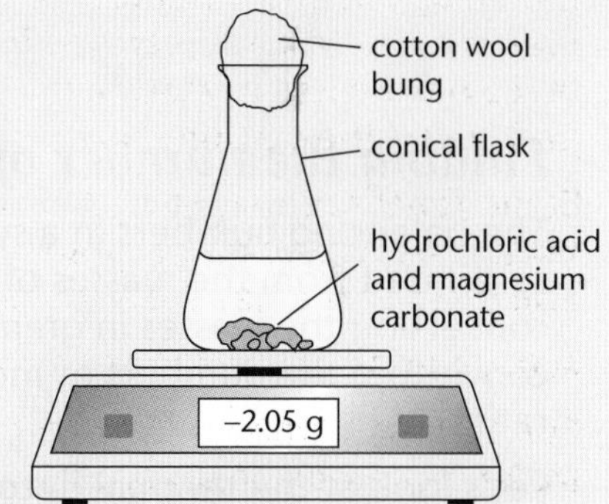

Figure 3.16 Measuring the mass of carbon dioxide given off

It is not only carbon dioxide or water vapour that will be produced in a reaction. Other reactions will produce other gases such as hydrogen. These reactions will produce a decrease in mass also.

$$Mg + 2HCl \rightarrow MgCl_2 + H_2$$

If the acid in this reaction is in excess, then the magnesium is the limiting reactant. Once the magnesium has been used up no more hydrogen can be made.

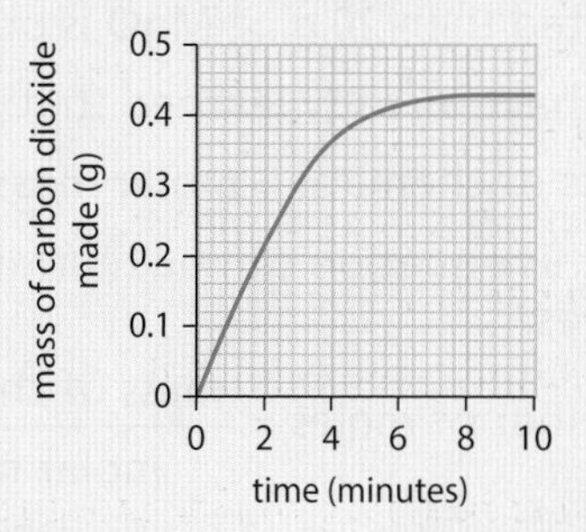

Figure 3.17 Graph showing mass of CO_2 given off

6. **Look at the graph in Figure 3.17. Explain when the reaction stopped and why it stopped.**
7. **The reaction between 8.4 g of $MgCO_3$ and excess hydrochloric acid solution stops after 7 minutes. The mass of the flask and contents has decreased by 4.4 g. Sketch the graph you might obtain when monitoring this reaction.**
8. **6.54 g of zinc was added to 6.97 g of nitric acid. At the end of the reaction, 2.92 g of zinc remained.**
   a Explain which of the reactants was the limiting reactant.
   b State the mass of zinc and mass of nitric acid that reacted.
9. **Explain whether mass will be gained or lost in the following reaction. CH_3COOH and H_2O are liquids and O_2 and CO_2 are gases.**

$$CH_3COOH + 2O_2 \rightarrow 2CO_2 + 2H_2O$$

DID YOU KNOW?

It is possible to work out exactly how much gas should be given off if we know the amount of limiting reactant we start with.

Using moles to balance equations

Learning objectives:

- convert masses in grams to amounts in moles
- balance an equation given the masses of reactants and products
- change the subject of a mathematical equation.

KEY WORDS

balanced equations
moles
thermal decomposition
molar mass

We have already learned how to calculate molar masses and use balanced equations **to calculate the mass of a product that should be made from a given mass of reactant. Can we turn these calculations around so that we can use the mass of substance to find the number of moles that react and so balance an equation?**

HIGHER TIER ONLY

Finding the number of moles

The balancing numbers in a symbol equation can be calculated from the masses of reactants and products by converting the masses in grams into amounts in moles and converting the numbers of **moles** to simple whole number ratios.

Let's look at the **thermal decomposition** of magnesium carbonate that produces magnesium oxide (MgO) and carbon dioxide (CO_2). We find that if 42 g of $MgCO_3$ is heated then 20 g of MgO is produced.

We have already learned that:

mass of chemical = **molar mass** × number of moles

So this equation can be rearranged to find the number of moles:

$$\text{number of moles} = \frac{\text{mass of chemical}}{\text{molar mass}}$$

So to calculate the number of moles in 20 g of MgO:

First calculate the molar mass M_r of MgO:

(A_r Mg = 24, A_r O = 16) M_r MgO = 24 + 16 = 40 g/mol.

Then use the equation: $\text{number of moles} = \frac{\text{mass of chemical}}{\text{molar mass}}$

$$\text{number of moles} = \frac{20}{40} = 0.5 \text{ moles}$$

If we heated 42 g of $MgCO_3$, then we would get 20 g of MgO.

Figure 3.18 The thermal decomposition of $MgCO_3$

KEY INFORMATION

In this case 22 g of CO_2 would have been driven off: remember the conservation of mass.

How many moles is 42 g of $MgCO_3$?

Using the same calculation as above we would calculate:

First the molar mass M_r of $MgCO_3$: (A_r Mg = 24, Ar C = 12 Ar O = 16)

$$M_r\ MgCO_3 = 24 + 12 + (3 \times 16) = 84\ g/mol.$$

Then use the equation: number of moles = $\frac{\text{mass of chemical}}{\text{molar mass}}$

number of moles = $\frac{42}{84}$ = 0.5 moles

1 What is the mass of MgO that would be produced by 84 tonnes of $MgCO_3$?

Balancing the equation

$$MgCO_3 \rightarrow MgO$$

So the previous calculation tells us that:

0.5 mole $MgCO_3 \rightarrow$ 0.5 mole MgO

so 1 mole of $MgCO_3$ makes 1 mole of MgO.

The ratio is 1:1

The ratio of moles in the equation is therefore:

$$MgCO_3 \rightarrow MgO + ?\ CO_2$$

But what about the number of moles of CO_2? Looking at the formula there must be 1 mole of CO_2 produced.

We need to check by mass. We saw that to conserve mass 22 g of CO_2 would have been produced. Molar mass of CO_2 is 44 g/mol.

number of moles = $\frac{\text{mass of chemical}}{\text{molar mass}} = \frac{22}{44}$

So again 0.5 moles of $MgCO_3$ makes 0.5 moles of MgO and 0.5 moles of CO_2, which is a 1:1:1 ratio.

The **balanced equation** is therefore:

$$MgCO_3 \rightarrow MgO + CO_2$$

Applying the calculations

Aluminium oxide, Al_2O_3 produces aluminium, Al and oxygen, O_2

If 204 g of Al_2O_3 produce 108 g of Al work out the number of moles of Al_2O_3, Al and O_2 involved and hence write out the full balanced equation.

2 Explain how you worked out the number of moles of Al_2O_3 and Al from the masses given.

3 Explain how you worked out the number of moles of O_2 and how you deduced the mole ratio of the three substances.

DID YOU KNOW?

In this example we found that 0.5 moles of $MgCO_3$ produced 0.5 moles of MgO. So we can say that 1 mole of $MgCO_3$ would produce 1 mole of MgO.

As these are ratios we can also say 17 tonnes of $MgCO_3$ would produce 17 tonnes of MgO.

KEY CONCEPT

Limiting reactants and molar masses

Learning objectives:

- identify which reactant is in excess
- explain the effect of a limiting quantity of a reactant on the amount of products
- calculate amount of products in moles or masses in grams.

KEY WORDS

limiting
excess
mole
directly proportional

When some sodium is put in water to show what happens, only a tiny amount is used. This is for safety. Only a tiny amount is used to limit the reaction. When it's gone, it's gone.

HIGHER TIER ONLY

Reactants in excess

In a reaction between magnesium and dilute hydrochloric acid a small amount of one of the reactants is used and a lot more of the other reactant is used. We need to decide:

- when to use a lot more of one reactant
- what 'a lot more' is.

For example, when making the salt magnesium chloride all the acid needs to be used up, as it is hard to separate the magnesium chloride as a solid from the acidic solution.

$Mg(s) + 2HCl(aq) \rightarrow MgCl_2(aq) + H_2(g)$

It is much easier to have too much magnesium metal, use up all the acid reacting with the metal, then filter off the solution from the **excess** magnesium solid.

In this case the **limiting** reagent is the hydrochloric acid and the excess is magnesium.

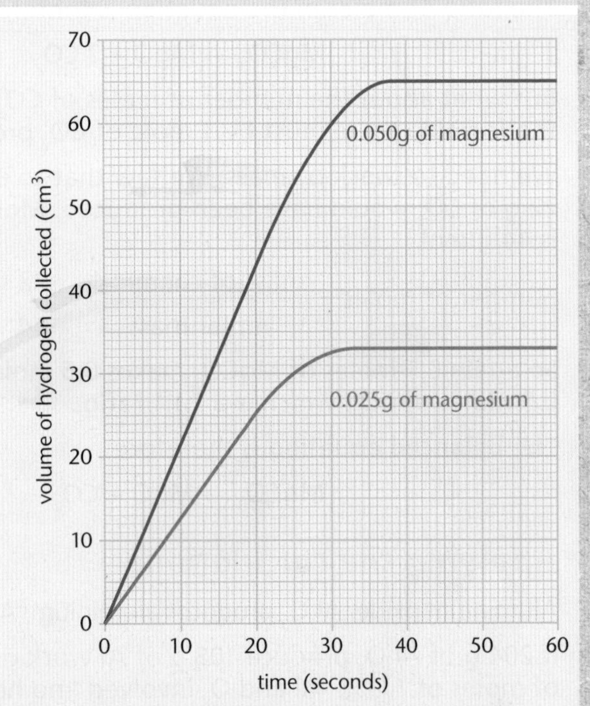

Figure 3.19 How much more hydrogen is formed if the amount of magnesium is doubled?

On the other hand if we are investigating the rate of reaction between magnesium and hydrochloric acid, we will want to use up all of the magnesium, as it is easier to measure the mass of the solid than adjust the concentration and volume of acid. So we use excess acid and a known amount of magnesium.

Magnesium is used up by the end of the reaction so it is known as the *limiting reactant*.

Look at the graph in Figure 3.19.

In this reaction, both Mg and HCl are the reactants.
To find the rate of reaction
we limit the amount of magnesium.

1. **Use construction lines to find the rate of reaction when 0.050 g of magnesium is used (blue line).**
2. **Determine the volume of hydrogen formed when:**
   a 0.025 g of magnesium is used.
   b 0.050 g of magnesium is used.
   Describe the relationship between the amount of reactant used and the amount of product formed.
3. **Predict the volume of hydrogen if 0.0125 g of magnesium is used. Sketch this reaction line on a copy of the graph.**

ADVICE

The limiting reactant *limits* the amount of product that can be made.

How much product?

The amount of product is **directly proportional** to the amount of limiting reactant used. So, if the amount of this reactant doubles, the amount of product doubles. The more reactant that is used, the more product it forms.

4. **Explain what is meant by 'limiting reactant'.**
5. **Predict what will happen to the amount of product if the limiting reactant is quadrupled.**

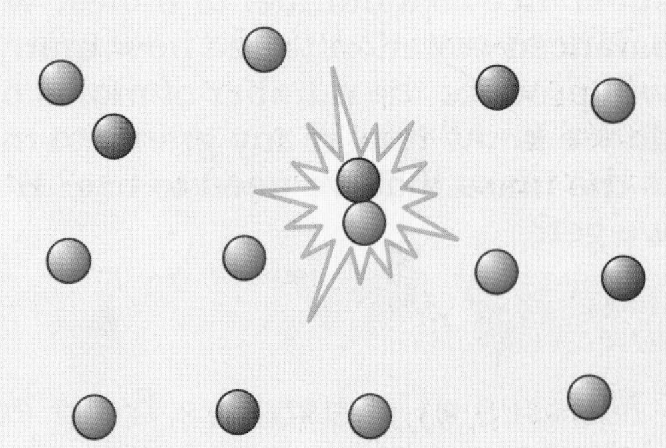

Figure 3.20 Which colour is the limiting reactant in this model of colliding particles?

Molar masses and moles again

$Mg(s) + 2HCl(aq) \rightarrow MgCl_2(aq) + H_2(g)$

Magnesium is the limiting reactant.

If 24 g of magnesium had been used in the reaction we should have made 95 g of $MgCl_2$

If 12 g of magnesium had been used in the reaction we should have made 47.5 g of $MgCl_2$

If 6 g of magnesium had been used in the reaction we should have made 23.75 g of $MgCl_2$

6. **Work out the mass of magnesium chloride that could be made from 3 g of magnesium.**
7. **Explain why the numbers in the first row above (24 g Mg and 95 g of $MgCl_2$) were chosen.**
8. **In the reaction:**

   $Mg(s) + 2HCl(aq) \rightarrow MgCl_2(aq) + H_2(g)$

   If 2 moles of $MgCl_2$ are needed,

   a determine how many moles of Mg must be used at the start.

   b determine the minimum number of moles of HCl that must be used for the reaction.

DID YOU KNOW?

If the mass of the atomic mass of an element is measured out in grams this will be the same as the mass of one mole (the molar mass).

Amounts of substances in equations

Learning objectives:

- calculate the masses of substances in a balanced symbol equation
- calculate the masses of reactants and products from balanced symbol equations
- calculate the mass of a given reactant or product.

KEY WORDS

mole
reactant mass
product mass
molar mass

We now know that chemists measure the amount of substance as a number of moles. We need to use a balanced equation to see how many moles of reactant will produce the number of moles of product. But how do we know how many grams to measure out? What is the mass that we need to use? How much mass will we get?

HIGHER TIER ONLY

Masses of substance from equation

The masses of reactants and products can be calculated from balanced symbol equations. For example, looking at the reaction between magnesium and oxygen to form magnesium oxide:

$$2Mg + O_2 \rightarrow 2MgO$$

we can see that:

2 moles of Mg react with 1 mole of O_2 to produce 2 moles of MgO.

We know that the relative formula mass of:

Mg is 24	O_2 is 16 × 2	MgO is 24 + 16

So if 2 ×24 g of Mg reacts with 32 g of O_2 then 2 × 40 g of MgO is made.

$$2Mg + O_2 \rightarrow 2MgO$$

$$48\text{ g} + 32\text{ g} \rightarrow 80\text{ g}$$

If 4.8 g of Mg is used then → 8.0 g of MgO is made.

Which can be written as:

$$\frac{\text{Molar mass of substance A}}{\text{Mass of A}} = \frac{\text{Molar mass of substance B}}{\text{Mass of B}}$$

$$\text{or Mass B} = \text{Mass of A} \times \frac{\text{Molar mass of B}}{\text{Molar mass of A}}$$

Figure 3.21 Magnesium reacting with oxygen to form magnesium oxide

1. **Calculate the mass of MgO made from 6.0 g of Mg.**
2. **Calculate the mass of Mg needed to make 2.0 g of MgO**

Measuring the number of moles in different ways

Chemical equations can be interpreted in terms of **moles**. Another example is:

$Mg + 2HCl \rightarrow MgCl_2 + H_2$

This shows that one mole of magnesium reacts with two moles of hydrochloric acid to produce one mole of magnesium chloride and one mole of hydrogen gas. These are the ratios in which reactants and products relate. The actual amount of moles made will be in these ratios but can be measured in terms of mass, concentration or volume.

3. **Determine the number of moles of H_2O that will be made from six moles of propane on combustion with O_2.**

   $C_3H_8 + 5O_2 \rightarrow 3CO_2 + 4H_2O$

Predicting masses

Moles can also be used to predict masses from equations.

This is the equation for burning a fuel called heptane, C_7H_{16}:

$$C_7H_{16} + 11O_2 \rightarrow 7CO_2 + 8H_2O$$

What mass of carbon dioxide, CO_2, is formed when 100 g of C_7H_{16} is burned?

Stage 1: find the **molar masses**

Molar mass of C_7H_{16} is $(7 \times 12) + (16 \times 1) = 100$ g

Molar mass of CO_2 is $(1 \times 12) + (2 \times 16) = 44$ g

Stage 2: use the number of moles from the equation to find the ratio

From the equation, 1 mole of heptane produces 7 moles of CO_2.

The mass of 7 moles of CO_2 is $7 \times 44 = 308$ g

So 100 g of heptane will produce 308 g of carbon dioxide.

Use the relative atomic masses in the periodic table to help you answer Questions 4 and 5.

4. **Find the mass of ZnO made when 1.25 g of $ZnCO_3$ are thermally decomposed**

   $ZnCO_3 \rightarrow ZnO + CO_2$
5. **Find the mass of $CuCO_3$ needed to make 7.95 g of CuO.**

KEY INFORMATION

Don't forget that as these are *ratios* the calculation can also be rearranged to:

$$\frac{\text{Mass of substance A}}{\text{Molar mass of A}} = \frac{\text{Mass of substance B}}{\text{Molar mass of B}}$$

DID YOU KNOW?

The number of moles of Mg that is actually used can be found from the mass of Mg used measured in grams. The number of moles of HCl can be found from the concentration of the acid in grams/dm³. The number of moles of H_2 made can be found from the volume collected compared with the molar volume.

KEY CONCEPT

Endothermic and exothermic reactions

Learning objectives:

- identify exothermic and endothermic reactions from temperature changes
- evaluate the energy transfer of a fuel
- investigate the variables that affect temperature changes in reacting solutions.

KEY WORDS

endothermic
energy transfer
exothermic
surroundings

How could you find out which fuel gives out most heat energy? How do you choose which heat pack to use to warm yourself? How do you choose which cool pack to use to treat an injury strain? To answer these questions, you need to investigate reactions and temperature changes.

Exothermic and endothermic reactions

Chemical reactions happen when reactants change into products.

Chemical reactions can either *give out* or *take in* energy.

Reactions that ***give out*** energy to the surroundings (release energy) are exothermic.	Reactions that ***take in*** energy from the surroundings (absorb energy) are endothermic.
Measuring temperature changes shows the type of reaction:	
if the temperature *goes up*, it is an *exothermic* reaction	if the temperature *drops*, it is an *endothermic* reaction
Exothermic reactions include combustion, many oxidation reactions and neutralisation.	Endothermic reactions include thermal decompositions, the reaction of citric acid and sodium hydrogencarbonate and photosynthesis.
Everyday uses of exothermic reactions include self-heating cans and hand warmers.	Some sports injury packs, which you use to cool a sprain, are based on endothermic reactions.
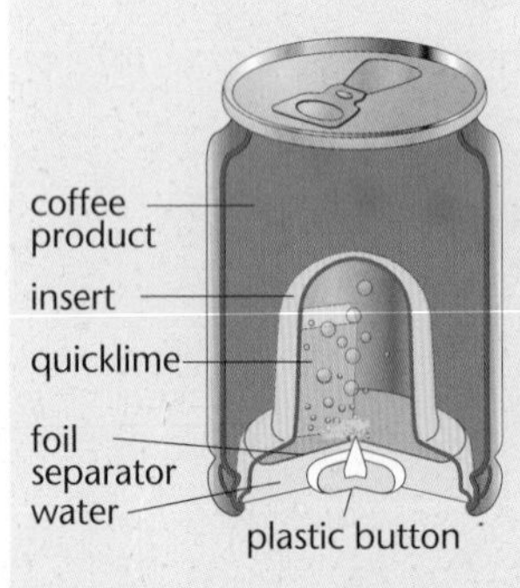 Figure 3.22 This uses an exothermic reaction	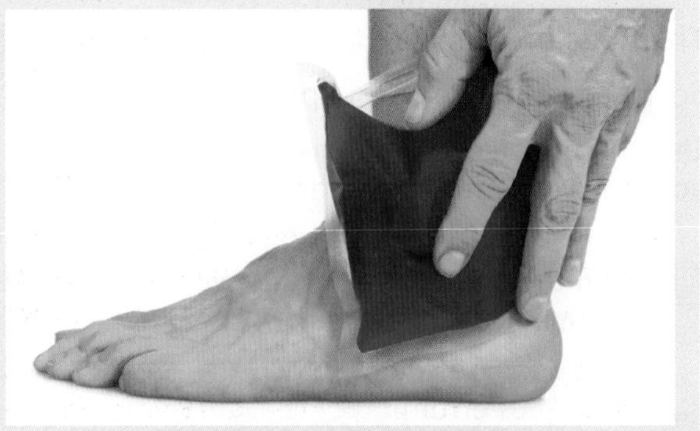 Figure 3.23 This uses an endothermic reaction

1. **Hydrochloric acid was added to sodium hydroxide and the temperature change measured. The temperature increased by 5°C. Explain whether this was an exothermic or an endothermic reaction.**
2. **Explain why photosynthesis is an endothermic reaction.**

Mending injuries

Sports injury packs rely on endothermic reactions.

	Mass of citric acid (g) added to 10 g of sodium hydrogencarbonate in water				
	6	7	8	9	10
Temperature at start °C	23	22	23	24	21
Temperature at end °C	19	16	15	14	10
Temperature difference °C					

3. **Look at the table above. Complete the data. A sports injury pack needs to lower the temperature by 7.5°C. Estimate the mass of citric acid that needs to be added to 10 g of sodium hydrogencarbonate in order to achieve this.**

Investigating and evaluating

Neutralisation is an exothermic reaction. A number of variables affect temperature changes of reacting solutions during neutralisations.

These can be investigated using apparatus such as in Figure 3.24.

Phil and Chris took these results when adding 5 cm^3 portions of acid to an alkali.

Volume of acid cm^3	0	5	10	15	20	25	30	35	40	45
Temperature °C	21	24	26	27	29	31	28	27	25	24

Table 5.1

4. **Draw a graph of the results from table 5.1. Explain when neutralisation took place. Suggest what was happening in the reaction at each stage, as indicated by the change in temperature.**
5. **Explain why a polystyrene cup was used during the experiment.**
6. **Phil and Chris want to investigate how the temperature change is affected by other factors. Suggest which variables could be changed and which should be kept constant.**
7. **Identify one source of error in the method and one possible source of error in the measurements. Suggest modifications to improve accuracy and minimise these errors.**

DID YOU KNOW?

You do not have to calculate the energy changes (as in Physics where you use $\Delta E = mc\theta$), but temperature change is a good way to indicate the energy change if you keep the substance the same (i.e. water, so *c* is constant) and use the same amount (so *m* is constant).

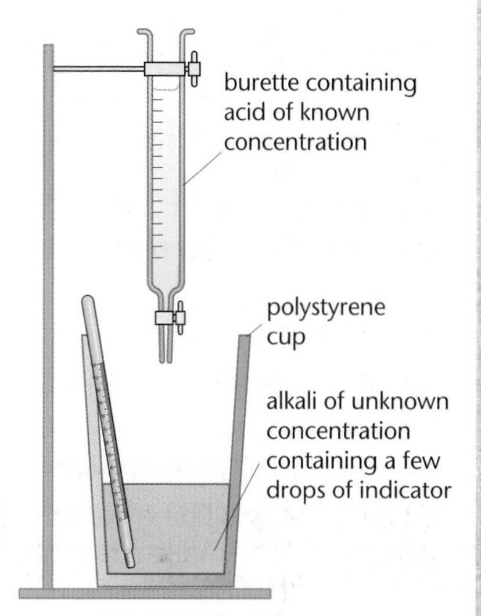

Figure 3.24 Measuring temperature rises during neutralisation

REMEMBER!

An exothermic reaction is one that transfers energy to the surroundings so the temperature of the surroundings increases.

An endothermic reaction is one that takes in energy from the surroundings so the temperature of the surroundings decreases.

Reaction profiles

Learning objectives:

- draw simple reaction profiles (energy level diagrams)
- use reaction profiles to identify reactions as exothermic or endothermic
- explain the energy needed for a reaction to occur and calculate energy changes.

KEY WORDS

exothermic
product
profile
reactant

All chemical reactions need enough energy to start. Sometimes energy needs to be taken in from the surroundings. We know that petrol burns easily in air. But you can have a beaker of petrol on the bench in your lab without anything happening. The reaction needs some energy to start. A lit match works well.

Colliding particles

Chemicals are made up from 'particles' that have their own energy.

When two chemicals react their particles have energy and are moving.

As the particles meet they collide.

Some collisions do not result in a reaction. They are unsuccessful. Some collisions do result in a reaction. They are *successful.* They have enough energy to react.

These successful colliding particles have at least the minimum energy needed to make a reaction. They have enough energy to activate the reaction.

This energy needed to start is called the *activation energy.*

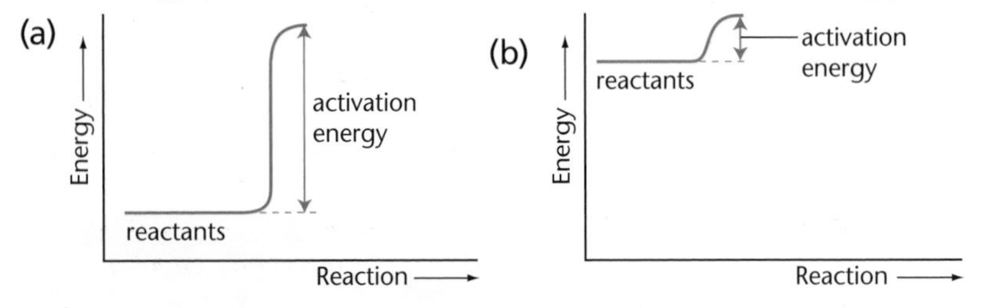

Figure 3.27 Activation energy for the reactions. Reactants (a) have lower energy than reactants (b) and a higher activation energy.

1 **Explain what provides the activation energy in a firework.**

Endothermic and exothermic reactions

Once the activation energy is reached a reaction can proceed until the products are made.

The products will have energy too, once they are made. Their energy level can be drawn onto the diagram of the reaction progress.

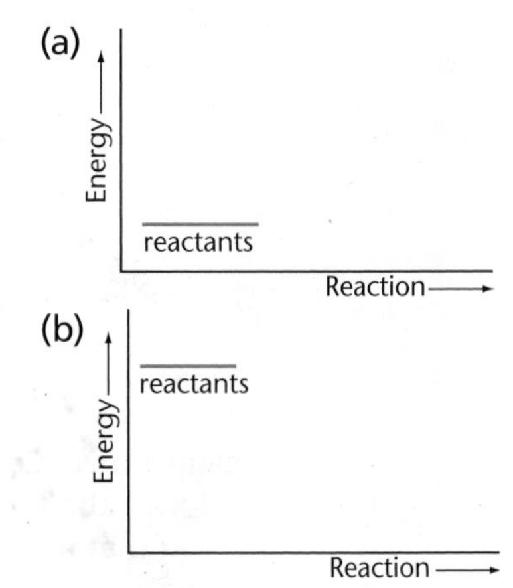

Figure 3.25 The reactants of two reactions. Reactants (a) have lower energy than reactants (b).

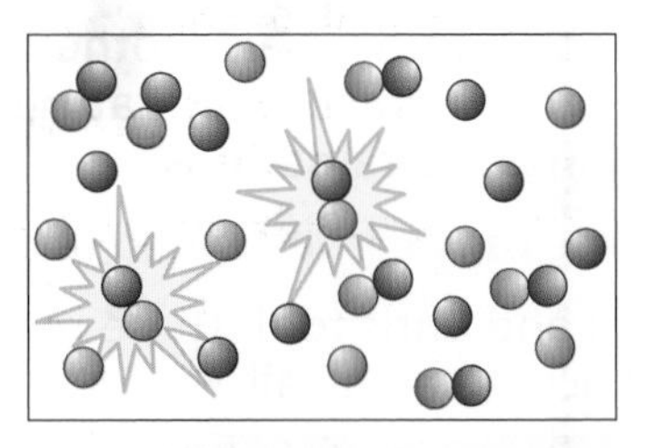

Figure 3.26 Some particles collide without sufficient energy but some particles collide with sufficient energy to react

DID YOU KNOW?

Energy is conserved in chemical reactions. The amount of energy in the universe at the end of a chemical reaction is the same as before the reaction takes place. If a reaction transfers energy to the surroundings the product molecules must have less energy than the reactants, by the amount transferred.

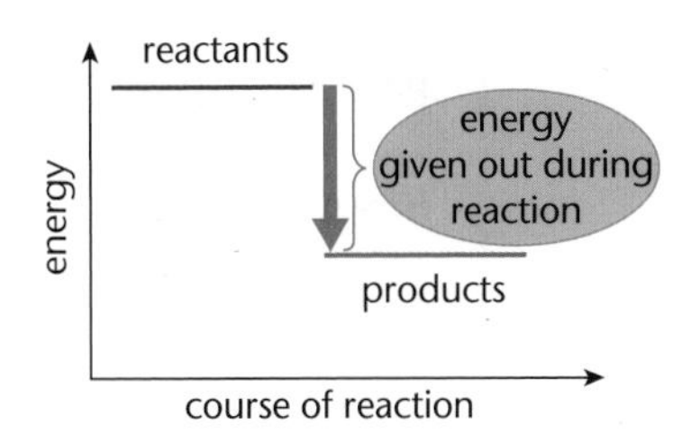

Figure 3.28 A reaction profile for an exothermic reaction

In Figure 3.28 you can see that the product energy level is lower than the reactant energy level. This means that energy has been *given out* (to the surroundings) in the reaction.

This is an *exothermic* reaction.

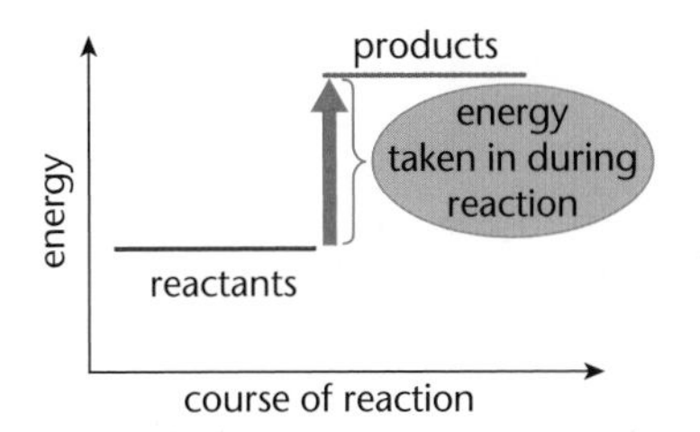

Figure 3.29 A reaction profile for an endothermic reaction

In Figure 3.29 you can see that the product energy level is higher than the reactant energy level. This means that energy has been *taken in* (from the surroundings) in the reaction.

This is an *endothermic* reaction.

REMEMBER!

Chemical reactions can occur only when reacting particles collide with each other and with sufficient energy. The minimum amount of energy that particles must have to react is called the activation energy. A lighted match heats petrol vapour enough to make the particles move with more energy and collide with oxygen particles successfully. The reaction that follows is usually obvious.

Now we have all the stages, we can draw complete reaction **profiles**.

Reaction profiles can be used to show three things:

1 the relative energies of **reactants** and **products**
2 the activation energy
3 the overall energy change of a reaction.

A reaction profile for an **exothermic** reaction can be drawn as shown in Figure 3.30.

2 **Draw a complete energy profile for an endothermic reaction.**

3 **Tom and Nabila reacted zinc with acid. They found that the temperature of the acid increased as the zinc was added. Draw a reaction profile for this reaction.**

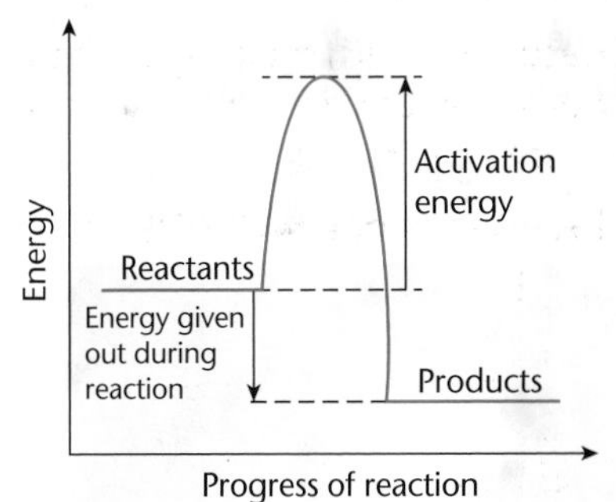

Figure 3.30 A complete reaction profile for an exothermic reaction

HIGHER TIER ONLY

Breaking and making bonds

Chemical reactions can be thought of as happening in two stages:

In stage 1 energy is needed to break the bonds in the reactants into separate atoms. This energy is called activation energy. It is an endothermic process.

In stage 2 the separated atoms combine to form the products, and energy is released. This is an exothermic process.

If less energy is needed to break bonds than released on making new bonds, then a reaction is exothermic overall.	*If more energy is needed to break bonds than released on making new bonds, then a reaction is endothermic overall.*

4 **The reaction between hydrogen and oxygen to form water is exothermic. Explain why in terms of the bond breaking and making.**

DID YOU KNOW?

Catalysts lower the activation energy. You will find out about this in topic 5.12

Energy change of reactions

Learning objectives:

- describe the energy changes in bond breaking and bond making
- explain how a reaction is endothermic or exothermic overall
- calculate the energy transferred in chemical reactions using bond energies.

KEY WORDS

bond breaking
bond making
endothermic
exothermic

The bond energy of bonds between different atoms can be measured. This data is widely available and can be used to calculate the overall energy of reactions. This is an example of using experimental data to provide data that can be applied theoretically to other reactions.

Breaking and making bonds

Bond *breaking* is an *endothermic* process	Bond *making* is an *exothermic* process

Burning fuels is a clear example where we can see that overall energy is given out. They are exothermic reactions.

However, the overall process is a combination of both **endothermic** and **exothermic** processes.

Bond breaking must take place first, followed by **bond making**.

The bonds of methane and of oxygen are broken. Energy is taken in.	The bonds of carbon dioxide and of water are made. Energy is given out.

In this reaction more energy is given out than taken in so the overall reaction is exothermic.

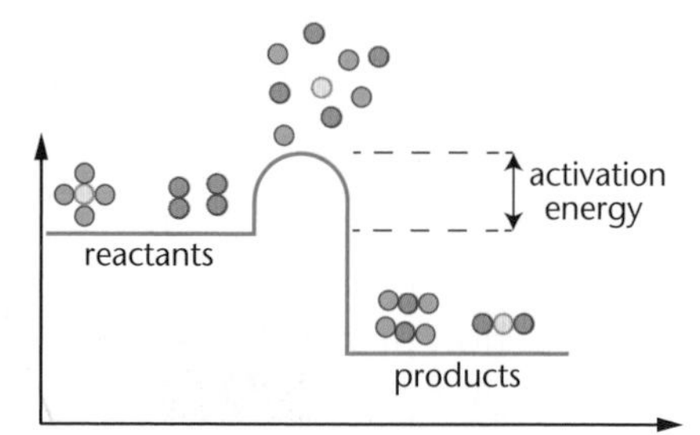

Figure 3.31 Energy is taken in to break bonds. Once the activation energy has been exceeded, energy is given out when making bonds.

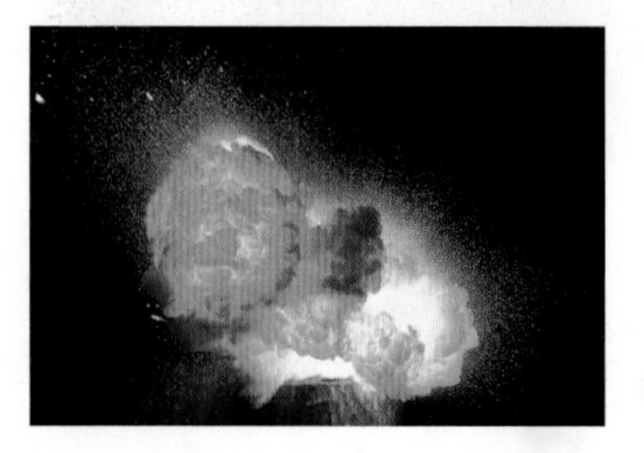

Figure 3.32 An exothermic reaction

Figure 3.33 Photosynthesis. Is it an endothermic or an exothermic reaction?

1 Draw an energy profile and explain what happens in terms of bond breaking and bond making in this reaction. During the reaction the temperature goes down.

$Na_2CO_3 + 2CH_3COOH \rightarrow 2CH_3COONa + H_2O + CO_2$

Bond energies

To decide if a chemical reaction is exothermic or endothermic, the amount of energy needed for bond breaking must be compared with the amount of energy released when new bonds are formed.

REMEMBER!

Energy must be supplied to break bonds in the reactants. Energy is released when bonds in the products are formed.

The energy needed to break bonds and the energy released when bonds are *formed* can be calculated from *bond energies*.

The difference between the		
sum of the energy needed to break bonds in the reactants	and	sum of the energy released when bonds in the products are formed
is the overall energy change of the reaction.		

In an endothermic reaction, the energy needed to break existing bonds is greater than the energy released from forming new bonds.	In an exothermic reaction, the energy released from forming new bonds is greater than the energy needed to break existing bonds.

Some common bond energies (in kJ/mol) are:

C—H	C—C	C—O	C=O	O=O	H—O
412	368	352	532	498	465

For example in the reaction:

$C_3H_8 + 5O_2 \longrightarrow 3CO_2 + 4H_2O$

$$H_3C{-}CH_2{-}CH_3 + 5O_2 \longrightarrow 3\ O{=}C{=}O + 4\ H{-}O{-}H$$

the sums of the bond energies (in kJ/mol) are:

Reactants	Products
2 × C—C = 2 × 368 = 736 8 × C—H = 8 × 412 = 3296 5 (1 × O=O) = 5 (1 × 498) = 2490	3 (2 × C=O) = 3 (2 × 532) = 3192 4 (2 × H—O) = 4 (2 × 465) = 3720
= 6522	= 6912

The difference is –390 kJ/mol. This means that the reaction is exothermic (energy released indicated by the – sign) and 390 kJ/mol is released.

KEY INFORMATION

It is a convention that an exothermic reaction has a minus (negative) sign before the number and an endothermic reaction has a plus (positive) sign.

2 **What is sum of the bond energies for 2 moles of the molecule below?**

$$H_3C{-}C({=}O){-}O{-}H$$

Calculations

Bond energies can be used to find the overall energy change in a reaction. Use the bond energies in the table above to complete the calculations in question 3.

3 **Pentane, C_5H_{12}, can be used as a fuel. Calculate the energy change for the reaction:**

$C_5H_{12} + 8O_2 \rightarrow 5CO_2 + 6H_2O$

4 **Find the energy released when $CH_2{=}CH_2$ reacts with H_2 to form C_2H_6. The bond energy for C=C is 614 kJ/mol and H–H is 432 kJ/mol.**

MATHS SKILLS

Recognise and use expressions in decimal form

KEY WORDS

decimals
integers
scale
fraction

Learning objectives:

- read scales in integers and using decimals
- calculate the energy change during a reaction
- calculate energy transferred for comparison.

It is important to be able to calculate the energy change during a reaction. If we want to make comparisons between fuels or foodstuffs, for example, we need to know the amount of energy transferred when they burn. Can you remember the unit for measuring energy changes?

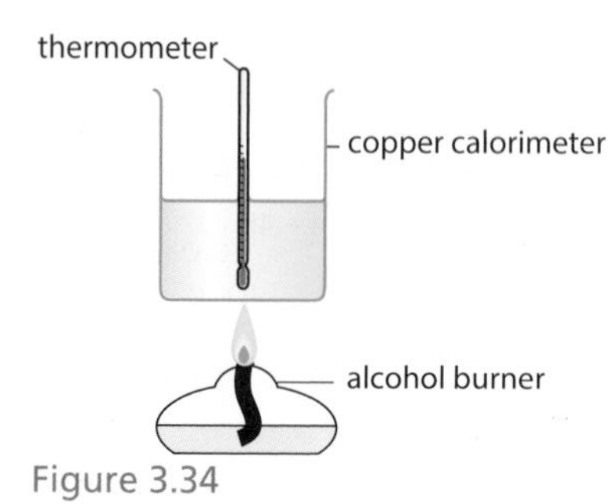

Figure 3.34

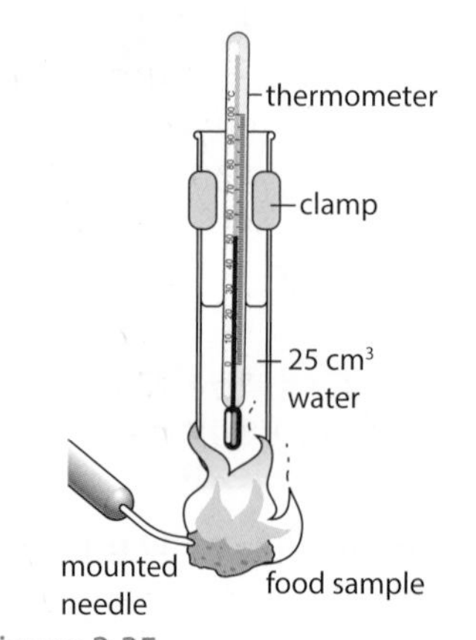

Figure 3.35

Temperature changes and energy changes

The easy part of any experiment is to measure the temperature.

We can read a thermometer with a **scale** that goes up in whole numbers (Figure 3.36). Whole numbers are called **integers**.

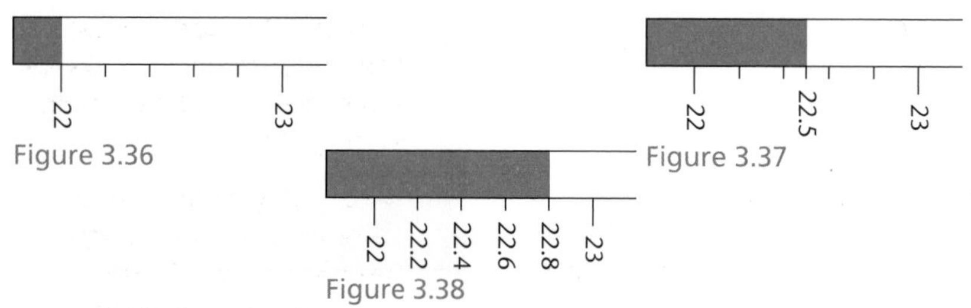

Figure 3.36

Figure 3.37

Figure 3.38

We can also read a thermometer with a scale that goes up in half degrees. Between each whole numbers is a 'half degree' mark. Figure 3.37 shows a reading of $22^1/_2$ °C which is also 22.5 °C.

$22^1/_2$ °C is a reading with a **fraction** form and 22.5 °C is a reading with a **decimal** form.

Figure 3.38 shows a thermometer with a scale that increases by 2/10th each time or in **decimal** form by 0.2 °C.

KEY INFORMATION

Energy change is *not* the same as temperature change. Temperature is *measured* but energy is *calculated*. Temperature is measured in °C (degrees Celsius). Energy is calculated in joules (J) or kilojoules (kJ).

1 A thermometer is used to measure a person's body temperature. The thermometer reads 36.8 °C. Which thermometer was used to make this recording?

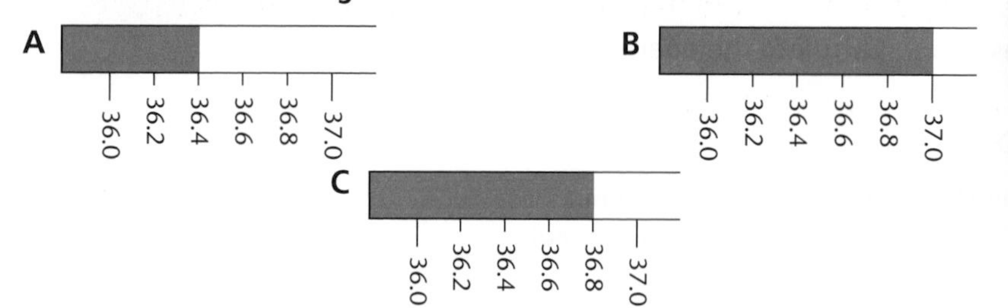

2 **Normal body temperature can be 0.4 °C higher than 36.8 °C. What is this higher temperature?**

Decimal form

It is possible to write numbers that are smaller than 1 either as fractions or in decimal form.

Decimal form begins when 1 is divided by 10. This is written as 0.1. The number 1 has moved down the column hierarchy to the right of the decimal point.

Can you think what 1 divided by 100 would be? The number 1 would move two places down to be 0.01

Fraction	Decimal	0	.	0	00	000	0000
one-tenth	zero point one	0	.	1			
one-hundredth	zero point zero one	0	.	0	1		
one-thousandth	zero point zero zero one	0	.	0	0	1	
one-ten thousandth	zero point zero zero zero one	0	.	0	0	0	1

3 **What would be the decimal if 1 was divided by 1000?**

4 **What would be the decimal if 1 was divided by 100 000?**

5 **What is 1 divided by to make the number 0.00000001?**

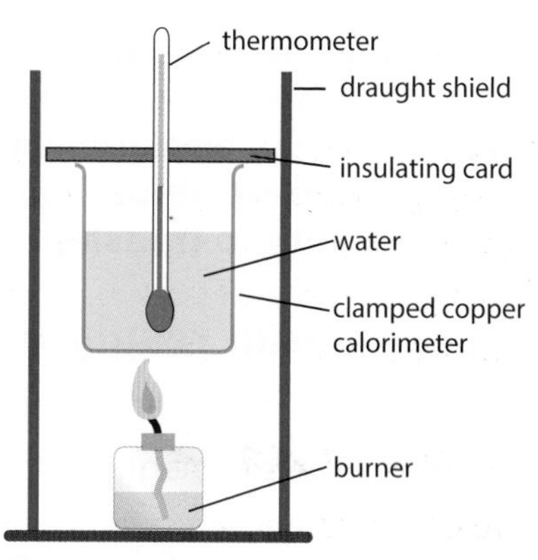

Figure 3.39

Comparing fuels

Alcohols are easily burned in oxygen and can be used as fuels. The amount of energy they transfer out *cannot* be measured, but this energy can be transferred to a body of water. The temperature increase of the water *can* be measured.

Imagine if the energy from some fuel is transferred either to 10 g of water or to 50 g of water. What would happen? The temperature of the 10 g of water would go higher. We can see that our calculation needs to take account of the *mass* of the water.

Now imagine that we transferred the energy to olive oil, not to water. Would the temperature go up by the same amount? Possibly not. So, we also need to take account of the *specific heat capacity* of the receiving liquid.

From these three quantities (temperature increase, mass of the receiving liquid, specific heat capacity of the receiving liquid) we can calculate the energy transferred from an alcohol (Figure 3.39) using the equation:

energy transferred by fuel = mass of water × specific heat capacity of water × temperature change of water

energy transferred by fuel = 100 × 4.2 × temperature change of water

$$E = m \times c \times (t_2 - t_1)$$

6 **Propanol burned to raise the temperature of 100 g of water by 24 °C. How much energy did it transfer?**

7 **When 3.8 g of propanol were burned, the energy transferred was 13 300 J. How many joules were transferred by 1 g of propanol?**

DID YOU KNOW?

In this example, we chose the mass of water to be 100g. The specific heat capacity of water is a fixed quantity, which is 4.2 J/g/°C.

Oxidation and reduction in terms of electrons

Learning objectives:

- use experimental results of displacement reactions to confirm the reactivity series
- write ionic equations for displacement reactions
- identify in a half equation which species are oxidised and which are reduced.

KEY WORDS

displacement
half equations
ionic equations
oxidation

As we have seen, a more reactive metal can displace a less reactive metal from a solution of its compound. This is called a displacement reaction. These reactions are used to save ships from rusting and rely on one metal's ability to transfer its electrons to another metal.

Experimental results

An example of a displacement reaction is:

iron sulfate + magnesium → magnesium sulfate + iron

It happens because magnesium is more reactive than iron. The iron has been pushed out or displaced. It forms a coating on the rest of the magnesium.

This table shows what happens when four different metals are used.

Solution used	Metal being added			
	magnesium	zinc	iron	copper
magnesium sulfate	X	X	X	X
zinc sulfate	✓	X	X	X
iron sulfate	✓	✓	X	X
copper sulfate	✓	✓	✓	X

Key:

X means that nothing happens

✓ means that the added metal gets coated with the metal from the solution

The table shows that copper does not react with any of the other solutions and is, therefore, the least reactive of these metals.

1. **Use the experimental results in the table to put the metals in order of reactivity.**
2. **When a piece of tin metal is placed in three different solutions this is seen:**

copper sulfate tin is coated	zinc sulfate nothing happens	iron(II) sulfate nothing happens

Explain the position of tin in the reactivity series.

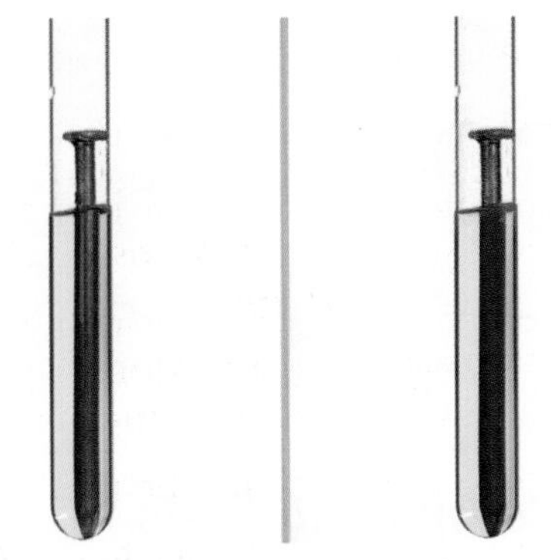

Figure 3.40 A displacement reaction. Iron is reacting with copper sulfate. What can you see is happening?

KEY SKILL

To work out what happens in a displacement reaction you do not need to remember every reaction. You need to know the order of reactivity: magnesium > zinc > iron > copper.

This means that:

- magnesium metal displaces zinc, iron and copper
- zinc displaces iron and copper
- iron displaces copper

Displacement equations

When writing an equation for a displacement reaction, the more reactive metal swaps places with the less reactive metal. (It does not matter whether the compound is a chloride, a nitrate or a sulfate).

For example: As zinc is more reactive than iron, iron is displaced from the solution.

$$Zn + FeCl_2 \rightarrow ZnCl_2 + Fe$$

3 Write an equation for the displacement reaction between nickel sulfate ($NiSO_4$) and the more reactive metal, magnesium.

DID YOU KNOW?

The half equation $Zn - 2e^- \rightarrow Zn^{2+}$ is conventionally written as

$$Zn \rightarrow Zn^{2+} + 2e^-$$

but with the first way it is easier to see what is happening.

HIGHER TIER ONLY

Ionic and half equations

All metals react by losing electrons and turning into ions. This is **oxidation**.

Displacement reactions depend on having a reactive metal element and the compound of a less reactive metal.

The more reactive the metal, the harder it pushes off electrons.

These electrons have to go somewhere; they are forced onto the ions of other metals that are not so reactive. The metal atoms that gain these electrons are reduced. This is *reduction*.

These **ionic equations** show what happens:

$$Mg + Zn^{2+} \rightarrow Mg^{2+} + Zn$$

$$Zn + Fe^{2+} \rightarrow Zn^{2+} + Fe$$

These equations can also be written as **half equations** to show the electron transfer happening.

$$Mg - 2e^- \rightarrow Mg^{2+} \qquad Zn^{2+} + 2e^- \rightarrow Zn$$

$$Zn - 2e^- \rightarrow Zn^{2+} \qquad Fe^{2+} + 2e^- \rightarrow Fe$$

oxidation is the loss of electrons and reduction is the gain of electrons.

KEY INFORMATION

Oxidation Is electron Loss Reduction Is electron Gain: OILRIG

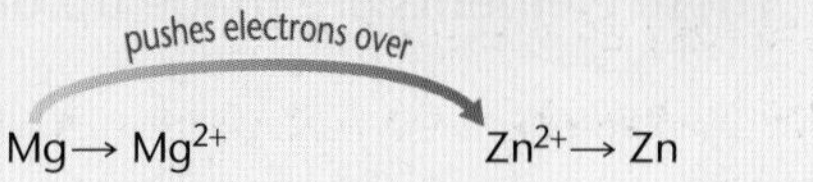

pushes electrons over

$Mg \rightarrow Mg^{2+}$ $\quad$ $Zn^{2+} \rightarrow Zn$

Figure 3.41

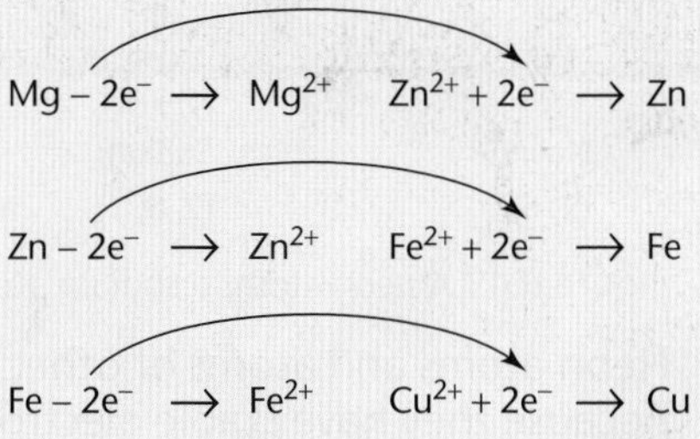

$Mg - 2e^- \rightarrow Mg^{2+}$ $\quad$ $Zn^{2+} + 2e^- \rightarrow Zn$

$Zn - 2e^- \rightarrow Zn^{2+}$ $\quad$ $Fe^{2+} + 2e^- \rightarrow Fe$

$Fe - 2e^- \rightarrow Fe^{2+}$ $\quad$ $Cu^{2+} + 2e^- \rightarrow Cu$

Figure 3.42

4 Write an ionic equation for reaction between magnesium metal and copper chloride solution.

5 Write two half equations for reaction between magnesium metal and iron(II) chloride solution.

6 Identify in these two half equations which species is oxidised and which is reduced.
$Al - 3e^- \rightarrow Al^{3+}$ $\qquad$ $Cr^{3+} + 3e^- \rightarrow Cr$

7 Magnesium is more reactive than silver. It can displace silver ions, Ag^+, from solution.

a Write the half equations for the displacement of silver ions by magnesium. Identify which is oxidised and which is reduced.

b Write the ionic equation for the displacement reaction. The number of electrons transferred must be equal.

KEY CONCEPT

Electron transfer, oxidation and reduction

Learning objectives:

- explain why atoms lose or gain electrons
- explain oxidation and reduction by electron transfer
- relate ease of losing electrons to reactivity.

KEY WORDS

oxidation
reduction
electron loss
electron gain

We have already seen that some elements react more easily than others. We also learned that atoms often need to lose electrons or gain electrons for reactions to take place and some atoms are able to do this more easily than others. How do all these observations and models relate?

KEY INFORMATION

Each element has the same *type* of atom but the atoms of different elements are different, so all sodium atoms have 11 electrons but all chlorine atoms have 17 electrons.

Losing and gaining electrons

Chemical reactions take place when elements rearrange. These elements are made up of atoms.

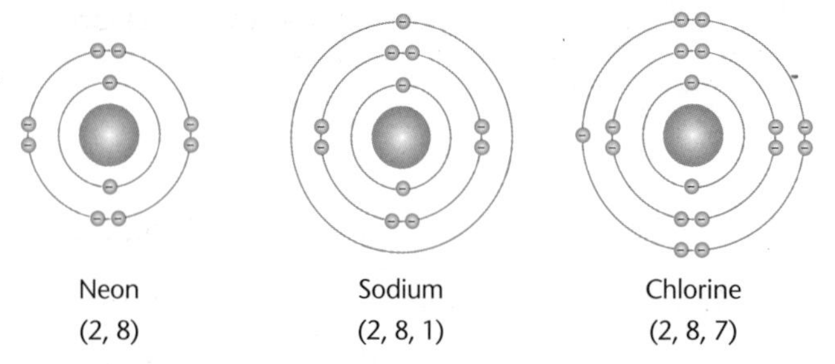

Figure 3.43 A neon atom, a sodium atom and a chlorine atom

Neon atoms and atoms of other noble gases are unreactive because they have stable electronic configurations. They do not need to gain or lose electrons to achieve stable configurations.

Sodium atoms and all other atoms of *metals* need to *lose* one or more electrons for the metals to react.

Chlorine atoms and atoms of most other *non-metals* need to *gain* one or more electrons to achieve stable configurations (or alternatively *share* electrons).

1. **How does an atom of sodium join with an atom of chlorine?**
2. **Explain why helium and argon are unreactive.**

HIGHER TIER ONLY

Oxidation and reduction

When magnesium burns in oxygen it becomes magnesium oxide. This is an **oxidation** reaction.

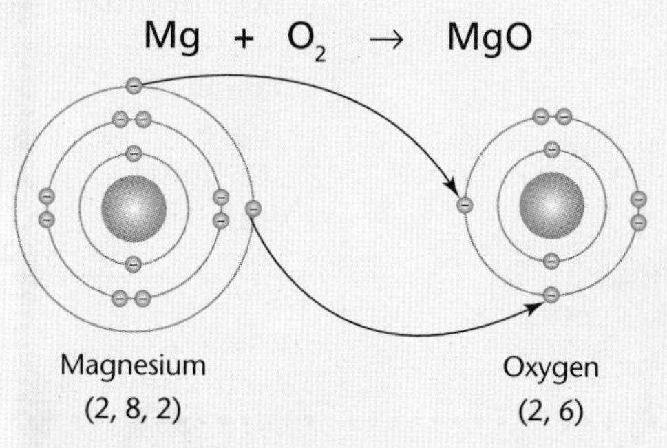

Figure 3.44 Magnesium loses two electrons during the oxidation process

When copper oxide is reacted with hydrogen the oxygen is removed. This is a **reduction** reaction.

Oxidation is the **loss of electrons.**

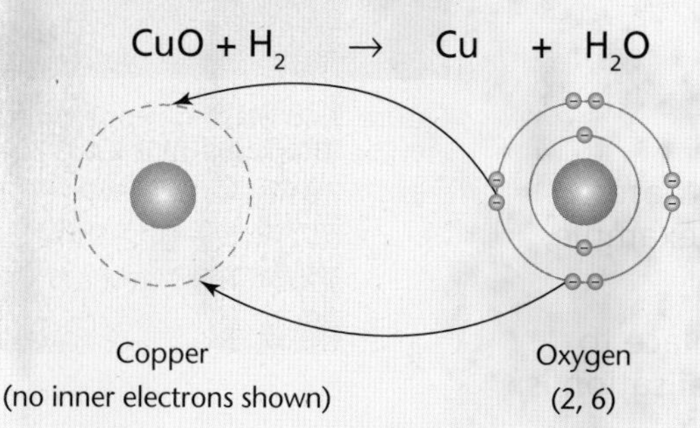

Figure 3.45 Copper atoms gain back two outer electrons when oxygen is removed

DID YOU KNOW?

Because two magnesium atoms are needed for balancing the equation there are $4e^-$ lost. That is $2Mg - 4e^- \rightarrow 2Mg^{2+}$

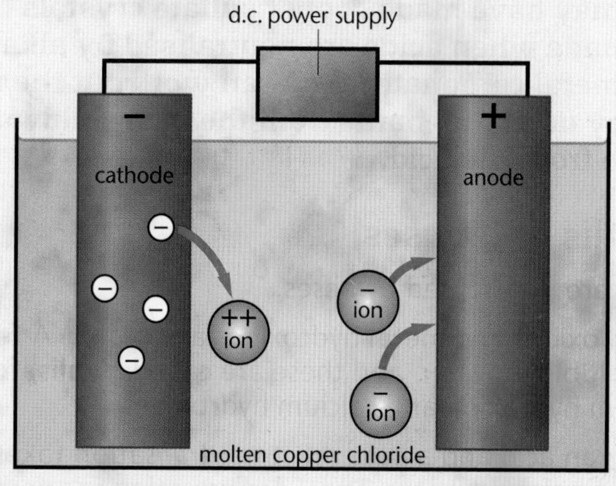

Figure 3.46 Reduction at the negative electrode (cathode), oxidation at the positive electrode (anode)

Reduction is the **gain of electrons.**

These processes also happen at electrodes during *electrolysis.*

3. **Explain these reactions in terms of oxidation and reduction:**
   a zinc to zinc oxide b Iron(II) oxide to iron
4. **Explain these reactions in terms of oxidation and reduction:**
   a silver depositing on a cathode b oxygen given off at an anode.

Redox reactions and ease of transfer

Some metals lose their electrons more easily than others. It depends on the 'pull' of the nucleus on the electrons leaving. Potassium atoms lose their one electron more easily than magnesium loses its two electrons. A reactivity series can be drawn up to show the order in which metals may lose their electrons more easily.

5. **Explain why the electrons from calcium are lost more easily than from magnesium.**
6. **Explain why sodium is more reactive than magnesium.**

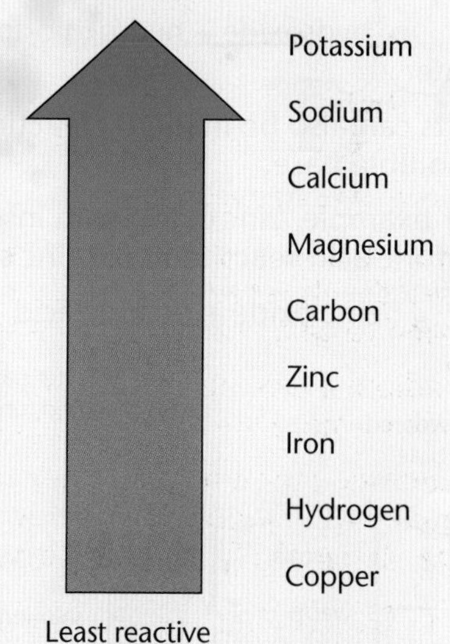

Figure 3.47 The reactivity series of metals

Neutralisation of acids and salt production

Learning objectives:

- describe ways that salts can be made
- predict products from given reactants
- deduce the formulae of salts from the formulae of common ions.

KEY WORDS

alkali
base
carbonate
salt

You may have made copper sulfate crystals before. Salts are made when acids are neutralised by alkalis, bases and metal carbonates. You can work out the name of the salt by using the name from the base followed by the name from the acid.

Figure 3.48 Making copper sulfate crystals. These are salts.

Alkalis and bases

Acids are neutralised by bases.

Metal oxides and metal hydroxides are **bases**. A few bases are soluble in water, and these are called **alkalis**, for example, sodium hydroxide and calcium hydroxide.

When an acid and a base react *neutralisation* takes place to make a **salt** and water. The word equation for neutralisation is:

acid + base → salt + water

For example, the reaction of sodium hydroxide, a soluble base, with hydrochloric acid produces sodium chloride.

sodium hydroxide + hydrochloric acid → sodium chloride + water

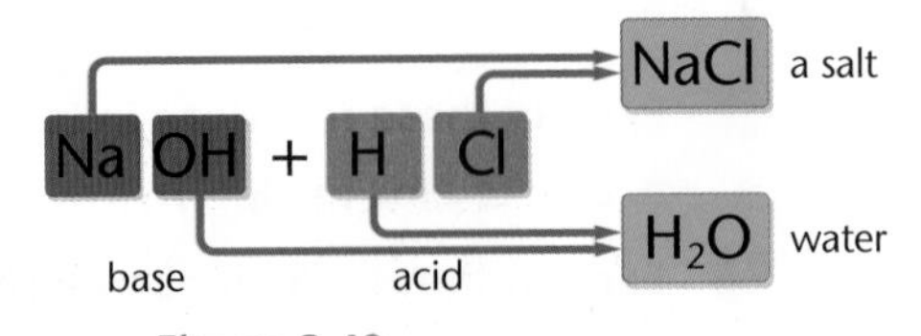

Figure 3.49

Salts can also be made by reacting acids with excess of an insoluble base.

For example, zinc oxide is an insoluble base. Zinc oxide and sulfuric acid react to make zinc sulfate and water.

zinc oxide + sulfuric acid → zinc sulfate + water

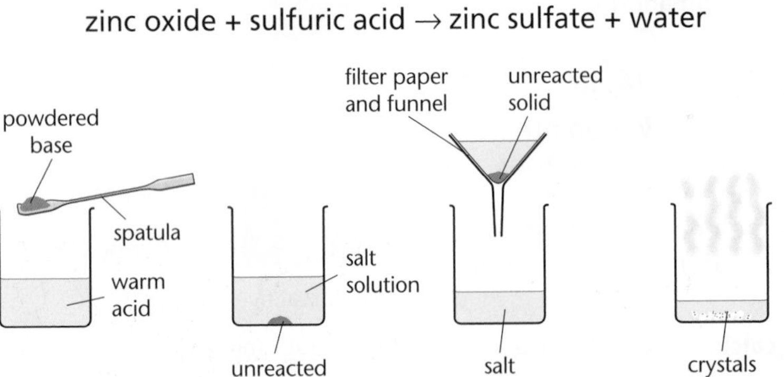

Figure 3.50 Making a salt with excess base or carbonate

Metal **carbonates** react with dilute acid to make a metal salt, but this time the gas carbon dioxide is produced. Water is also formed.

KEY INFORMATION

'Common salt' is sodium chloride, NaCl, but other chlorides, sulfates and nitrates are also salts.

acid + metal carbonate → salt + water + carbon dioxide

Magnesium carbonate reacts with hydrochloric acid to make magnesium chloride, carbon dioxide and water.

magnesium carbonate + hydrochloric acid → magnesium chloride + carbon dioxide + water

1. **Write the word equation for the reaction between potassium hydroxide and hydrochloric acid.**
2. **Write the word equation for the reaction between copper carbonate and nitric acid.**

Predicting the names of salts

Acids react with alkalis to make salt and water. The second part of the name of the salt is taken from the acid.

The first part of the name of the salt is taken from the base.

3. **Predict the name of the salt formed when nitric acid reacts with zinc oxide.**
4. **Predict the name of the salt made from the reaction between sulfuric acid and copper carbonate.**

Common acid	The salt made
nitric	nitrate
sulfuric	sulfate
hydrochloric	chloride

DID YOU KNOW?

The names of salts from other acids can be predicted. For example, citric acid (the acid in lemons) makes sodium citr*ate*.

Formulae of salts from ions

The formula of a salt can be deduced from the formulae of the ions.

This is a table of common ions showing their charges:

Ions from alkalis and carbonates		Ions from bases and carbonates		Ions from acids	
Sodium	Na^+	Magnesium	Mg^{2+}	chloride	Cl^-
Potassium	K^+	Zinc	Zn^{2+}	nitrate	NO_3^-
Calcium	Ca^{2+}	Copper	Cu^{2+}	sulfate	SO_4^{2-}

For example, potassium hydroxide and nitric acid make potassium nitrate. The formula of the salt can be written by looking at the ions in the table.

K^+ single charge, NO_3^- single charge, formula KNO_3

Magnesium sulfate:

Mg^{2+} double charge, SO_4^{2-} double charge, formula $Mg SO_4$

Copper chloride:

Cu^{2+} double charge, Cl^- single charge, formula $CuCl_2$

Sodium sulfate:

Na^+ single charge, SO_4^{2-} double charge, formula Na_2SO_4

5. **Deduce the formula of zinc chloride.**
6. **Deduce the formula of copper nitrate.**
7. **Write a balanced equation for the reaction between calcium carbonate and nitric acid.**

KEY INFORMATION

The nitrate ion will need a bracket around it in this formula. For example, calcium nitrate is $Ca(NO_3)_2$

Soluble salts

Learning objectives:

- describe how to make pure, dry samples of soluble salts
- explain how to name a salt
- derive a formula for a salt from its ions.

KEY WORDS

concentrate
crystallise
filter
filtrate

We have seen how crystals can be made from acids and insoluble substances. Soluble salts, such as bath salts and some fertilisers, can be made very simply in the lab by reacting, filtering, evaporating and crystallising.

Soluble salts

A soluble salt can be made by reacting an acid with an insoluble solid substance.

- The solid is added to the acid until no more of the solid reacts and it is in *excess*.
- The excess solid is **filtered** off leaving a solution of the salt.

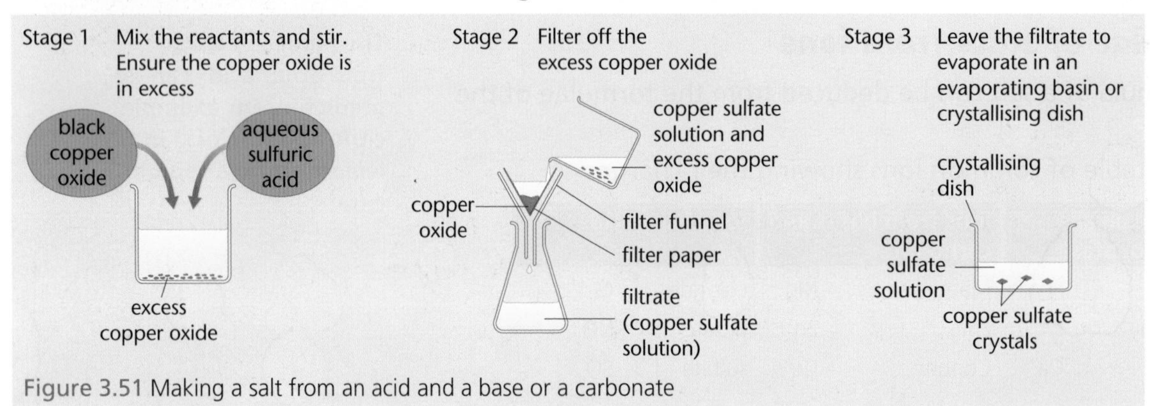

Figure 3.51 Making a salt from an acid and a base or a carbonate

- The salt solution is then heated a little to **concentrate** it.
- The concentrate is left to cool and **crystallise** to produce a solid salt.

DID YOU KNOW?

Crystals are larger if the solution is cooled more slowly.

1. **Explain why an excess of the solid is added to the acid.**
2. **Explain why the solution that has been filtered needs to be concentrated by heating a little.**

Which salts can be made?

Acids can make salts by reacting with insoluble substances such as metals, metal oxides, metal hydroxides or metal carbonates.

metal + acid → metal salt + hydrogen

metal oxide + acid → metal salt + water

metal hydroxide + acid → metal salt + water

metal carbonate + acid → metal salt + water + carbon dioxide

Some common chemical equations for the neutralisation of acids by a base or metal carbonate can be constructed using these formulae:

Acids		Bases and alkalis		Carbonates		Metals	
sulfuric acid	H_2SO_4	potassium hydroxide	KOH	sodium carbonate	Na_2CO_3	Magnesium	Mg
nitric acid	HNO_3	sodium hydroxide	NaOH	calcium carbonate	$CaCO_3$	Zinc	Zn
hydrochloric acid	HCl	calcium hydroxide	$Ca(OH)_2$	zinc carbonate	$ZnCO_3$	Iron	Fe
		copper oxide	CuO	copper carbonate	$CuCO_3$		

The equation for the reaction:

- between iron and sulfuric acid is $Fe + H_2SO_4 \rightarrow FeSO_4 + H_2$
- between zinc oxide and sulfuric acid is $ZnO + H_2SO_4 \rightarrow ZnSO_4 + H_2O$
- between sodium hydroxide and nitric acid is $NaOH + HNO_3 \rightarrow NaNO_3 + H_2O$
- between copper carbonate and sulfuric acid is $CuCO_3 + H_2SO_4 \rightarrow CuSO_4 + H_2O + CO_2$

3 Write the equation for the reaction between copper oxide and sulfuric acid.

4 Write the equation for the reaction between potassium hydroxide and hydrochloric acid.

KEY INFORMATION

Remember the nitrate ion will need a bracket around it in this formula, e.g. $X(NO_3)_y$

Charges on ions

The ions used in the examples above are:

Single charge	Cl^-	NO_3^-	OH^-	Na^+	K^+	H^+
Double charge	SO_4^{2-}	Ca^{2+}	Mg^{2+}	Zn^{2+}	Fe^{2+}	Cu^{2+}

The equations underneath the table are all balanced. This is because all the ions in the reactants and products are matched for charges.

If the salt is made using ions that are +2 and –1 (for example, copper chloride $CuCl_2$) or using ions that are +1 and –2 (for example, sodium sulfate Na_2SO_4) then the equations will need to be balanced.

$$CuO + 2HCl \rightarrow CuCl_2 + H_2O$$

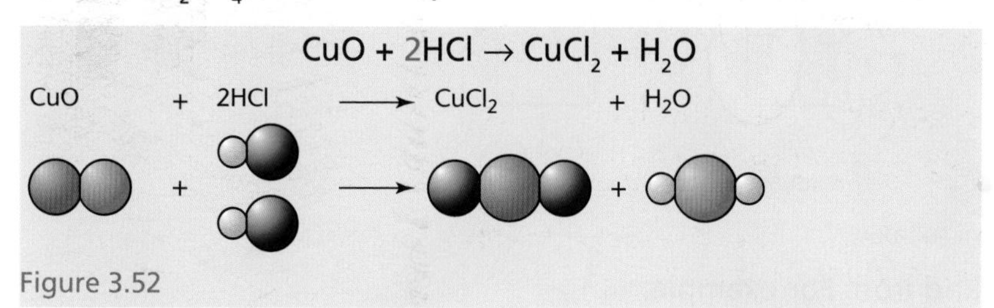

Figure 3.52

$$2NaOH + H_2SO_4 \rightarrow Na_2SO_4 + H_2O$$

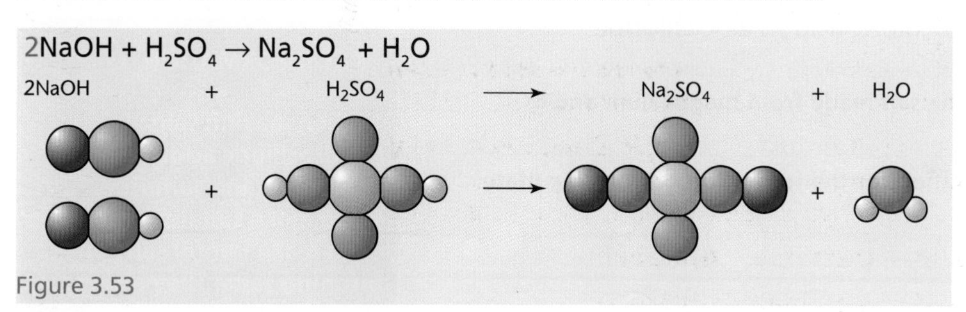

Figure 3.53

5 Write a balanced equation for the reaction between zinc carbonate and hydrochloric acid.

6 Write a balanced equation for the reaction between magnesium oxide and nitric acid.

Reaction of metals with acids

Learning objectives:

- describe how to make salts from metals and acids
- write full balanced symbol equations for making salts
- use half equations to describe oxidation and reduction.

KEY WORDS

acid
half equation
metal
salt

Crystals of salts can be made in many ways, including reacting metals with acids. To make a different salt, you use a different acid.

KEY INFORMATION

Don't forget that a salt is the neutral product of an acid added to a base or a carbonate. It is made of two ions, the positive ion from the base or carbonate and the negative ion from the acid. Sodium chloride is only one example, there are many, many others.

Making salts

Acids react with some metals to produce salts and hydrogen.

Magnesium metal reacts fairly vigorously with dilute sulfuric acid to make a solution of magnesium sulfate and hydrogen.

magnesium + sulfuric acid → magnesium sulfate + hydrogen

To make a crystallised salt:

a excess magnesium needs to be added to the acid
b the solution needs to be filtered into a crystallising dish
c the solution needs to be concentrated by evaporation
d and the solution then left to evaporate, to crystallise.

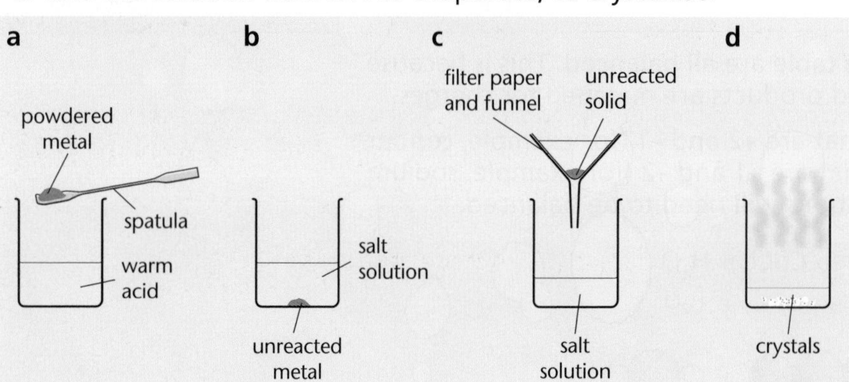

Figure 3.54 Making crystals of a magnesium sulfate

Salts can also be made using zinc and iron. For example,

zinc + hydrochloric acid → zinc chloride

KEY INFORMATION

When making salts, hydrochloric acid makes chlorides and sulfuric acid make sulfates. Iron can make iron(II) salts or iron(III) salts. They are usually different colours.

1. **Write the name of the salt made from magnesium and hydrochloric acid.**
2. **Write the word equation for the formation of iron(II) sulfate.**

Forming ions

Chemical equations for the reaction of metals with acids can be constructed using symbols of metals and formulae of acids:

Metals		Acids			
magnesium	Mg^{2+}	sulfuric acid	H_2SO_4	$2H^+$	SO_4^{2-}
iron	Fe^{2+}	hydrochloric acid	HCl	H^+	Cl^-
zinc	Zn^{2+}				

DID YOU KNOW?

Nitric acid is a common acid that you would use to make nitrate salts.

nitric acid	HNO_3	H^+	NO_3^-

In making magnesium sulfate a 2+ ion is joining with a 2– ion, so the formula is a 1:1 relationship, $MgSO_4$.

magnesium + sulfuric acid → magnesium sulfate + hydrogen

$$Mg + H_2SO_4 \rightarrow MgSO_4 + H_2$$

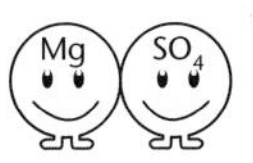

Figure 3.55

In making magnesium chloride a 2+ ion is joining with a 1– ion. This means that two of the 1– ions are needed, so the formula is a 1:2 relationship, $MgCl_2$.

magnesium + hydrochloric acid → magnesium chloride + hydrogen

$$Mg + 2HCl \rightarrow MgCl_2 + H_2$$

Figure 3.56

As two chloride ions (Cl^-) are needed to join with one Mg^{2+} ion, they must both be provided by the HCl. Therefore, two amounts of HCl are needed. The equation is balanced.

3. **Write a full balanced symbol equation for the production of zinc chloride from zinc.**
4. **Write a full balanced symbol equation for the production of iron(III) chloride, $FeCl_3$, from iron.**

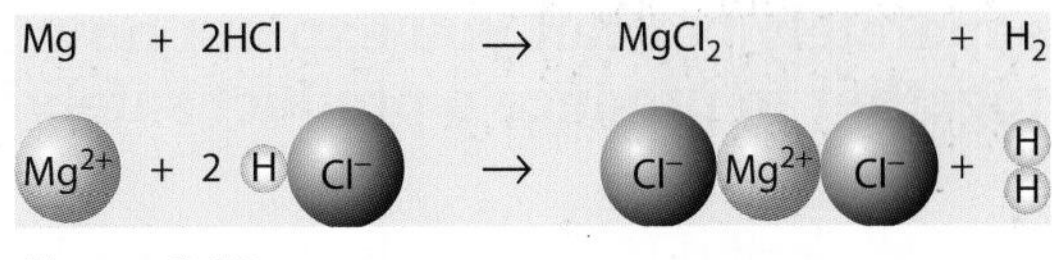

Figure 3.57

HIGHER TIER ONLY

Using half equations

When making a magnesium salt from magnesium this reaction happens:

$$Mg + 2H^+ \rightarrow Mg^{2+} + H_2$$

The **half equations** are:

$Mg - 2e^- \rightarrow Mg^{2+}$	$2H^+ + 2e^- \rightarrow H_2$
Electron loss is oxidation	Electron gain is reduction
As both these reactions take place at the same time this is a *redox* reaction	

In the equation:

Magnesium loses electrons so oxidation takes place to produce magnesium ions, (Mg^{2+}).

Hydrogen ions (H^+) gain electrons so reduction takes place.

5. **Write two half equations for the reaction between zinc and an acid.**
6. **In the reaction between iron and sulfuric acid, which component is oxidised and which is reduced?**
7. **Write a balanced chemical equation for the reaction of aluminium with sulfuric acid. Write half equations for this reaction. Aluminium forms Al^{3+} ions.**

PRACTICAL

Preparing a pure, dry sample of a soluble salt from an insoluble oxide or carbonate

Learning objectives:

- describe a practical procedure for producing a salt from a solid and an acid
- explain the apparatus, materials and techniques used for making the salt
- describe how to safely manipulate apparatus and accurately measure melting points.

KEY WORDS

insoluble
oxide
carbonate
filter
crystallisation

Being able to choose the correct techniques and carry out a specified procedure to produce a pure product is an important skill for scientists. You will probably know how to complete each of the necessary techniques separately but can you explain how to use them together to produce a product safely?

These pages are designed to help you think about aspects of the investigation rather than to guide you through it step by step.

A number of different skills are needed to carry out the production of a sample of a chemical. This topic looks at the skills of selecting techniques, apparatus and materials and managing safety.

DID YOU KNOW?

Most salts have a very high melting point, so a special heater would be needed.

Carrying out procedures

The making of magnesium sulfate from magnesium carbonate can be done in several stages using different pieces of apparatus.

This diagram has some stages of an experimental plan depicted but not all stages.

In Figure 3.58(a) the quantities of two chemicals have been measured to make the salt:

1. **Identify the two substances needed to make magnesium sulfate.**
2. **State how the quantities are measured.**
3. **State the units of measurement for each substance.**
4. **Describe the safety measures that need to be taken when adding substances.**
5. **After the two substances are mixed and stirred as in Figure 3.58(b), explain why there is some solid left in the bottom of the solution.**
6. **The excess solid is removed as in Figure 3.58(c). Name the process of removal of the solid.**

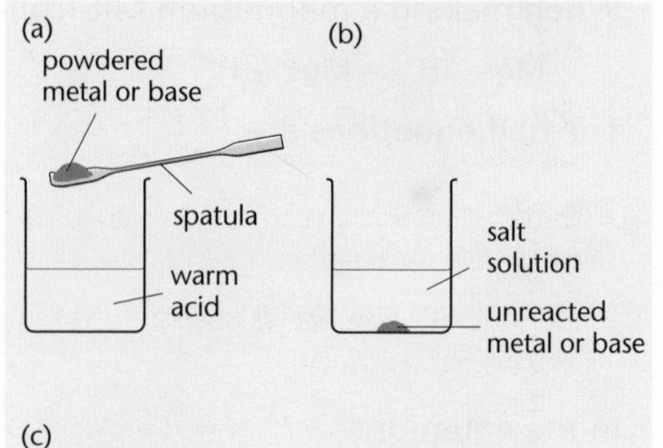

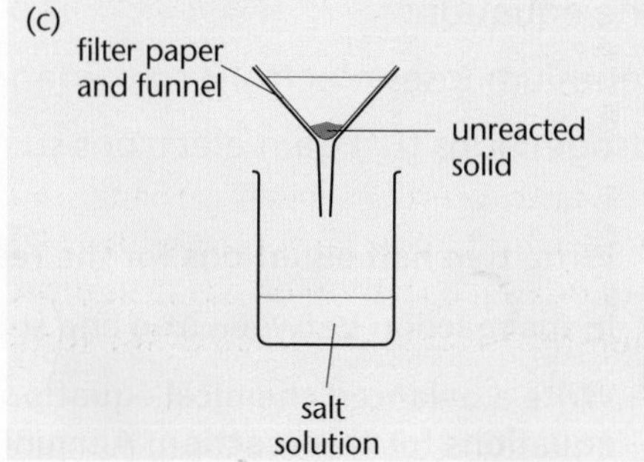

Figure 3.58

Metals		Acids			
magnesium	Mg^{2+}	sulfuric acid	H_2SO_4	$2H^+$	SO_4^{2-}
iron	Fe^{2+}	hydrochloric acid	HCl	H^+	Cl^-
zinc	Zn^{2+}				

DID YOU KNOW?

Nitric acid is a common acid that you would use to make nitrate salts.

nitric acid	HNO_3	H^+	NO_3^-

In making magnesium sulfate a 2+ ion is joining with a 2– ion, so the formula is a 1:1 relationship, $MgSO_4$.

magnesium + sulfuric acid → magnesium sulfate + hydrogen

$$Mg + H_2SO_4 \rightarrow MgSO_4 + H_2$$

Figure 3.55

In making magnesium chloride a 2+ ion is joining with a 1– ion. This means that two of the 1– ions are needed, so the formula is a 1:2 relationship, $MgCl_2$.

magnesium + hydrochloric acid → magnesium chloride + hydrogen

$$Mg + 2HCl \rightarrow MgCl_2 + H_2$$

Figure 3.56

As two chloride ions (Cl^-) are needed to join with one Mg^{2+} ion, they must both be provided by the HCl. Therefore, two amounts of HCl are needed. The equation is balanced.

3. **Write a full balanced symbol equation for the production of zinc chloride from zinc.**
4. **Write a full balanced symbol equation for the production of iron(III) chloride, $FeCl_3$, from iron.**

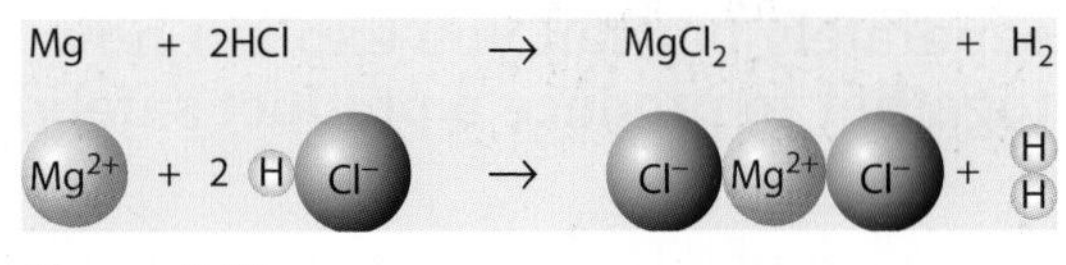

Figure 3.57

HIGHER TIER ONLY

Using half equations

When making a magnesium salt from magnesium this reaction happens:

$$\mathbf{Mg + 2H^+ \rightarrow Mg^{2+} + H_2}$$

The **half equations** are:

$Mg - 2e^- \rightarrow Mg^{2+}$	$2H^+ + 2e^- \rightarrow H_2$
Electron loss is oxidation	Electron gain is reduction
As both these reactions take place at the same time this is a *redox* reaction	

In the equation:

Magnesium loses electrons so oxidation takes place to produce magnesium ions, (Mg^{2+}).

Hydrogen ions (H^+) gain electrons so reduction takes place.

5. **Write two half equations for the reaction between zinc and an acid.**
6. **In the reaction between iron and sulfuric acid, which component is oxidised and which is reduced?**
7. **Write a balanced chemical equation for the reaction of aluminium with sulfuric acid. Write half equations for this reaction. Aluminium forms Al^{3+} ions.**

PRACTICAL

Preparing a pure, dry sample of a soluble salt from an insoluble oxide or carbonate

Learning objectives:

- describe a practical procedure for producing a salt from a solid and an acid
- explain the apparatus, materials and techniques used for making the salt
- describe how to safely manipulate apparatus and accurately measure melting points.

KEY WORDS

insoluble
oxide
carbonate
filter
crystallisation

Being able to choose the correct techniques and carry out a specified procedure to produce a pure product is an important skill for scientists. You will probably know how to complete each of the necessary techniques separately but can you explain how to use them together to produce a product safely?

These pages are designed to help you think about aspects of the investigation rather than to guide you through it step by step.

A number of different skills are needed to carry out the production of a sample of a chemical. This topic looks at the skills of selecting techniques, apparatus and materials and managing safety.

DID YOU KNOW?

Most salts have a very high melting point, so a special heater would be needed.

Carrying out procedures

The making of magnesium sulfate from magnesium carbonate can be done in several stages using different pieces of apparatus.

This diagram has some stages of an experimental plan depicted but not all stages.

In Figure 3.58(a) the quantities of two chemicals have been measured to make the salt:

1. **Identify the two substances needed to make magnesium sulfate.**
2. **State how the quantities are measured.**
3. **State the units of measurement for each substance.**
4. **Describe the safety measures that need to be taken when adding substances.**
5. **After the two substances are mixed and stirred as in Figure 3.58(b), explain why there is some solid left in the bottom of the solution.**
6. **The excess solid is removed as in Figure 3.58(c). Name the process of removal of the solid.**

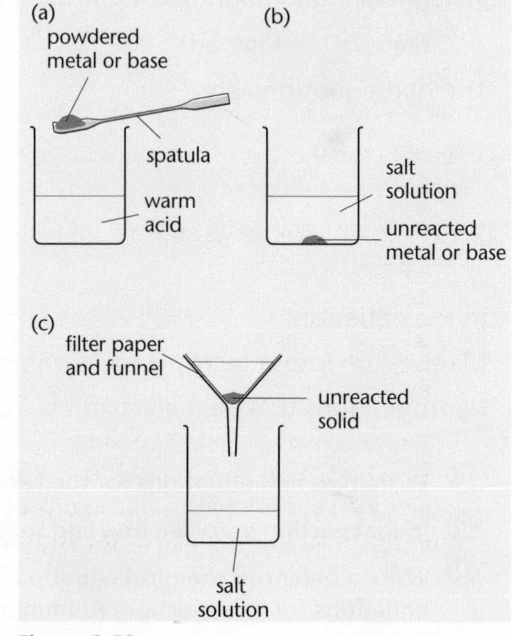

Figure 3.58

Using apparatus and techniques

The solution is collected after the removal of the excess solid.

To obtain the product from the solution two more processes are needed.

- evaporation
- **crystallisation**

To make a better product a pure sample is needed. The crystals formed will need to be.

- washed and dried
- recrystallised

Think about these further questions for making pure magnesium sulfate.

7 Describe how the first sample of dry impure salt is obtained from the solution.

8 Explain how the purity of the salt can be improved.

9 Explain how the purity of the final sample is tested.

REMEMBER!

It is helpful to imagine setting up the apparatus needed and draw each part of the procedure by stages.

Measuring accurately

When a sample of product has been made, it needs to be tested for purity. Sam and Alex made samples of calcium nitrate crystals. They looked up the melting point for these crystals and found it was 42.7 °C. These are their results. Who made the purest sample?

Sam	MP °C	40.3	40.5
Alex	MP °C	41.6	41.4

10 Explain the affect of impurities on the melting point.

11 Suggest why it might have been advisable for Sam and Alex to take another temperature measurement.

12 Give the equation for the reaction between magnesium carbonate, $MgCO_3$ and the acid?

13 Calculate the mass of the two substances needed to produce 6 g of $MgSO_4$

14 Explain why the solid $MgCO_3$ needs to be added *in excess* of the amount calculated.

15 What may happen to the yield if the aim is to increase purity?

HIGHER TIER ONLY

16 The uncertainty in reading the thermometer was plus or minus 0.2°C. Discuss whether the difference in melting point between the data book value and the measurements is significant. Explain which of the results are most accurate.

pH and neutralisation

Learning objectives:

- describe the use of universal indicator to measure pH
- use the pH scale to identify acidic or alkaline solutions
- investigate pH changes when a strong acid neutralises a strong alkali.

KEY WORDS

hydroxide ion
neutralisation
pH
universal indicator

We use acids in our normal lives every day. Stomach acid is essential for digesting food. The acidic food we eat is sometimes delicious and sometimes sour. Why are citric acid and ethanoic acid available in supermarkets but hydrochloric acid in the laboratory needs to be used with care and safety glasses need to be worn?

Figure 3.59 Which of these contains sulfuric acid, citric acid, nitric acid, ethanoic acid?

Acids and alkalis

Acids are substances that produce hydrogen ions in aqueous solution.

Examples of acids are: Examples of alkalis are:

hydrochloric acid	HCl	sodium hydroxide	$NaOH$
nitric acid	HNO_3	potassium hydroxide	KOH
sulfuric acid	H_2SO_4	ammonia	$NH_3(aq)$
ethanoic acid	CH_3COOH		
citric acid	$C_6H_8O_7$		
The hydrogen ions they produce have the symbol H^+. They are ions with a positive charge.		Alkalis are substances that make **hydroxide ions** in aqueous solution. These hydroxide ions have the symbol OH^-. They are ions with a negative charge.	

1 Identify the ion in sulfuric acid that makes it acidic.

2 Explain the difference between an acid and an alkali.

The pH scale

When **universal indicator** (UI) is added to solutions it changes colour.

This is because UI is a different colour when the number of hydrogen ions in the solution changes.

The number of hydrogen ions in a solution is related to a scale called the pH scale.

Type of solution	Colour of UI	pH
acidic	red/ orange/yellow	1–6
neutral	green	7
alkaline	blue/purple	8–14

Universal indicator can be used to estimate the **pH** of a solution, by matching to a colour/pH chart.

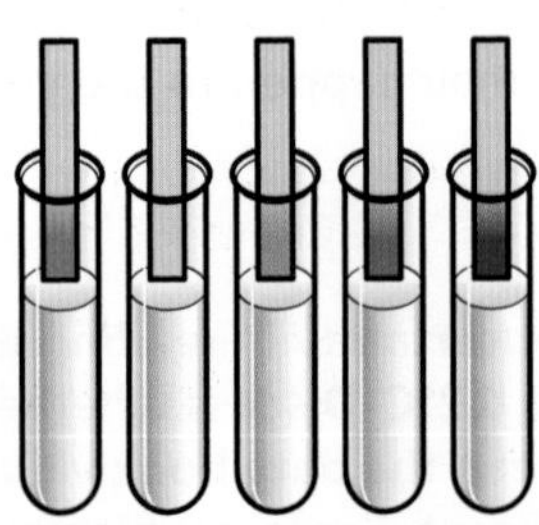

Figure 3.60 Universal indicator colour changes in strong acid, weak acid, neutral, weak alkali and strong alkali

1	2	3	4	5	6	7	8	9	10	11	12	13	14

1 = very acidic, 7 = neutral, 14 = very alkaline

Figure 3.61 pH colour match chart

A more accurate way of measuring pH is to use a pH probe.

3. **Universal indicator solution was added to HCl in a conical flask. It was then exactly neutralised by NaOH. Excess NaOH was then added. Describe the colour change and estimate the pH at each stage.**

4. **What would be the pH of a solution if UI turned green?**

> **DID YOU KNOW?**
>
> Some indicators show a sudden colour change at one pH value. Universal indicator shows a gradual range of colour changes, as it contains a mixture of different indicators.

Neutralisation

If an acid is added to an alkali **neutralisation** takes place.

An acid solution has a low pH.
If an alkali is added slowly to an acid, the pH number of the acid will gradually increase.
When it gets to pH 7 the acid is neutralised.

An alkaline solution has a high pH. If acid is slowly added to an alkali, the pH number will gradually decrease.
When it gets to pH 7 the alkali has been neutralised.

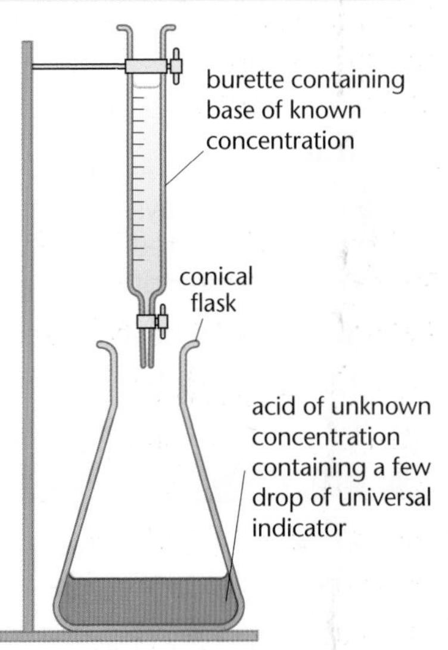

Figure 3.62 Investigating the changes of pH when a base is added to an acid

The higher the concentration of H^+ ions the lower the pH.

Alkalis contain OH^- ions (hydroxide ions).

Neutralisation involves this reaction:

$$H^+_{(aq)} + OH^-_{(aq)} \rightarrow H_2O_{(l)}$$

Neutralisation leaves no free H^+ ions.

The neutralisation reaction takes place with all common acids and common alkalis.

Acid		
Hydrochloric acid	$H^+ \rightarrow$	Cl^-
Nitric acid	$H^+ \rightarrow$	NO_3^-
Sulfuric acid	$H^+ \rightarrow$	SO_4^{2-}
Ethanoic acid	$H^+ \rightarrow$	CH_3COO^-

Alkali		
Sodium hydroxide	Na^+	$OH^- \rightarrow$
Potassium hydroxide	K^+	$OH^- \rightarrow$
Calcium hydroxide	Ca^{2+}	$OH^- \rightarrow$

$$H^+_{(aq)} + OH^-_{(aq)} \rightarrow H_2O_{(l)}$$

5. **Phosphoric acid has the formula H_3PO_4**
   a **Identify the ions in phosphoric acid.**
   b **Compare the concentration of H^+ and OH^- ions in phosphoric acid and sodium hydroxide.**

6. **Write an *ionic* equation for the reaction between hydrochloric acid and potassium hydroxide.**

7. **Indigestion is caused by the overproduction of hydrochloric acid in the stomach. Indigestion remedies often contain magnesium hydroxide.**
   a **Estimate the pH of an indigestion remedy and explain in terms of the ions present.**
   b **Explain how the indigestion remedy works in terms of the ions present.**

> **DID YOU KNOW?**
>
> You can plot a graph of the pH change against every 1 cm^3 of alkali you add to acid. You will get a surprising curve.

> **KEY INFORMATION**
>
> Remember: you know that the process of neutralisation creates salts.
>
> acid + alkali → salt + water

Strong and weak acids

KEY WORDS

concentration
dilution
ionisation
strength

Learning objectives:

- explain weak and strong acids by the degree of ionisation
- describe neutralisation by the effect on hydrogen ions and pH
- explain dilute and concentrated as amounts of substance.

Sulfuric acid, a strong acid, is used in car batteries. Vinegar and citric acid are weak acids that are used in cooking. The terms 'strong' and 'weak' are not the same as 'concentrated' and 'dilute'. You will need to consider the number of particles and the amount they ionise.

HIGHER TIER ONLY

Ionisation

There are two ways to describe any solution of an acid. It can be either concentrated or dilute and it can also be either strong or weak. These do not mean the same thing.

It is possible to have an acid that is one of four combinations of these terms. Look at Figure 3.65 to see these.

Examples of strong and weak acids are:

strong acids	weak acids
hydrochloric nitric sulfuric	ethanoic citric carbonic

Strong and weak acids both have the same chemical reactions.

Acid reactions are caused by hydrogen ions, H^+

Strong and weak acids react with:

- metals to make hydrogen
  $Mg + 2H^+ \rightarrow Mg^{2+} + H_2$
- metal carbonates to make carbon dioxide
  $CaCO_3 + 2H^+ \rightarrow Ca^{2+} + H_2O + CO_2$

So why are some acids (H^+A^-) strong and some weak?

Strong: In water all of the acid molecules, HA, become ions (H^+ and A^-).	Weak: In water only a few of the acid molecules, HA, become ions (H^+ and A^-), most stay as molecules.
Strong acids ionise completely in water. $HCl \rightarrow H^+ + Cl^-$	Weak acids do not ionise fully. The equilibrium lies to the left. $CH_3COOH \rightleftharpoons H^+ + CH_3COO^-$
A high concentration of H^+ means that the pH is low.	A low concentration of H^+ means that the pH is higher.

1 **Name the ions produced when sulfuric acid ionises.**

2 **Explain why citric acid, $H^+(citrate)^-$ is a weak acid.**

strong and concentrated

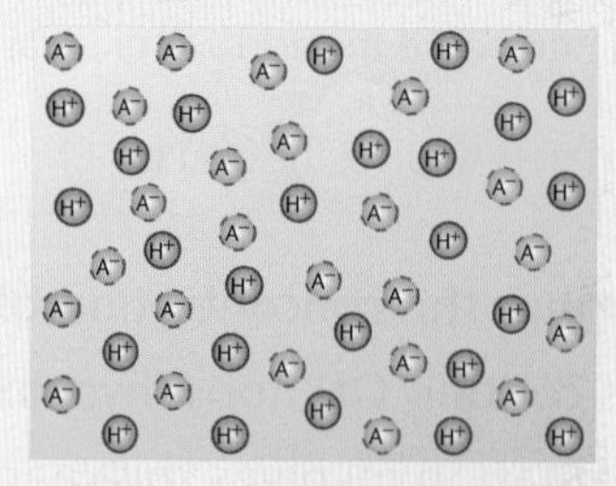

strong and dilute

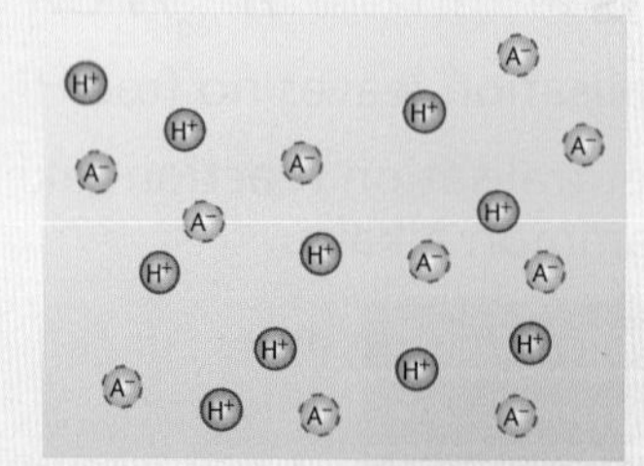

weak and concentrated

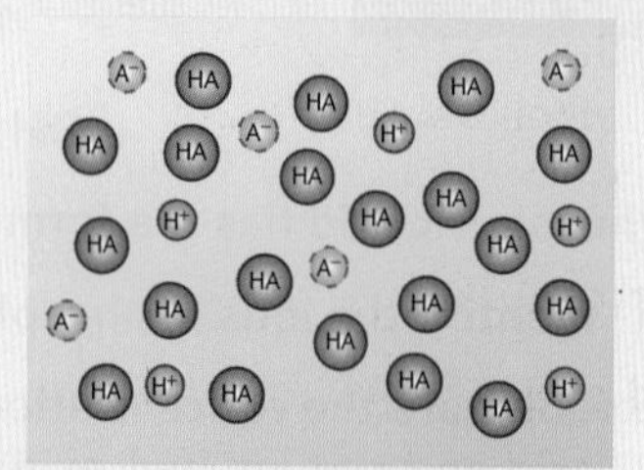

weak and dilute

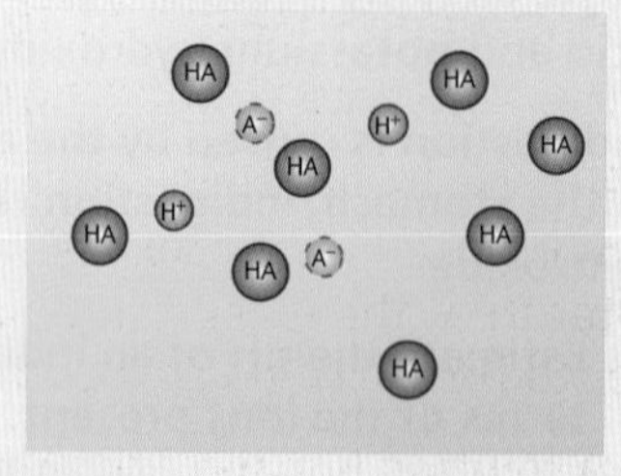

Figure 3.63

pH, neutralisation and titration curves

The pH scale is related to the concentration of H^+ ions.

Strong acids have a lower pH than weak acids.

low pH number = high concentration of H^+	higher pH number = lower concentration of H^+

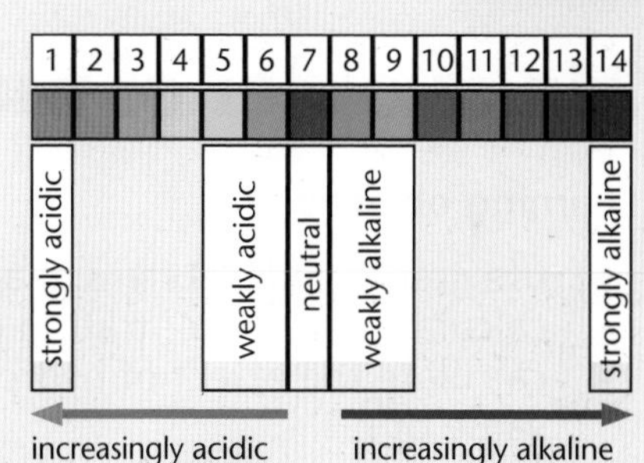

Figure 3.64 The pH scale

For a given **concentration** of aqueous solutions, the stronger an acid, the lower the pH. As the pH decreases by one unit, the hydrogen ion concentration of the solution increases by a factor of 10. This because pH is based on a logarithmic scale.

During neutralisation of an alkali, if acid is added to the solution the pH will decrease. This can be seen in a titration curve.

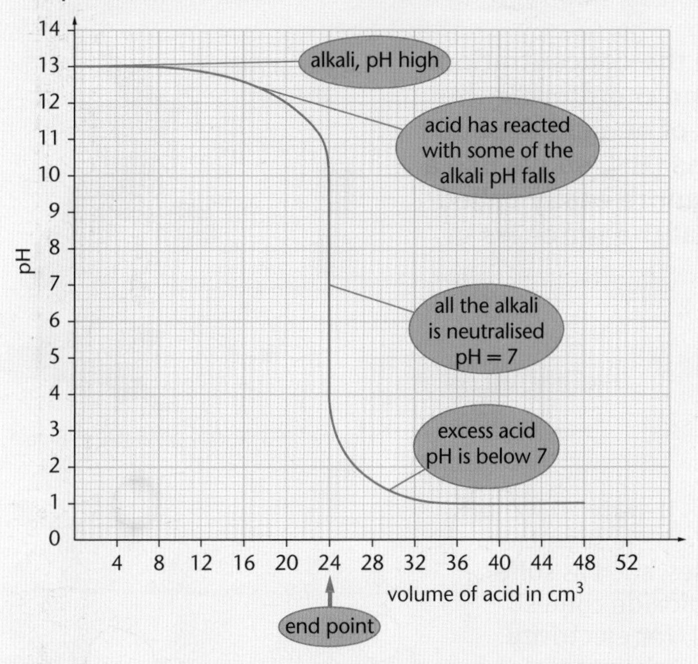

Figure 3.65 Titration curve of a strong alkali pH 14 neutralised by a strong acid

DID YOU KNOW?

Titration curves for weak acids and weak alkalis do not give a very clear neutralisation point.

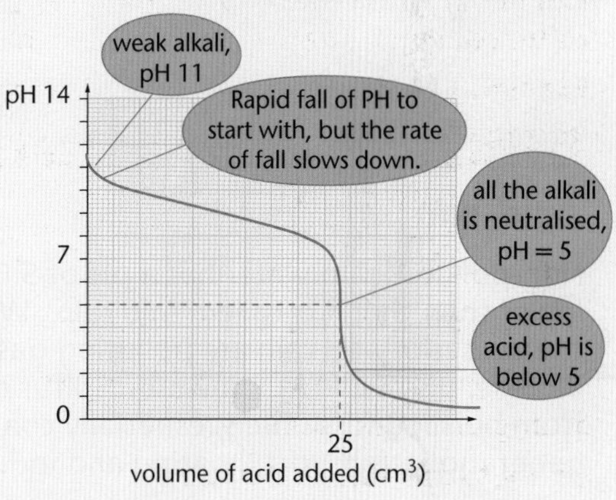

Figure 3.66 Titration curve of a weak alkali pH 11 neutralised by a strong acid

3 **Suggest a sketch of a titration curve for a weak alkali being added to a strong acid. Take the end point as 25 cm³.**

Concentration

Acid concentration is not the same as H^+ ion concentration. A concentrated solution of a weak acid still has a low concentration of H^+ ions.

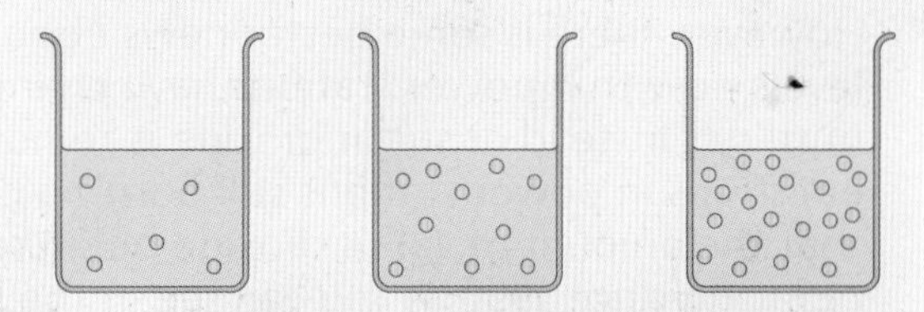

Figure 3.67 Solutions of different concentrations

Concentration is measured in mol/dm³ or g/dm³.

1 dm³ is the same as 1000 cm³.

100 cm³ is one-tenth of 1000 cm³.

The mass of a solute added to 100 cm³ of water needs to be one-tenth of the mass that would be added to 1000 cm³ of water to be the *same concentration*.

4 **A solution is made from 6 g of citric acid added to 100 cm³ of water. What is the concentration of the solution in g/dm³?**

KEY INFORMATION

The concentration of an acid tells us how many moles of acid there are in 1 dm³.

The strength of an acid tells us how much an acid ionises.

MATHS SKILLS

Make order of magnitude calculations

Learning objectives:

- use graphs and diagrams to apply the pH scale to acid rain distribution
- calculate the concentration of acids
- calculate the effect of hydrogen ion concentration on the numerical value of pH.

KEY WORDS

magnitude
logarithmic scale
derived
relative

It is important to be able to calculate the concentrations of acid for using them in reactions and monitoring the acidity of rainfall. The concentration of an acid depends on the concentration of hydrogen ions [H+], so we use a logarithmic scale to describe acidity more easily called the pH scale. Can you remember what the numbers of the scale represents?

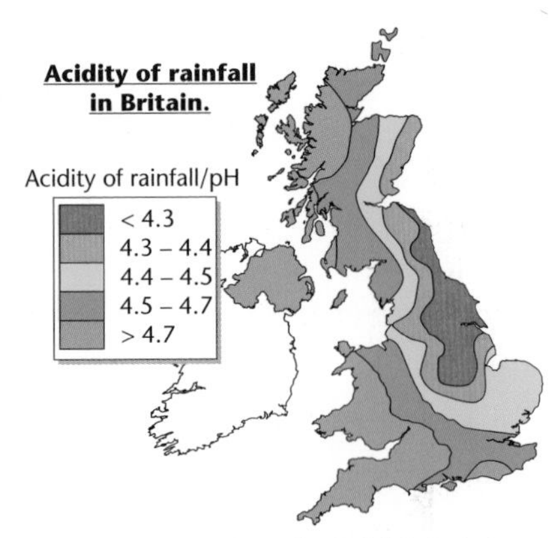

Figure 3.68 Acidity of rainfall in Britain

HIGHER TIER ONLY

Acid rain

Non-polluted rain water has a pH of 5.6. If water has a pH lower than that it is called 'acid rain'. We have seen that the causes of acid rain are polluting gases, such as sulfur dioxide, dissolving in rainwater. Sulfur dioxide comes from burning fossil fuels, especially coal, for electricity generation, transport, heating and industrial processes.

Figure 3.68 shows a map where the rainfall has a pH of less than 5.6 and so is more acidic than normal.

However, there is some better news about sulfur dioxide levels now that power stations have changed. They either burn gas instead or sulfur dioxide is removed from the emissions in power stations still burning coal. This means that the amount of sulfur dioxide being released has been reducing since 1990 as you can see in the bar chart, Figure 3.69.

1. **Which side of the UK experiences rainfall that is the most acidic? Explain how you can tell from the data.**
2. **Estimate the decrease in sulfur dioxide in thousand tonnes equivalent from 1990 to 2006 from the graph.**
3. **Calculate the approximate ratio of sulfur dioxide equivalents for 1990 and 2011.**

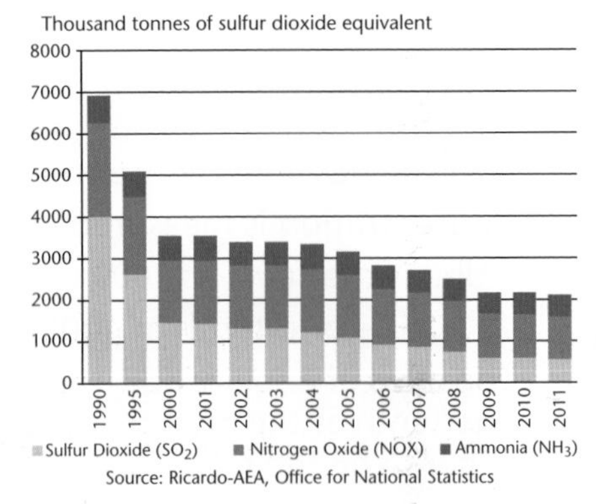

Figure 3.69 Acidity of rainfall in Britain

Concentrations of acid

The concentration of acids is measured in g/dm^3 or mol/dm^3. The shorthand is to write a square bracket [HA] to mean concentration.

The **relative** *formula mass* (M_r) of nitric acid HNO_3 is calculated from the *relative atomic masses* (A_r)

(A_r: H is 1, N is 14, O is 16) M_r: $HNO_3 = 1 + 14 + (3 \times 16) = 63$

If a solution of nitric acid contains 63 g of acid dissolved in 1 dm^3 of water the $[HNO_3] = 63\,g/dm^3$ The relative formula mass is also the mass of 1 mole or the *molar mass*, so $[HNO_3] = 1\,mol/dm^3$

A 100 cm^3 solution that contains 6.3 g has decreased both volume and mass by a factor of 10, so the concentration remains the same, $[HNO_3] = 63\,g/dm^3$ and also $[HNO_3] = 1\,mol/dm^3$

A 1000 cm^3 solution that contains 6.3 g has decreased only the mass by a factor of 10 not the volume, so the concentration reduces by a factor of 10, so $[HNO_3] = 6.3\,g/dm^3$ and $[HNO_3] = 0.1\,mol/dm^3$

Finding what happens to the pH as acid is diluted or neutralised:

What does pH mean? It is a scale that is **derived** from the concentration of hydrogen ions in an acid solution. If the concentration of a strong acid is 0.1 mol/dm^3, then the concentration of hydrogen ions, $[H^+]$, is also 0.1 mol/dm^3. In standard form this is 10^{-1} mol/dm^3.

The pH number is the 'index' number but positive. So the pH of this solution is pH 1. If the concentration decreases by a factor of 10, the standard form decreases by the index of –1 and the pH increases by positive 1.

Concentration of acid in mol/dm^3	Concentration of acid in standard form in mol/dm^3	pH
0.1	10^{-1}	1
0.01	10^{-2}	2
0.001	10^{-3}	3
0.0001	10^{-4}	4
0.00001	10^{-5}	5
0.000001	10^{-6}	6
0.0000001	10^{-7}	7

So you can see from the table that as the concentration of the acid decreases by a factor of 10, the pH increases by one unit.

8 Determine the pH of a solution with $[H^+]$= 0.000001 mol/dm^3.

9 Estimate the concentration of hydrogen ions for a solution that turned universal indicator green.

DID YOU KNOW?

Robert Angus Smith first showed a relationship between atmospheric pollution and acid rain in 1852 and used the term acid rain in 1872. It took us 100 years to take the issue seriously and do something about it!

4 What mass of sulfuric acid, H_2SO_4, is needed to make a solution $[H_2SO_4] = 1\,mol/dm^3$? (A_r: S is 32)

5 What mass of sulfuric acid is needed to make a solution $[H_2SO_4] = 0.05\,mol/dm^3$?

6 9.13 g of hydrochloric acid was dissolved in 250 cm^3 of water. Calculate the concentration of hydrochloric acid in mol/dm^3.

7 0.46 g of methanoic acid, HCOOH, was dissolved in 200 cm^3 of water. Calculate the concentration of methanoic acid in mol/dm^3.

DID YOU KNOW?

The pH value is actually a value on a logarithm scale. The pH equals the negative log to the base 10 of the hydrogen ion concentration. This is more elegantly written as: $pH = -\log_{10}[H^+{}_{aq}]$.

PRACTICAL

Investigate the variables that affect temperature changes in reacting solutions such as, acid plus metals, acid plus carbonates, neutralisations, displacement of metals

KEY WORDS

temperature
carbonate
thermometer
neutralisation

Learning objectives:

- use scientific theories and explanations to develop hypotheses
- plan experiments to make observations and test hypotheses
- evaluate methods to suggest possible improvements and further investigations.

Using previous knowledge to develop a hypothesis for further investigation is an important skill for all scientists. Temperature change is measurable when iron is added to an acid. If the iron is changed to another metal, is the effect the same? Some metals are more reactive than others, does this pattern allow us to predict the change?

These pages are designed to help you think about aspects of the investigation rather than to guide you through it step by step.

This topic looks at a few of the skills you need to use to explore the factors that affect a **temperature** change.

DID YOU KNOW?

Temperature changes during reactions are used in heat packs for pain relief and for heating instant coffee drink packs.

Developing a hypothesis

We need to ask a question first, such as 'what are temperature changes affected by'?

They may be affected by any of the variables that can be changed in a reaction.

Here are the four reactions that could be investigated and some of the variables that could be changed:

Which variables affect the temperature change when	Variables that could be changed			
Acids react with metals	type of metal	type of acid	concentration of acid	size of pieces of metal
Acids react with **carbonates**	type of carbonate	type of acid	concentration of acid	size of pieces of carbonate
Acids react with alkalis	type of alkali	type of acid	concentration of acid	concentration of alkali
A metal is added to a solution of ions of another metal				

You might be able to think of some more.

Do we have previous knowledge of reactions or patterns that might help us focus?

1. **Suggest whether different reactants and changing concentration would affect the temperature change.**
2. **Fill in the bottom row for investigating temperature changes in a displacement reaction.**

Planning an investigation

What is done in an investigation into any of these reactions is that one of the variables is selected and altered. All the other factors are kept the same.

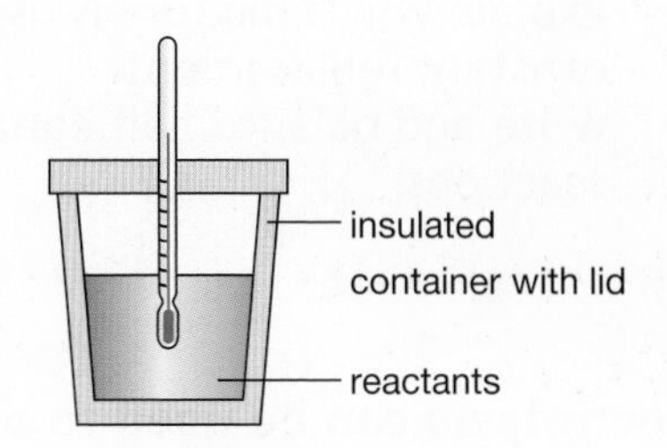

Figure 3.70

3. **Explain why it is important to just change one variable at a time.**
4. **A hypothesis might be: The smaller the piece of metal the bigger the temperature change when reacting with acid. Suggest which variables you would keep the same.**
5. **Some students developed a hypothesis: the more concentrated the acid the greater the temperature change when reacting with carbonates. They gathered these results:**

Concentration of acid in mol/dm^3	Temperature in °C					
	1st trial			2nd trial		
	Start temp	Final temp	Temp change	Start temp	Final temp	Temp change
2.0	20.0	27.3	7.3	21.0	28.2	
1.5	22.0	27.2	5.2	22.4	27.8	
1.0	18.5	23.3	4.8	20.6	31.7	
0.5	21.6	25.5	3.9	21.2	24.9	

a Calculate the temperature changes for the 2nd trial.

b Identify the anomaly in the data. Suggest how the students should deal with this result.

c Suggest what conclusion the students came to with this data.

KEY INFORMATION

Patterns such as the reactivity series or evidence from rate investigations may be a starting point for developing ideas for hypotheses.

Evaluating the experiment

When experiments have been completed a discussion should take place. Could the techniques have been improved. How well does the data gathered support the predictions and conclusions made. Were the observations reasonable or could they have another explanation, for example, 'did particle size affect the final temperature change or just the rate at which it achieved the highest temperature?'

HIGHER TIER ONLY

6. **Explain how the evidence supports or refutes the original hypothesis developed by the students.**
7. **Explain any problems with the solid carbonates that may affect the repeatability of the experiment.**
8. **Explain why heat loss needs to be prevented.**

The process of electrolysis

Learning objectives:

- identify reactions at electrodes during electrolysis
- explain why a mixture is used and the anode needs constant replacement
- write and balance half equations for the electrode reactions.

KEY WORDS

electrode
electrolysis
electrolyte
ion migration

Electrolysis can be used to obtain useful products such as oxygen and hydrogen from sulfuric acid and copper or aluminium from their ores.

Electrolytes

Electrolysis is the process of passing direct current (d.c.) through a solution or melted ionic compound to move the ions apart and so break the compound down and discharge some of the elements at the **electrodes**.

The solution or molten compound is called the electrolyte. The ions need to be free to move about within the solution or liquid to be an **electrolyte**. An electrolyte is a liquid that conducts electricity and is decomposed during the electrolysis.

Figure 3.71 Electrolysis of sulfuric acid to produce hydrogen and oxygen

Two **electrodes** are used. They are called the *cathode* and the *anode*. These 'dip into' the solution or 'melt'.

The cathode is the negative electrode.	The anode is the positive electrode.

Electrolytes are made from *ions*. During electrolysis the passing of an electric current through electrolytes causes the ions to move to the electrodes:

positive ions are attracted to the cathode and are called ***cations***	negative ions are attracted to the anode and are called ***anions***

DID YOU KNOW?

A molten ionic compound involved in electrolysis can be referred to as a 'melt'.

1 Explain why electrolytes need to be molten or in solution.

Positive ions and negative ions

Ions that have a formula with a positive charge, such as H^+ and Na^+ are called cations.

Ions that have a formula with a negative charge, such as OH^- and Cl^- are called anions.

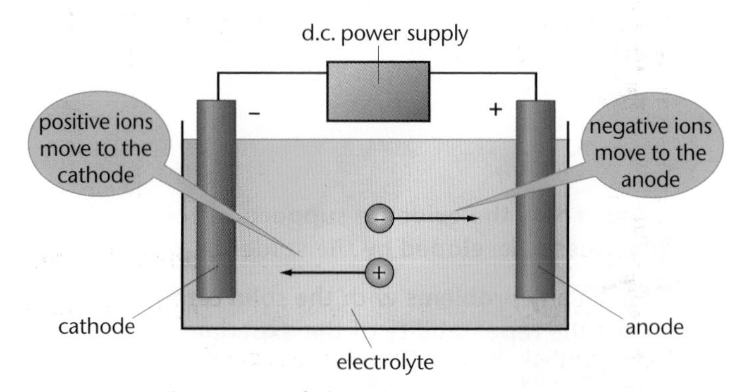

Figure 3.72 The process of electrolysis

Positive ions (cations)		Negative ions (anions)	
H^+	hydrogen	OH^-	hydroxide
Na^+	sodium	Cl^-	chloride
Cu^{2+}	copper	SO_4^{2-}	Sulfate
Al^{3+}	aluminium	O^{2-}	Oxide

Ions are discharged at the electrodes producing elements. Positive ions move towards negative electrode. Negative ions move towards the positive electrode. Opposite charges attract.

KEY INFORMATION

Check the charges on the ions when explaining the migration of the ions to the electrodes.

2. **Identify the two products that would form if molten copper chloride was electrolysed.**

3. **Why do aluminium ions move towards the cathode?**

HIGHER TIER ONLY

Ions discharging

The **ions migrate** towards the two electrodes.

All the positive ions are attracted to the negatively charged electrode.

All the negative ions are attracted to the positively charged electrode.

When the negative ions reach the anode (the positively charged electrode) they lose electrons to the electrode.

When the positive ions reach the cathode (the negatively charged electrode) they gain electrons at the electrode.

These reactions can be written as half equations.

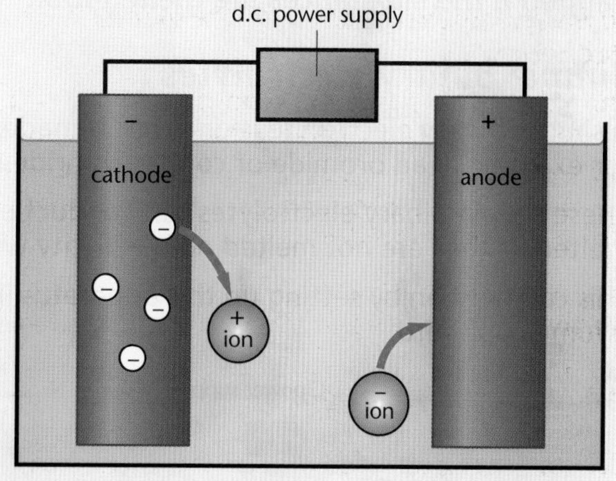

Figure 3.73 Ions discharging at the cathode and anode

The half equation for the discharge of the sodium ion would be:

$Na^+ + e^- \rightarrow Na$

The half equation for the discharge of the chloride ion would be:

$Cl^- - e^- \rightarrow Cl$

However, chlorine exists as molecules of two atoms so two ions need to be discharged.

$2Cl^- - 2e^- \rightarrow Cl_2$

DID YOU KNOW?

The discharge of chlorine at the anode is more correctly written as:
$2Cl^- \rightarrow Cl_2 + 2e^-$

4. **Write the two half equations for the electrolysis of molten copper chloride.**

5. **Aqueous magnesium bromide solution can be electrolysed.**
   - **a** **Write the half equation for the reaction at the anode.**
   - **b** **Predict the product at the cathode.**
   - **c** **Explain why hydrogen is actually produced at the cathode and where the hydrogen comes from.**

Electrolysis of molten ionic compounds

Learning objectives:

- identify which ions migrate to the cathode and anode
- explain how the ions of a molten electrolyte are discharged
- predict the products of electrolysis of molten binary compounds
- understand how to write half equations to explain ion activity at electrodes.

KEY WORDS

anode
cathode
discharged
molten

We have already learned that ions need to be free to move in an electrolyte if electrolysis is to take place. Ions that are not free to move cannot conduct electricity. Discharging the elements from compounds of highly reactive metals can happen if the substances are melted first.

Simple binary electrolytes

A simple binary electrolyte is one that is made up of two ions, for example, lead bromide or copper chloride.

These simple binary electrolytes can conduct electricity if melted. If they are not melted no electricity will flow.

This can be seen by setting up this apparatus in a laboratory, in a fume cupboard.

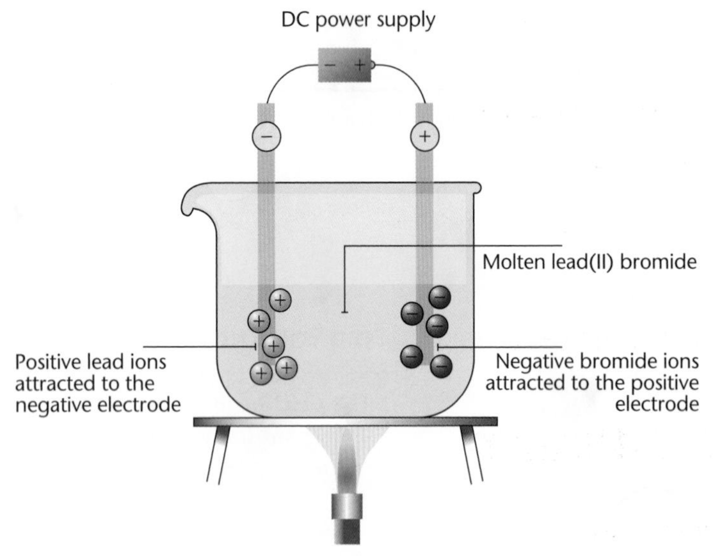

Figure 3.74 Electrolysis of molten lead bromide in the lab

When the lead bromide is cold and solid no electricity will flow and the lamp will not light.

If the crucible is heated and the solid begins to melt a current will flow and the lamp lights.

The ions, because they are now free to move, will migrate towards the electrodes.

1 **Lead makes a positive ion. To which electrode will it migrate?**

Positive and negative electrodes

When **molten** lead bromide is electrolysed using inert electrodes the ions will migrate towards them.

The lead ion, Pb^{2+}, is positive so will migrate towards the negative electrode.

The negative electrode is the cathode so a lead ion is a cation.

The ion is **discharged** and the metal lead, Pb, is produced at the **cathode**.

The non-metal bromide ion, Br^- is negative so will migrate towards the positive electrode.

The positive electrode is the anode so a bromide ion is an anion.

Two ions are discharged and the non-metal gas bromine, Br_2, is produced at the **anode**.

Another molten binary electrode that can be used in the lab is molten copper chloride.

The ions produced are Cu^{2+} and Cl^-.

2 **To which electrode will a chloride ion migrate?**

3 **Predict the products of the electrolysis of molten sodium bromide at each electrode.**

KEY INFORMATION

The electrodes used are inert. That means they do not react with any element that is discharged during electrolysis. They are normally made of carbon/graphite.

DID YOU KNOW?

Bromine is a brown gas that is a toxic irritant that causes burns. It has an unpleasant smell and bleaching action. That is why you should use a fume cupboard if it is produced.

HIGHER TIER ONLY

Electrode half equations

The ions migrate towards the electrodes and transfer electrons.

At the cathode	At the anode
Electrons are gained	Electrons are given up

For $PbBr_2$ the half equations are:

$$Pb^{2+} + 2e^- \rightarrow Pb \qquad 2\,Br^- - 2e^- \rightarrow Br_2$$

Other decompositions can be written as half equations if the formulae of the ions are known.

Electrolyte	Half equation at cathode	Half equation at anode
KCl	$K^+ + e^- \rightarrow K$	$2Cl^- - 2e^- \rightarrow Cl_2$
$CuCl_2$	$Cu^{2+} + 2e^- \rightarrow Cu$	$2Cl^- - 2e^- \rightarrow Cl_2$
PbI_2	$Pb^{2+} + 2e^- \rightarrow Pb$	$2I^- - 2e^- \rightarrow I_2$
Al_2O_3	$2Al^{3+} + 6e^- \rightarrow 2Al$	$6O^{2-} - 12e^- \rightarrow 3O_2$

4 **Write the half equations at the cathode and anode for the electrolysis of molten copper bromide.**

Electrolysis of aqueous solutions

Learning objectives:

- explain the electrolysis of copper sulfate using inert electrodes
- predict the products of the electrolysis of aqueous solutions
- represent reactions at electrodes by half equations.

KEY WORDS

electrode reaction
half equation
preferential discharge
relative reactivity

We have already looked at the electrolysis of binary electrolytes in aqueous solution. We have seen that that when more than one ion migrates towards an electrode that it is the most reactive that stays in solution and the *least reactive* that is discharged. The discharge of the least reactive ion is called preferential discharge.

DID YOU KNOW?

There is a list of 'ease of discharge' for negative ions as well: as positive ions. This is SO_4^{2-} NO_3^- Cl^- Br^- I^- OH^-

Preferential discharge of ions

The preferential discharge is based on the reactivity series.

The relatively least reactive ions are preferentially discharged as elements.

At the cathode	At the anode
H^+ is discharged in preference to Zn^{2+} Al^{3+} Mg^{2+} Ca^{2+} Na^+ and K^+	OH^- is discharged in preference to SO_4^{2-}
Cu^{2+} and Ag^+ are discharged in preference to H^+ (as Cu^{2+} and Ag^+ are less reactive than H^+)	Cl^- and Br^- are discharged in preference to OH^-
H^+ produces hydrogen gas	OH^- produces oxygen gas

1 Explain whether silver ions or hydrogen ions would be discharged first during the electrolysis of aqueous silver nitrate.

The electrolysis of copper sulfate

The electrolysis of copper(II) sulfate using inert carbon electrodes follows the rules of preferential discharge. Copper is less reactive than hydrogen. What we *see* is that this electrolyte is a blue solution. As the electrolysis happens the blue colour of the solution fades.

We see the cathode being coated with copper and bubbles at the anode.

The copper ions (+) from the solution migrate to the cathode (–) and can be seen to deposit. Let's look at the four ions involved, Cu^{2+} H^+ SO_4^{2-} and OH^-.

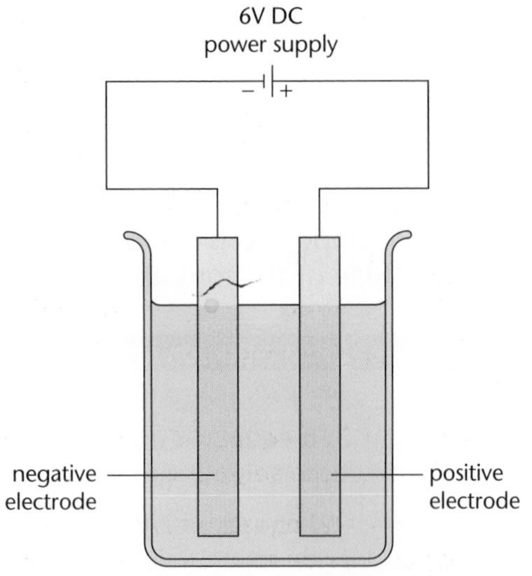

Figure 3.75 Which ions are discharged in the electrolysis of copper sulfate solution, $CuSO_4(aq)$ with inert electrodes?

To the cathode	To the anode
Copper ions (Cu^{2+}) Hydrogen ions (H^+) (from the water)	Sulfate ions (SO_4^{2-}) Hydroxide ions (OH^-) (from the water)
Hydrogen ions stay in the electrolyte. Copper ions are discharged *in preference* and make copper metal.	Sulfate ions stay in the electrolyte as hydroxide ions are more easily discharged. Hydroxide ions discharge *in preference* and form oxygen gas.

So, if $CuSO_4$ solution is electrolysed with carbon electrodes, copper is formed at the cathode and oxygen is formed at the anode.

2 Sodium chloride was added to a solution of copper(II) sulfate and the mixture electrolysed. Compare the discharge products to the electrolysis of pure copper(II) sulfate solution. Explain your answer.

Oxidation, reduction and half equations

During electrolysis, positively charged hydrogen **ions migrate** to the cathode and gain electrons.

$$2H^+ + 2e^- \rightarrow H_2$$

Because the ions gain electrons the reactions are *reductions.*

Negatively charged ions migrate to the anode and lose electrons.

Because these reactions lose electron they are *oxidations.*

> **KEY INFORMATION**
>
> **Oxidation Is electron Loss**
> **Reduction Is electron Gain:**
> **OILRIG**

The hydroxide ions migrate to the anode but the half equation is not as simple as that for H^+ (where only two electrons were involved). This time four electrons are involved.

$$4OH^- - 4e^- \rightarrow O_2 + 2H_2O$$

(or more correctly $4OH^- \rightarrow O_2 + 2H_2O + 4e^-$)

So for other aqueous solutions the electrode reactions are also the same, for example, in the electrolysis of NaOH(aq) or H_2SO_4(aq) they are:

At the cathode:	At the anode:
$2H^+ + 2e^- \rightarrow H_2$	$4OH^- - 4e^- \rightarrow 2H_2O + O_2$

In the electrolysis of NaOH, H_2 is made and not Na, because Na is much higher up in the reactivity series so H_2 is discharged *in preference.*

However, the electrode reactions in the electrolysis of $CuSO_4$(aq) with carbon electrodes are:

At the cathode:	At the anode:
$Cu^{2+} + 2e^- \rightarrow Cu$	$4OH^- - 4e^- \rightarrow 2H_2O + O_2$

because copper is less reactive than hydrogen and is preferentially discharged.

3 Write the half equations for the reaction at the cathode and the anode in the electrolysis of dilute sodium sulfate solution.

4 Write the half equations for the reaction at the cathode and the anode in the electrolysis of dilute copper chloride solution. Explain which is an oxidation and which is a reduction reaction.

5 Predict the half equations at the electrodes for the electrolysis of phosphoric acid, H_3PO_4, solution.

PRACTICAL

Investigating what happens when aqueous solutions are electrolysed using inert electrodes

Learning objectives:

- use scientific theories and explanations to develop hypotheses
- plan experiments to make observations and test hypotheses
- apply a knowledge of the apparatus needed for electrolysis including use of inert electrodes and varying electrolytes
- make and record observations.

KEY WORDS

electrolysis
electrode
inert
electrolyte
cathode
anode

Using patterns in reactions to form hypotheses about what may happen in other reactions is a very important skill for scientists. Patterns in other types of reactions, such as the reactions of metals with water and acids, may be linked to observations noted in electrolysis. Will you be able to find a link?

These pages are designed to help you think about aspects of the investigation rather than to guide you through it step by step.

Developing hypotheses

You already know the pattern of the reactivity series and that copper can electroplate other metals. Can potassium electroplate other metals?

From observation you already know that copper is less reactive than potassium and that copper from solution will deposit on a graphite cathode.

From scientific theory you know that potassium ions tend to form more easily than copper ions.

So one question could be 'does the ability to deposit on an electrode link to the reactivity of the metal ion in solution'? From this you could develop more than one hypothesis.

Let's consider one: '*The more reactive the metal the easier it is to deposit on an electrode*'. Is this reasonable given what you already know or should it be the other way round?

To plan the experiments to test this hypothesis:

What set-up of apparatus and what variables need to be considered when testing this hypothesis?

Here are some ideas:

You might be able to think of some more variables that may need to be kept the same.

DID YOU KNOW?

The reactivity series is closely linked to the electrochemical series.

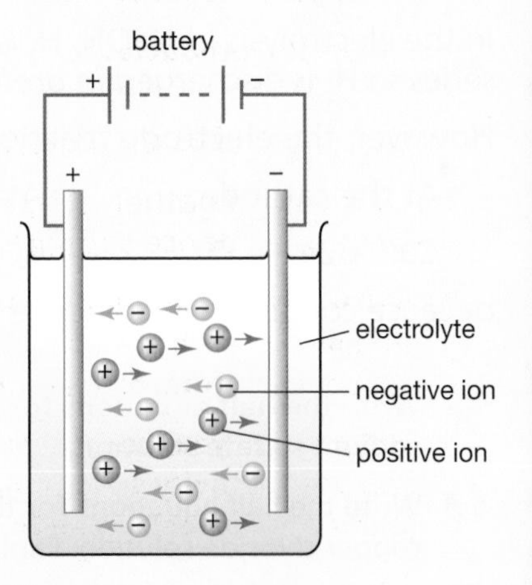

Figure 3.76

Between copper sulfate and potassium sulfate, think about which other 5 metal sulfate solutions may give a range of results to test the hypothesis.

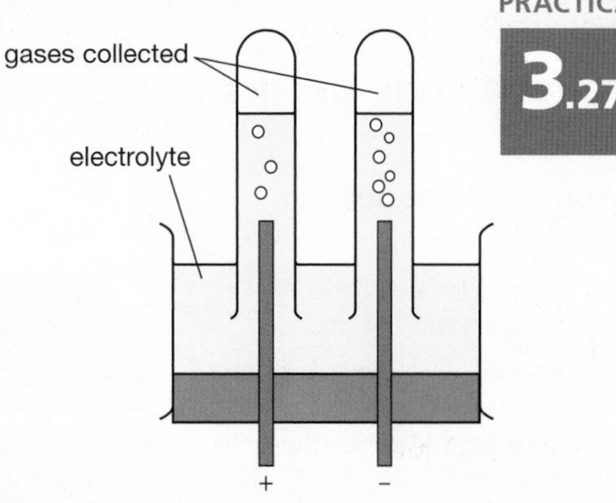

Figure 3.77

1. **Identify which electrode the metal will deposit on to.**
2. **Identify which gas may be given off instead of a metal. Describe how you would test for it and what you would see.**

Recording and analysing the results

When observations have been gathered, they will need to be interpreted. Sam and Alex tried these **electrolytes** and have started to record these results:

Electrolyte	Prediction for product at cathode	Appearance at cathode or test for gas	Product at cathode	Prediction at cathode supported?
zinc sulfate	zinc	gas popped with lighted spill		
copper sulfate	copper	red/brown metal coat		
sodium chloride	sodium	gas popped with lighted spill		
silver nitrate	silver	silvery/black metal coat		

KEY INFORMATION

A similar set of observations can be made for the reactions at the anode.

3. **Explain if the observations at the cathode support their predictions. Identify the products formed.**
4. **Explain which pattern they used to make these predictions.**
5. **Explain why their prediction was correct for copper and silver but incorrect for sodium and zinc. Predict what formed instead of sodium and zinc.**
6. **Draw a model, using ions, of what happens in the electrolysis of copper sulfate solution.**
7. **Predict the products that form at the anode.**

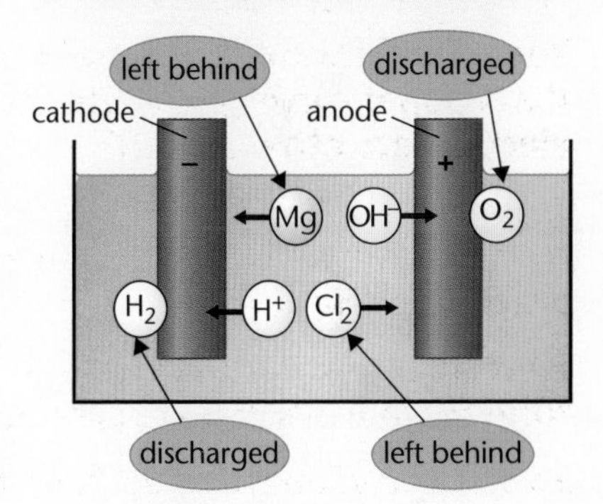

Figure 3.78

Evaluating the experiment

Look at the data gathered by Sam and Alex and reflect on whether the evidence supported the original hypotheses.

8. **Explain *the reasons why* these particular products were obtained at each electrode. What is the overall conclusion that can be made from Sam and Alex's results?**
9. **Explain how your conclusion supports or refutes the hypothesis that the reactivity series is linked to ability of a metal ion in solution to deposit on an inert electrode.**

Check your progress

You should be able to:

state the law of the conservation of mass →	explain how to balance equations in terms of numbers of atoms on both sides of the equation →	explain the meaning of subscripts within a formula and multipliers before a formula in a balanced equation
explain that when there is a mass change in a reaction it may be because a gas is being given off →	explain why there appears to be a mass change when metal carbonates are heated or metals are heated in oxygen →	explain observed changes in mass in non-enclosed systems and explain the changes in terms of the particle model
describe the measurement of amounts of substance in moles →	calculate the number of moles in a given mass →	calculate the mass of a given number of moles
calculate the masses of substances in a balanced symbol equation →	calculate the masses of reactants and products from balanced symbol equations →	calculate the mass of a given reactant or product
identify exothermic and endothermic reactions from temperature changes →	identify exothermic reactions as causing a temperature rise →	identify endothermic reactions as causing a temperature decrease
identify examples of exothermic reactions →	identify examples of endothermic reactions →	explain and evaluate the uses of some exothermic and endothermic reactions
investigate changes in temperature of different reactions →	investigate the variables that affect temperature changes in reacting solutions →	explain how the variables investigated affect temperature changes
recognise that energy transfer during a reaction is due to bonds being broken and then new bonds being made →	describe the energy changes in bond breaking as endothermic and bond making as exothermic and explain how the energy of a reaction is calculated overall →	calculate the energy transferred in chemical reactions using bond energies
describe how to make pure, dry samples of soluble salts →	explain how to name a salt →	derive a formula for a salt from its ions
describe the use of universal indicator to measure pH →	use the pH scale to identify acidic or alkaline solutions →	investigate pH changes when a strong acid neutralises a strong alkali.
explain weak and strong acids by the degree of ionisation →	describe neutralisation through the effect on hydrogen ions and pH →	explain the terms dilute and concentrated as the amounts of substances dissolved
explain why some metals need to be extracted by electrolysis →	explain the process of the electrolysis of aluminium oxide →	explain which non-metals are formed at the anode in preference
use apparatus to electrolyse aqueous solutions in the laboratory →	explain which metals (or hydrogen) are formed at the cathode in preference →	predict the products of the electrolysis of aqueous solutions containing a single ionic compound
explain the electrolysis of copper sulfate using inert electrodes →	predict the products of the electrolysis of aqueous solutions →	represent reactions at electrodes by half equations

Worked example

Kim and Jo are electrolysing dilute sulfuric acid.

1. **Identify the substance seen at the anode.**

   a nitrogen **b** hydrogen **c** sulfur **(d oxygen)**

The answer oxygen is correct.

2. **Describe how they will test for hydrogen gas.**

   It pops with a lighted splint.

This test is correct.

3. **Construct the half equation for the discharge of hydrogen at the electrode.**

   $H^+ + e^- \rightarrow H_2$

The charge on the ion and the gain of an electron are correct. The molecule H_2 is correct. The equation needs to be balanced, $2H^+ + 2e^- \rightarrow H_2$

4. **Next Kim and Jo want to electrolyse copper sulfate. Jo says that they cannot use solid copper sulfate. Explain why.**

   The ions need to be free to move.

This answer is partly correct. The ions need be free to move to conduct electricity, so need to be molten or in solution.

5. **They choose to use a solution. They pass a current through the solution of copper sulfate, using carbon electrodes.**

   a Describe what will happen at the cathode.

   It will get a coat of pink/brown copper.

The answer is correct.

   b Describe what they will see at the anode.

   It will disintegrate.

The student needs to be clear that this happens with copper electrodes, not carbon, and is used in the purification process. The answer should be that bubbles of oxygen will appear.

   c What will they see happening to the copper sulfate solution?

   It will stay blue.

The answer is again confused with the purification process. The correct answer is that the blue colour disappears. This is due to the copper depositing from the solution on to the electrode.

6. **Construct the half equation for the discharge of copper at an electrode and explain whether this reaction is oxidation or reduction. Explain why copper is deposited and hydrogen is not evolved.**

   $Cu^{2+} + 2e^- \rightarrow Cu$ This is reduction as electrons are gained (RIG). Copper is deposited because it is more reactive than hydrogen.

The half equation is correct. Reduction is correct – no need for the memory aid. Copper is less reactive than hydrogen, (which is why it is deposited).

End of chapter questions

Getting started

1. Hydrogen and oxygen will not react until a lighted splint is used. Suggest why. **1 Mark**
2. Which of these statements about a fuel cell is false?
   a Produces water as the only product. **b** Produces electricity.
   c Operates continuously. **d** Converts electricity into heat. **1 Mark**
3. Describe two uses of exothermic or endothermic reactions. **2 Marks**
4. Two solutions are added together. The temperature goes down. What type of reaction is this?
   a addition **b** endothermic **c** thermal decomposition **d** exothermic **1 Mark**
5. 1g of an insoluble metal carbonate reacts completely with 25cm^3 of acid, but 2g does not react completely. In an experiment to make a salt how many grams of the metal carbonate would you add to 25cm^3 of this acid? **1 Mark**
6. Sodium carbonate is added to a flask containing hydrochloric acid and a few drops of universal indicator. A small excess of sodium carbonate was added. Explain what happens to the colour in the flask. **2 Marks**
7. Match the result readings to the experiment. **2 Marks**

°C	20	28	34	36	37
cm^3	0	15	20	24	26
°C	21	18	16	15	14

rate of reaction by gas volume
endothermic reaction
exothermic reaction

Going further

8. Which oxide is a base? **1 Mark**
   a carbon dioxide **b** magnesium oxide **c** sulfur dioxide **d** silicon dioxide
9. Zinc is reacted with sulfuric acid. Bubbles form. What gas is this? **1 Mark**
   a oxygen **b** carbon dioxide **c** hydrogen **d** chlorine
10. Predict the products formed in the electrolysis of molten potassium iodide. **2 Marks**
11. Describe the stages of making pure zinc sulfate salt crystals from sulfuric acid. Give equations where appropriate. **4 Marks**
12. In an experiment a student reacted different metals with solutions of sulfates of the other metals. One metal, X, was unknown. These are the results:

	Mg	Fe	Cu	X
$MgSO_4$	✗	✗	✗	
$FeSO_4$	✓		✗	✓
$CuSO_4$	✓	✓		✓
XSO_4	✓	✗	✗	

Describe where X lies in the reactivity series, giving reasons. **2 Marks**

More challenging

13. What is the number of particles in one mole of a substance? **1 Mark**
    a 6.02×10^{23} **b** 6.02×10^{4} **c** 6.02×10^{-9} **d** 6.02×10^{-23}

14 Hydrogen peroxide, H_2O_2, decomposes when warmed. If oxygen is the desired product, the atom economy is not 100%. Suggest one reason why this is not important for this reaction.

$2\ H_2O_2\ (aq) \rightarrow 2\ H_2O\ (l) + O_2\ (g)$

1 Mark

15 Calculate the number of moles in 10 g of MgO.

2 Marks

16 Salts can be made by reacting acids with a variety of other substances. Show how salts are made using sulfuric acid and calcium and its compounds as an example.

- You should use equations to illustrate your answer
- Describe any observations you might make during the reactions.

2 Marks

17 The table below gives some data on three compounds.

4 Marks

a Work out values A to E.

b Work out the molecular formula of C_nH_{2n+2}.

Formula	Mass in grams	Molar mass	Number of moles
$CaCO_3$	A	B	0.2
$Ca(OH)_2$	148	C	D
C_nH_{2n+2}	25	E	0.25

Most demanding

18 Tin is present as tin(IV) oxide, SnO_2, in cassiterite ore. The ore is heated with carbon and other materials such as limestone and sand.

a Write a balanced equation for the reaction of tin(IV) oxide with carbon.

b Explain what is happening in terms of oxidation and reduction.

c Part of the reactivity series is (in decreasing order) as follows: K, Na, Mg, Al, C, Zn, Fe, Sn, Pb, H_2, Cu, Ag, Au. Explain why it is not necessary to use electrolysis to extract tin.

2 Marks

19 Explain why a fuel cell does not produce polluting gases.

4 Marks

20 Some bond energies in kJ/mol are given in the table below.

C–H	C=O	O=O	H–O	C–C
414	806	498	465	368

- Explain the overall energy change for the combustion of methane.
- Determine whether propane, C_3H_8, gives out more or less energy than methane when combusted.

4 Marks

40 Marks

PREDICTING AND IDENTIFYING REACTIONS AND PRODUCTS

IDEAS YOU HAVE MET BEFORE:

METALS AND NON-METALS

- Gold, iron, copper and lead are metals known for centuries.
- Oxygen and nitrogen are gases of the air.
- Sulfur is a yellow non-metal.

REACTIVITY SERIES

- Magnesium reacts more readily than zinc with acid.
- Copper does not react with dilute acid.
- If a piece of zinc is put into copper sulfate, copper coats the zinc.

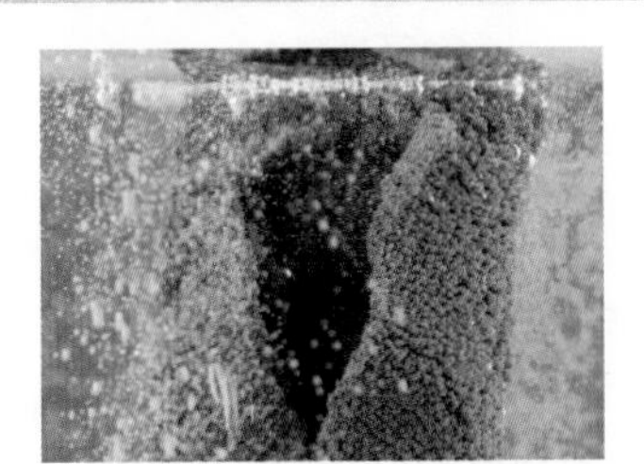

TESTING GASES

- Oxygen relights a glowing splint.
- Hydrogen pops with a burning splint.
- Carbon dioxide turns limewater milky.

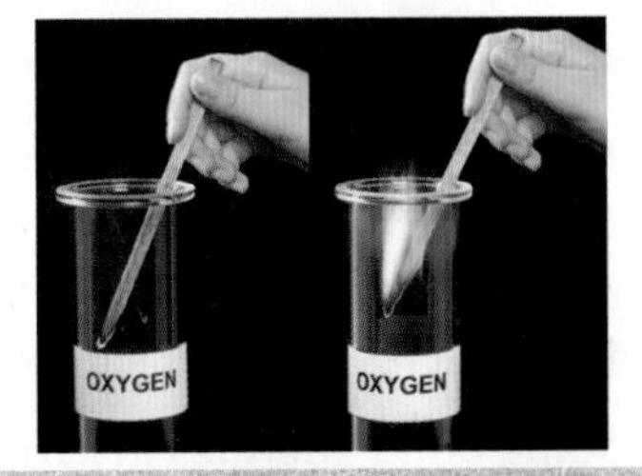

METALS CAN REACT TO MAKE SALTS

- Sodium chloride is a salt.
- Iron can react with acid to make a salt.
- Copper compounds react to make blue salts.

TESTING WITH INSTRUMENTS

- Thermometers measure melting and boiling points.
- Voltmeters measure voltage.
- Ammeters measure current.

4

IN THIS CHAPTER YOU WILL FIND OUT ABOUT:

WHAT DO TRANSITION METAL COMPOUND SOLUTIONS LOOK LIKE?

- Transition metals are harder and stronger than Group 1 metals.
- Transition metals are often used as catalysts.
- Transition metal compounds often form coloured solutions.

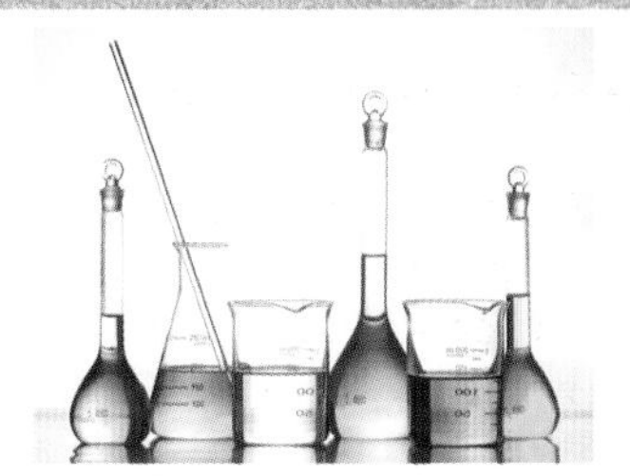

WHY ARE SOME METALS MORE REACTIVE THAN OTHERS?

- Metals from Group 1 show reactivity with water.
- Other metals react with acids with varying degrees of activity.
- More reactive metals displace less reactive metals.

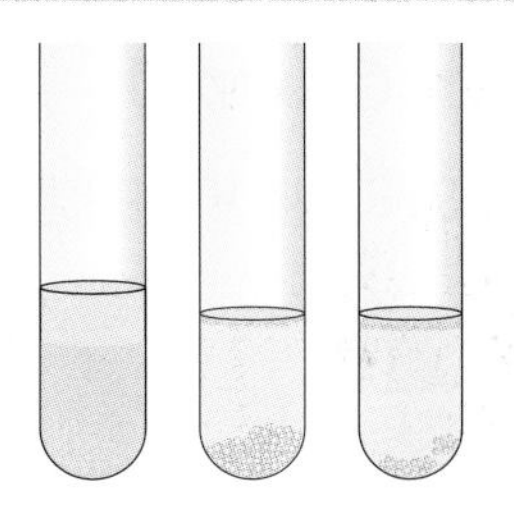

HOW CAN WE ANALYSE POSITIVE IONS?

- Flame tests are used to analyse Group 1 metals.
- Some metal ions produce coloured hydroxide precipitates.
- Aluminium hydroxide redissolves with excess addition of hydroxide.

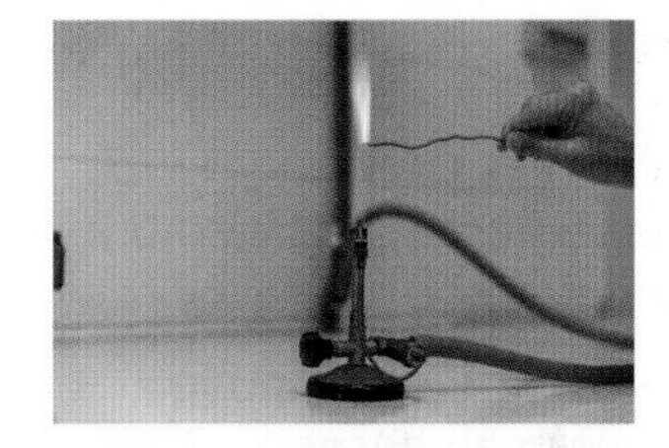

HOW CAN WE ANALYSE NEGATIVE IONS?

- Anions form precipitates with testing solutions.
- Halides produce different coloured precipitates with silver ions.
- Sulfates produce a precipitate with barium chloride.

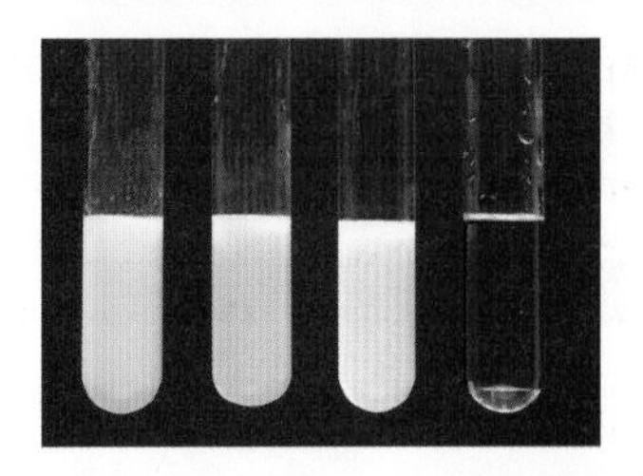

HOW CAN WE USE INSTRUMENTAL TECHNIQUES FOR ANALYSIS?

- Thin layer chromatography analyses drugs and medicines.
- Flame emission spectroscopy analyses metal ions in solution.
- Instrumental methods are accurate, sensitive and rapid.

Exploring Group 0

Learning objectives:

- describe the unreactivity of the noble gases
- predict and explain the trend in boiling point of the noble gases (going down the group)
- explain how properties of the elements in Group 0 depend on the outer shell of electrons of their atoms.

KEY WORDS

elements
helium
neon
argon
density
unreactive

Ever wondered why helium rather than hydrogen is used in party balloons and for weather balloons? Hydrogen is less dense than helium and so a hydrogen balloon would 'float' better. However, hydrogen is highly flammable whereas helium is unreactive. The decision comes down to safety.

Figure 4.1 Helium balloons

Patterns in Group 0

If you find Group 0 in the periodic table, you will see these **elements** in order.

Group 0
Helium (He)
Neon (Ne)
Argon (Ar)
Krypton (Kr)
Xenon (Xe)

All the elements of Group 0 in the periodic table have two things in common.

- They are all **unreactive**.
- They are all gases.

The boiling points of the elements in Group 0 show a trend.

Helium has the lowest boiling point. This means the atoms of the element keep moving rapidly (as a gas) at lower temperatures than the atoms of xenon.

Boiling points (°C)	
He	–268
Ne	–246
Ar	–186
Kr	–153
Xe	–108

The trend is that the boiling points of the gases increase down the group.

EXTENSION

Mendeleev did not predict the noble gases. They were discovered much later by William Ramsey. You need to be able to explain that if the noble gases had not fitted into Mendeleev's pattern then the evidence would have *refuted* his predictions but instead they did fit his pattern so the new evidence *supported* his predictions.

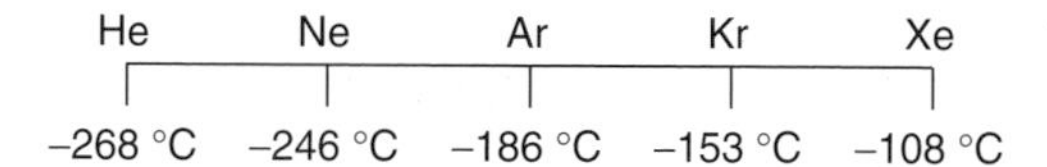

Figure 4.2

1. **Find the relative atomic mass of neon and argon. Determine which has the higher relative atomic mass.**
2. **There is a noble gas with a bigger relative atomic mass than Xe. Predict its boiling point.**

Why does helium stay as a gas at lower temperatures?

All the elements in Group 0 are gases.

These gases exist as single atoms, not molecules.

The smaller the atom the 'easier' it is for it to keep moving around rapidly.

So at lower temperatures the small atoms of helium move 'more easily' than the larger atoms of krypton. So krypton has a higher boiling point than helium.

3. **Describe how boiling point varies with relative atomic mass for Group 0 elements.**
4. **Suggest a relationship between the diameter of Group 0 atoms and their boiling point. Explain your answer.**

Why do elements in Group 0 exist as single atoms?

The elements of Group 0 do not generally make compounds with other elements and are unreactive.

They do not make compounds because the atoms have 8 electrons in their outer shell.

This is a very stable configuration. So there is no electron movement from one atom to another.

5. **Draw the electronic structure for helium. Explain why this atom does not join with any other atom.**
6. **Write out the electronic structure (in numbers) for argon. Explain why argon is unreactive but has a higher boiling point than helium.**
7. **One of the first noble gas compounds to be made was xenon tetrafluoride, made in 1962. It is a stable crystalline solid at room temperature.**

   a **Complete and balance the following equation:**

   __ Xe + __ $F_2 \rightarrow$ ____

   b **Suggest why it was a surprise that a noble gas compound had been made.**

DID YOU KNOW?

Helium is also used with neon in the lasers that scan supermarket barcodes.

Street lights used in long tunnels contain both neon and sodium. Argon is used for specialist welding and to fill the space between double glazed windows.

Find out what krypton and xenon are used for.

Figure 4.3 Neon lights were first used years ago

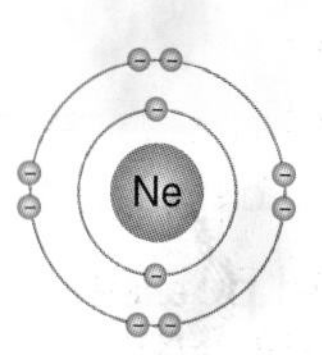

Figure 4.4 The electronic structure of neon. The outer shell is full.

Exploring Group 1

Learning objectives:

- explain why Group 1 metals are known as the alkali metals
- predict the properties of other Group 1 metals from trends down the group
- relate the properties of the alkali metals to the number of electrons in their outer shell.

KEY WORDS

alkali
density
indicator
ion
reactivity
stable electronic structure

Ever wondered how fireworks are made to have such stunning colours? It is because they have specific compounds added to them. Sodium compounds, for instance, produce a bright yellow colour against the night sky. Sodium is an element in the periodic table in the first column.

Figure 4.5 Which metal gives the firework this lilac colour?

Properties of Group 1 elements

Figure 4.6 This is sodium on water. Why does it float?

Lithium, sodium and potassium are Group 1 elements that are less **dense** than water.

Group 1 metals react vigorously with water and make hydrogen. Group 1 metals also burn in oxygen to form oxides. Sodium burns to make sodium oxide.

Figure 4.7 Potassium burning in oxygen

1. **Explain why sodium floats on water.**
2. **Identify the gas is given off when potassium reacts with water.**

Reaction trends of alkali metals

When lithium, sodium and potassium react with water, hydrogen is given off. A solution is also made in the reaction. It is an **alkali**. The alkali is the hydroxide of the metal. Sodium forms sodium hydroxide.

Lithium reacts vigorously with water.

Sodium reacts very vigorously with water.

Potassium reacts extremely vigorously with water and produces a lilac flame.

The word equation for the reaction between sodium and water is:

sodium + water → sodium hydroxide + hydrogen

The balanced symbol equation for the reaction is:

$2Na + 2H_2O \rightarrow 2NaOH + H_2$

3 **Suggest a way to show that the solution is an alkali.**

4 **Describe the trend in reactivity down Group 1.**

5 **Predict the reactivity of rubidium with water. Justify your answer.**

Making ions

Alkali metals have similar chemical properties. This is because when they react their atoms need to lose one electron to form the electronic structure of a noble gas. This is then a **stable electronic structure.**

When the atom loses one electron it forms an **ion**. The atom becomes charged. It has one more positive charge in its nucleus than the negative electrons surrounding it. So it is now a positive ion that carries a charge of +1.

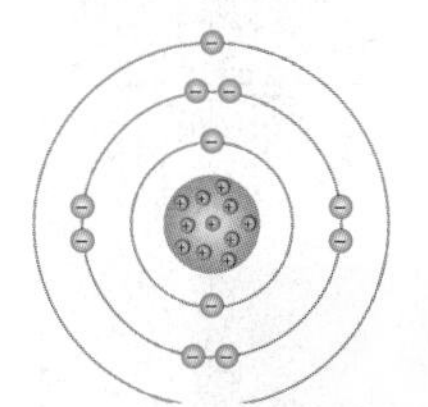

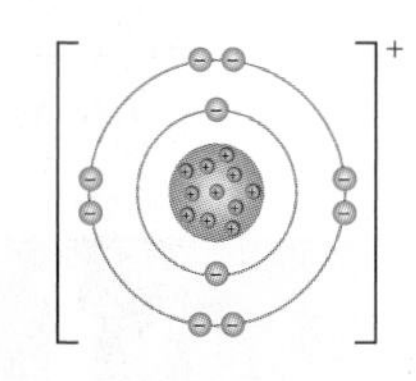

Figure 4.9 How alkali metals achieve a stable electronic structure

Sodium reacts with chlorine to make sodium chloride. The sodium makes an ion that carries a +1 charge. It makes ionic compound. The compound is a white solid that dissolves in water to form a colourless solution. The other alkali metals make these compounds too.

6 **Explain why Group 1 atoms lose electrons.**

7 **Draw a 'dot and cross' diagram to show a Li ion.**

8 **Potassium reacts with oxygen to form an oxide.**

a **Write a balanced equation for the reaction.**

b **State the electron configurations of potassium and oxygen in the oxide.**

c **The oxide dissolves in water to form an alkaline solution. Identify this solution.**

REMEMBER!

You should try to remember the order of reactivity of the alkali metals.

DID YOU KNOW?

Sodium hydroxide can be used as oven cleaner. This picture shows you what can happen if you get sodium hydroxide on your skin.

Figure 4.8 Why do the instructions on the bottle tell you to use gloves?

Sodium hydroxide is more dangerous to get into the eyes than acids. The hydroxide ions travel to the back of the eyes and can irreversibly damage the retina.

That is why your teacher will always tell you to wear safety glasses when handling chemicals, especially alkalis.

Exploring Group 7

Learning objectives:

- recall that fluorine, chlorine, bromine and iodine are non-metals called halogens
- describe that they react vigorously with alkali metals
- construct balanced symbol equations for the reactions of metals with halogens.

KEY WORDS

bromine
chlorine
halogen
iodine

Group 7 elements are known as the halogens. The first use of chlorine was to bleach textiles and it is still used in chemicals for bleaching toilets. Chlorine is used to sterilise water which prevents diseases such as cholera from spreading. Chlorine was also used as a weapon in the First World War. The effects were devastating.

The halogens

Group 7 elements, fluorine, **chlorine**, **bromine** and **iodine**, are called the **halogens**.

Halogens are non-metals. Halogens exist as pairs of atoms (molecules), so their symbols are F_2, Cl_2, Br_2 and I_2.

They react vigorously with **metals** such as sodium, potassium and magnesium.

Potassium reacts with chlorine to make potassium chloride. The word equation is:

potassium + chlorine → potassium chloride

It is possible to construct a balanced symbol equation for the reaction

$$K + Cl_2 \rightarrow KCl \text{ (unbalanced)}$$

The halogens react with metals to make salts.

The halogens react with non-metals to make gases or liquids such as acids.

$$2K + Cl_2 \rightarrow 2KCl \text{ (balanced)}$$

1. **Give the balanced symbol equation for the reaction between sodium and bromine.**
2. **Identify the product of the reaction between lithium and fluorine.**

13	14	15	16	17	18
					2 **He** helium 4.0
5 **B** boron 10.8	6 **C** carbon 12.0	7 **N** nitrogen 14.0	8 **O** oxygen 16.0	9 **F** fluorine 19.0	10 **Ne** neon 20.2
13 **Al** aluminium 27.0	14 **Si** silicon 28.1	15 **P** phosphorus 31.0	16 **S** sulfur 32.1	17 **Cl** chlorine 35.5	18 **Ar** argon 39.9
31 **Ga** gallium 69.7	32 **Ge** germanium 72.6	33 **As** arsenic 74.9	34 **Se** selenium 79.0	35 **Br** bromine 79.9	36 **Kr** krypton 83.8
49 **In** indium 114.8	50 **Sn** tin 118.7	51 **Sb** antimony 121.8	52 **Te** tellurium 127.6	53 **I** iodine 126.9	54 **Xe** xenon 131.3
81 **Tl** thallium 204.4	82 **Pb** lead 207.2	83 **Bi** bismuth 209.0	84 **Po** polonium	85 **At** astatine	86 **Rn** radon

Figure 4.10 Group 7 elements in the periodic table

Figure 4.11 Chlorine reacting with potassium to make potassium chloride

DID YOU KNOW?

Silver bromide is a compound of bromine with silver. It was used in early photography as it turned from cream to a purplish colour when exposed to light.

Group 7 trends

There is a **trend** in the physical appearance of the halogens at room temperature. Chlorine is a gas and iodine is a solid.

Figure 4.12 At room temperature chlorine is a green gas and iodine is a grey solid. What colour is the toxic and volatile bromine?

> **KEY INFORMATION**
>
> **Bromine displaces iodine from iodide solutions. It is seen as a red-brown solution.**
>
> $Br_2 + 2KI \rightarrow 2KBr + I_2$
> **(red-brown solution)**

Halogen	Atomic mass	Relative molecular mass	Melting point (°C)	Boiling point (°C)	State at room temperature
Chlorine		71	–101	–34	gas
Bromine	80	160	–7	59	
Iodine	127		114	184	

3 **Complete the data table.**

4 **Explain how the appearance of iodine in Figure 4.12 is confirmed by data column 5 of the table.**

5 **Describe the trend in boiling point as molecular mass changes.**

Displacement reactions of halogens

The **reactivity** of the halogens decreases down the group.

If halogens are bubbled through solutions of metal halides there are two possibilities: no reaction, or a **displacement** reaction.

If chlorine is bubbled through potassium bromide solution a displacement reaction occurs. The orange colour of bromine is seen. This is because chlorine is more reactive than bromine.

16	17	18
		2 **He** helium 4.0
8 **O** oxygen 16.0	9 **F** fluorine 19.0	10 **Ne** neon 20.2
16 **S** sulfur 32.1	17 **Cl** chlorine 35.5	18 **Ar** argon 39.9
34 **Se** selenium 79.0	35 **Br** bromine 79.9	36 **Kr** krypton 83.8
52 **Te** tellurium 127.6	53 **I** iodine 126.9	54 **Xe** xenon 131.3
84 **Po** polonium	85 **At** astatine	86 **Rn** radon

decreasing reactivity

Figure 4.13

6 **Identify which halogens will displace iodine from a solution of potassium iodide. Justify your answer.**

7 **Write a balanced symbol equation for the reaction between chlorine and potassium iodide.**

8 **Astatine is at the bottom of Group 7. It is radioactive and extremely rare so its chemistry has not been well studied.**

a **Predict the state of astatine at room temperature.**

b **Write a balanced equation for the reaction between chlorine and sodium astatide, NaAt.**

c **Explain whether astatine, At_2, will react with sodium iodide, NaI.**

Transition metals

Learning objectives:

- compare the properties of transition metals with Group 1 metals
- describe the formation of coloured ions with different charges
- explore the properties of Cr, Mn, Fe, Co, Ni and Cu.

KEY WORDS

catalyst
chromium
cobalt
manganese
nickel
transition element

Transition metals have been known since ancient times. Copper and iron are two metals with symbols from their old Latin name. The symbol for copper is Cu, which was known as cuprum and the symbol for iron is Fe which was known as ferrum.

DID YOU KNOW?

Gold (Au) had the Latin name aurum (the root of a word we use now – aurora) and silver (Ag) had the name argentum (which is still used in French for money, argent).

Comparing properties

All **transition elements** are metals and have typical metallic properties.

Transition elements are found in the middle part of the **periodic table**.

key: atomic number / **symbol** / name / relative atomic mass

(1) 1	(2) 2	3	4	5	6	7	8	9	10	11	12	(3) 13	(4) 14	(5) 15	(6) 16	(7) 17	(0) 18
1 **H** hydrogen 1.0																	2 **He** helium 4.0
3 **Li** lithium 6.9	4 **Be** beryllium 9.0											5 **B** boron 10.8	6 **C** carbon 12.0	7 **N** nitrogen 14.0	8 **O** oxygen 16.0	9 **F** fluorine 19.0	10 **Ne** neon 20.2
11 **Na** sodium 23.0	12 **Mg** magnesium 24.3											13 **Al** aluminium 27.0	14 **Si** silicon 28.1	15 **P** phosphorus 31.0	16 **S** sulfur 32.1	17 **Cl** chlorine 35.5	18 **Ar** argon 39.9
19 **K** potassium 39.1	20 **Ca** calcium 40.1	21 **Sc** scandium 45.0	22 **Ti** titanium 47.9	23 **V** vanadium 50.9	24 **Cr** chromium 52.0	25 **Mn** manganese 54.9	26 **Fe** iron 55.8	27 **Co** cobalt 58.9	28 **Ni** nickel 58.7	29 **Cu** copper 63.5	30 **Zn** zinc 65.4	31 **Ga** gallium 69.7	32 **Ge** germanium 72.6	33 **As** arsenic 74.9	34 **Se** selenium 79.0	35 **Br** bromine 79.9	36 **Kr** krypton 83.8
37 **Rb** rubidium 85.5	38 **Sr** strontium 87.6	39 **Y** yttrium 88.9	40 **Zr** zirconium 91.2	41 **Nb** niobium 92.9	42 **Mo** molybdenum 95.9	43 **Tc** technetium	44 **Ru** ruthenium 101.1	45 **Rh** rhodium 102.9	46 **Pd** palladium 106.4	47 **Ag** silver 107.9	48 **Cd** cadmium 112.4	49 **In** indium 114.8	50 **Sn** tin 118.7	51 **Sb** antimony 121.8	52 **Te** tellurium 127.6	53 **I** iodine 126.9	54 **Xe** xenon 131.3
55 **Cs** caesium 132.9	56 **Ba** barium 137.3	57-71 lanthanides	72 **Hf** hafnium 178.5	73 **Ta** tantalum 180.9	74 **W** tungsten 183.8	75 **Re** rhenium 186.2	76 **Os** osmium 190.2	77 **Ir** iridium 192.2	78 **Pt** platinum 195.1	79 **Au** gold 197.0	80 **Hg** mercury 200.6	81 **Tl** thallium 204.4	82 **Pb** lead 207.2	83 **Bi** bismuth 209.0	84 **Po** polonium	85 **At** astatine	86 **Rn** radon
87 **Fr** francium	88 **Ra** radium	89-103 actinides	104 **Rf** rutherfordium	105 **Db** dubnium	106 **Sg** seaborgium	107 **Bh** bohrium	108 **Hs** hassium	109 **Mt** meitnerium	110 **Ds** darmstadtium	111 **Rg** roentgenium	112 **Cn** copernicium		114 **Fl** flerovium		116 **Lv** livermorium		

Figure 4.14 The periodic table with the transition elements coloured green. Can you find the symbol for the transition metal called nickel?

Look at the data table and compare the properties of the transition metals with those of Group 1. What comparisons do you see?

Symbol	Li	Na	K	Cr	Fe	Cu
Melting point (°C)	180.5	98	63	1907	1538	1084
Density (g/cm³)	0.53	0.97	0.89	7.15	7.87	8.96
Hardness (Moh scale)	0.6	0.5	0.4	8.5	4	3
Tensile strength	Weak	Weak	Weak	n/a	n/a	n/a

Predict these properties for Mn, Co and Ni. Look them up in the Royal Society of Chemistry Periodic Table and check if you were correct.

Transition elements have different reactivities. They are less reactive than elements in Group 1.

Copper, iron and manganese do not react quickly with water or with oxygen.

Figure 4.15 Sodium in Group 1 is such a soft metal it can be cut with a knife. Iron cannot be cut like this as it is much harder.

1. **Use the data table above and the data for Mn, Co and Ni from the RSC Periodic Table to draw graphs of (a) melting point against atomic number and (b) density against atomic number. Explain the trends shown.**

2. **Use the graphs in Q1 to predict the properties of zinc.**

Catalysts

A transition element and its compounds are often **catalysts**.

- Iron is used in the Haber process to make ammonia.
- Nickel is used in the manufacture of margarine to harden oils.

A catalyst is an element or compound that changes the rate of a chemical reaction without taking part in the reaction as a reactant. Catalysts are unchanged by the reaction.

3. **Explain why catalysts such as iron in the Haber process can be reused.**

Ions and coloured compounds

A compound of a Group 1 metal is usually white.

A compound that contains a transition element is often coloured:

Compound	copper	iron(II)	iron(III)	nickel
Colour	blue	pale green	orange/brown	green

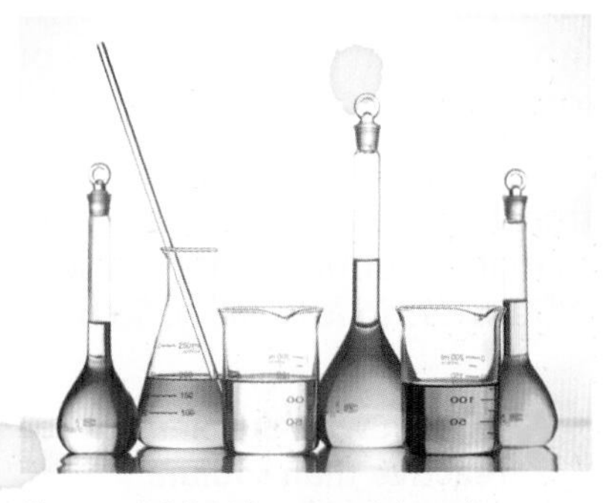

Figure 4.16 What transition elements do these compounds contain?

The reason that iron forms compounds that are different colours is that it can form two different types of positive **ion**.

Transition metals are able to lose different numbers of electrons. Iron can lose two electrons or three electrons. So it can make a 2+ ion or a 3+ ion.

From the table above you can see that compounds of iron making a 2+ ion make a pale green solution.

4. **Zinc reacted with a solution of an iron compound. The colour changed from brown to pale green. Explain what happened.**

5. **How many electrons will nickel need to lose to make a Ni^{3+} ion?**

6. **Copper can form two different ions. Suggest the charges on the ions given the following formulae: CuO, Cu_2O.**

EXTENSION

Iron(II) compounds were called ferrous compounds and the iron(III) ions were called ferric ions. These names are from the name ferrum. You may still need to use these names if you are looking up information. Nowadays the names used are iron(II) and iron(III).

Reaction trends and predicting reactions

Learning objectives:

- explain why the trends down the group in Group 1 and in Group 7 are different
- explain the changes across a period
- predict the reactions of elements with water, dilute acid or oxygen from their position in the periodic table.

KEY WORDS

electron arrangement
reactivity
trend

Fluorine, at the top of Group 7, is more reactive than iodine lower down the group. Yet why is caesium, lower down Group 1, so much more reactive with water than lithium at the top of the group? It is all to do with the electronic arrangement and the outer electrons.

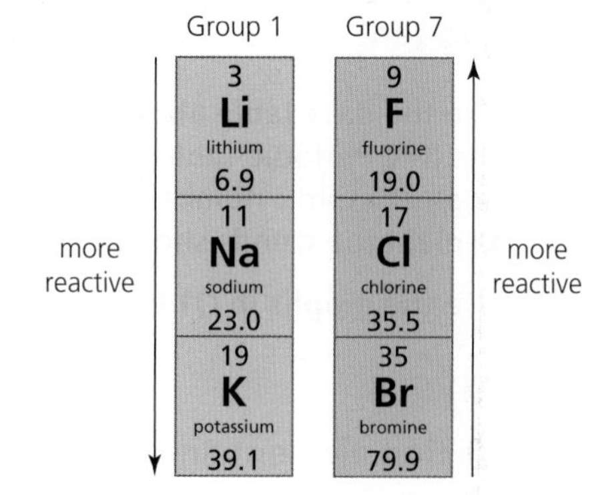

Figure 4.17 (a) Group 1; (b) Group 7

Opposite trends

The **trends** in **reactivity** in Group 1 and Group 7 are in the opposite directions.

Reactivity increases down Group 1 as the outer electron in a potassium atom is further away from the nucleus than in a lithium atom so there is 'less pull' on it by the nucleus so it is lost ***more*** easily. So potassium is more reactive than lithium.

Reactivity decreases down Group 7 as the electron trying to transfer into a bromine atom is further away from the nucleus than in a fluorine atom so there is 'less pull' on it by the nucleus so it transfers in ***less*** easily. So bromine is less reactive than fluorine.

1. **Explain why the element caesium (also in Group 1) is much more reactive than sodium.**
2. **Explain why astatine is less reactive than chlorine.**

DID YOU KNOW?

Caesium was used to make the first atomic clock in 1955.

Trends across the periodic table

The trend across the table is from metallic to non-metallic.

The trends across the periodic table also depend on atomic structure and the electronic configurations.

In Periods 2 and 3 across the table the outer electrons will increase by one. Group 1 has one outer electron and Group 2 has two outer electrons up to Group 7 has seven outer electrons.

Group 1 elements lose one electron to form positive ions easily. Fig 4.18 shows Na in Group 1. Na is more reactive than Mg in

Group 2. It is more difficult for elements to lose two or three electrons so they are less reactive.

Group 7 elements need to gain one electron to form ions.

Group 1	2											Group 3	4	5	6	7	0
1 **H** hydrogen 1.0																	2 **He** helium 4.0
3 **Li** lithium 6.9	4 **Be** beryllium 9.0											5 **B** boron 10.8	6 **C** carbon 12.0	7 **N** nitrogen 14.0	8 **O** oxygen 16.0	9 **F** fluorine 19.0	10 **Ne** neon 20.2
11 **Na** sodium 23.0	12 **Mg** magnesium 24.3											13 **Al** aluminium 27.0	14 **Si** silicon 28.1	15 **P** phosphorus 31.0	16 **S** sulfur 32.1	17 **Cl** chlorine 35.5	18 **Ar** argon 39.9
19 **K** potassium 39.1	20 **Ca** calcium 40.1	21 **Sc** scandium 45.0	22 **Ti** titanium 47.9	23 **V** vanadium 50.9	24 **Cr** chromium 52.0	25 **Mn** manganese 54.9	26 **Fe** iron 55.8	27 **Co** cobalt 58.9	28 **Ni** nickel 58.7	29 **Cu** copper 63.5	30 **Zn** zinc 65.4	31 **Ga** gallium 69.7	32 **Ge** germanium 72.6	33 **As** arsenic 74.9	34 **Se** selenium 79.0	35 **Br** bromine 79.9	36 **Kr** krypton 83.8

metals — non-metals

Figure 4.18 Period 3 is highlighted.

3 **Explain why neon is an unreactive element.**

4 **Explain why sodium is a more reactive metal than aluminium.**

Predicting reactions

Knowing the position of an element in the periodic table will allow us to predict its behaviour with water, acid or oxygen.

From the trends that you know, make some predictions.

- Will barium be more reactive with dilute acid than magnesium?
- Will sulfur react more like phosphorus with oxygen or more like sodium?
- If bromine is bubbled through potassium chloride solution what reaction will there be?

Barium is lower down Group 2 than magnesium so will lose electrons more easily so is more reactive.

Sulfur is more like phosphorus as they are both non-metals, so they will react with oxygen in a similar way.

There will be no reaction because chlorine is more reactive than bromine and so will not be displaced.

5 **Predict how rubidium will behave with water, acid and oxygen, giving reasons for your predictions.**

6 **Predict how strontium will react with dilute acid and give reasons for your predictions.**

7 **Hydrogen reacts with halogens to form hydrogen halides.**

a **Give the balanced chemical equation for the reaction of chlorine with hydrogen.**

b **A mixture of chlorine and hydrogen explode in sunlight. Predict the reactivity of fluorine with hydrogen. Explain your answer.**

KEY INFORMATION

An atom with one or two electrons will tend to lose them to make positive ions. For example Na loses one electron to become Na^+.

Reactivity series

Learning objectives:

- describe the reactions, if any, of metals with water or dilute acids
- deduce an order of reactivity of metals based on experimental results
- explain how the reactivity is related to the tendency of the metal to form its positive ion.

KEY WORDS

displacement
positive ion
reactivity
tendency

You have already seen that Group 1 metals, sodium and potassium, react vigorously with water. Potassium is more reactive than sodium. You will have seen a tiny amount of potassium skimming on the surface of water and bursting into a lilac flame. Caesium is even more reactive than potassium.

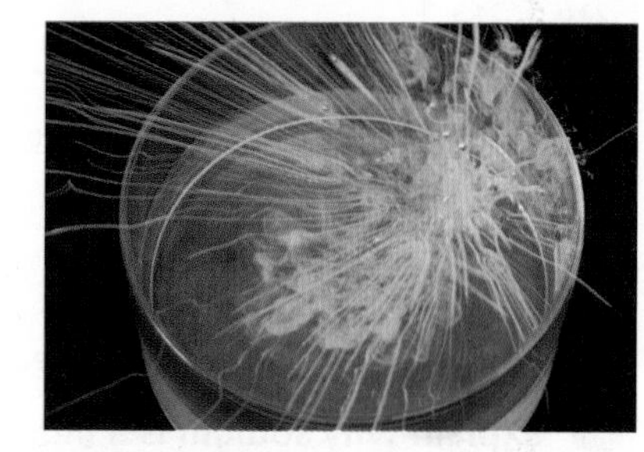

Figure 4.19 Potassium on water. A highly reactive metal. Don't try this at home.

Reactions of metals with water

- Potassium, sodium and lithium react vigorously with water.
- Calcium reacts with water less vigorously than lithium.
- Magnesium reacts with steam but not with cold water.

We can put these five metals into decreasing order of their reactivity with water.

potassium → sodium → lithium → calcium → magnesium

Other, less reactive, metals do not react with water but do react with acid, so the order of reactivity can be continued.

Magnesium, zinc, iron and copper can be added to solutions of acid and their reactivity observed. Copper does not react with dilute acids but magnesium does.

The decreasing order of reactivity of metals with acids is:

magnesium → zinc → iron → copper

So from the two sets of observations the **reactivity** series so far is: potassium → sodium → lithium → calcium → magnesium → zinc → iron → copper

DID YOU KNOW?

Gold is so unreactive that it does not react with oxygen. It is always beautifully shiny.

Figure 4.20 Gold can remain untarnished and shiny for millennia, as it is so unreactive

1 Explain how you know that sodium is more reactive than zinc.

Positive ions

The metals potassium, sodium, lithium, calcium, magnesium, zinc, iron and copper all form positive ions.

They can be put in order of their reactivity by observation of their reactions with water and dilute acids. The reason that there is a difference in reactivity is because of the different **tendency** of each metal to form positive ions.

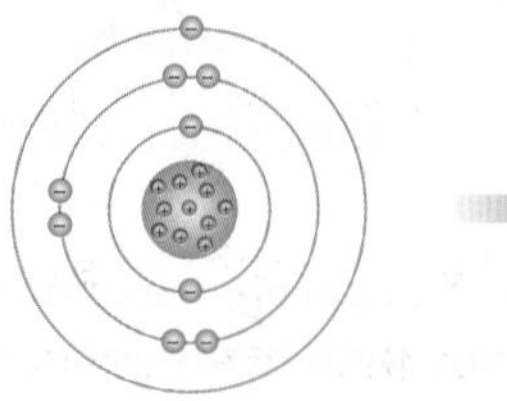

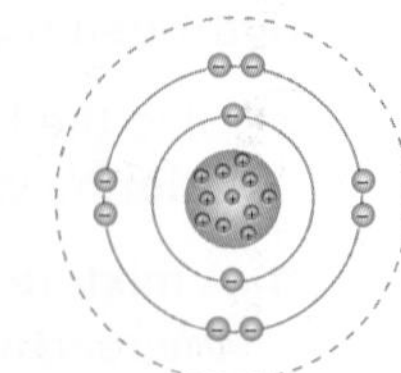

Figure 4.21 Sodium atom forming a sodium ion by losing an electron

Sodium forms **positive ions** much more easily than copper and more easily than lithium.

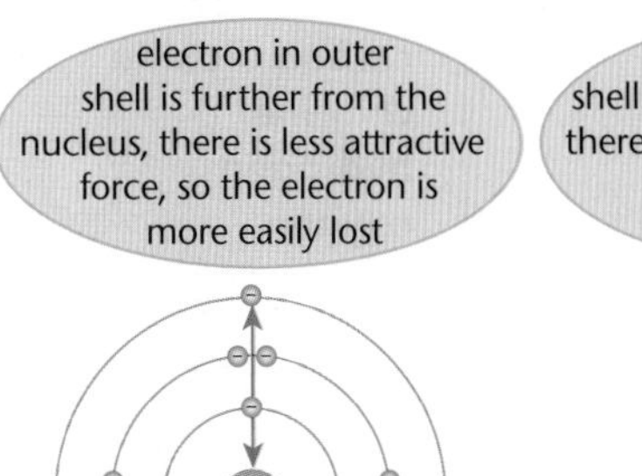

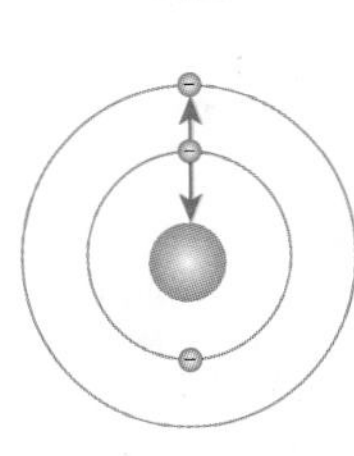

Figure 4.22 Sodium forms ions more easily than lithium

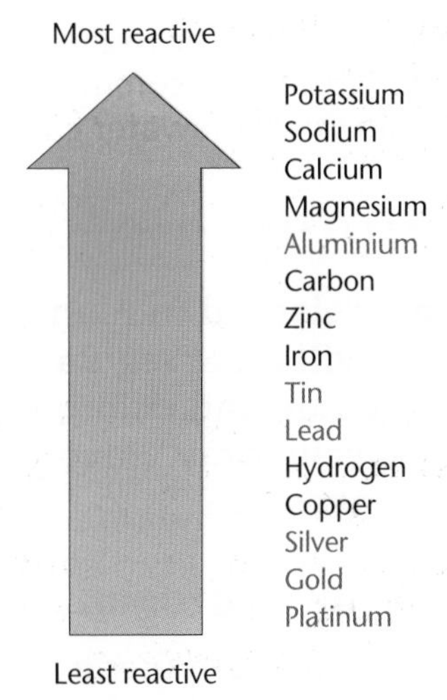

Figure 4.23 The reactivity series

The non-metals hydrogen and carbon are often included in the reactivity series. Hydrogen is placed between iron and copper. Carbon lies between magnesium and zinc.

A more reactive metal can displace a less reactive metal from a solution of a compound of the less reactive metal.

For example, if magnesium metal is put into iron sulfate solution, the magnesium 'pushes the iron out' of the iron sulfate. Solid iron and a solution of magnesium sulfate are made.

This is called a **displacement** reaction.

iron sulfate + magnesium → magnesium sulfate + iron

It happens because magnesium is more reactive than iron. The iron has been pushed out or displaced. The iron forms a coating on the piece of magnesium.

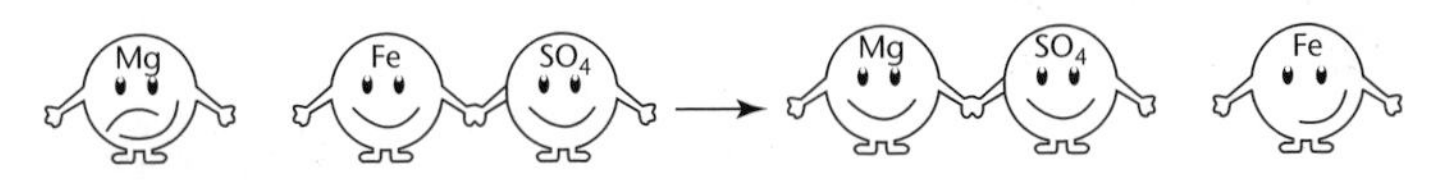

Figure 4.24

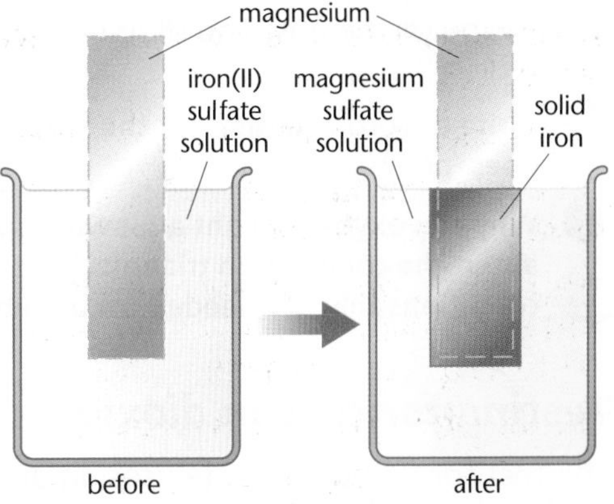

Figure 4.25 Why can magnesium displace iron?

2 **Explain why potassium forms positive ions more easily than sodium.**

HIGHER TIER ONLY

Ion formation

Positive ions are formed by the loss of electrons.

$$K - e^- \rightarrow K^+ \text{ [or } K \rightarrow K^+ + e^- \text{]}$$

$$Mg - 2e^- \rightarrow Mg^{2+} \text{ [or } Mg \rightarrow Mg^{2+} + 2e^- \text{]}$$

Displacement of a less reactive metal is represented by:

$$Mg + Fe^{2+} \rightarrow Mg^{2+} + Fe$$

3 **Write an ionic equation to show the displacement of copper ions by zinc metal. Both copper ions and zinc ions have a 2+ charge.**

KEY SKILLS

Before you start to write equations find out how many electrons the metals lose to make an ion.

Test for gases

Learning objectives:

- recall the tests for four common gases
- identify the four common gases using these tests
- explain why limewater can be used for testing CO_2.

KEY WORDS

carbon dioxide
chlorine
hydrogen
oxygen

Gases are released in many chemical reactions. The gases we will meet regularly are hydrogen, oxygen, carbon dioxide and chlorine. Each one can be tested for in a specific way. Have you tried testing for any of these gases yet?

Testing for hydrogen and oxygen

The test for hydrogen uses a burning splint held at the open end of a test tube of the gas. Hydrogen burns rapidly with a 'pop' sound.

The basis for this test is that hydrogen is reacting rapidly as a fuel by burning in oxygen. The product is water.

The test for oxygen uses a glowing splint inserted into a test tube of the gas. The glowing splint relights in oxygen.

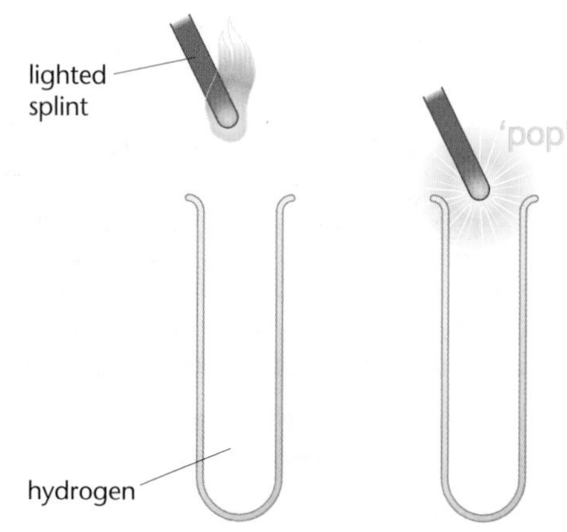

Figure 4.26 Testing for hydrogen

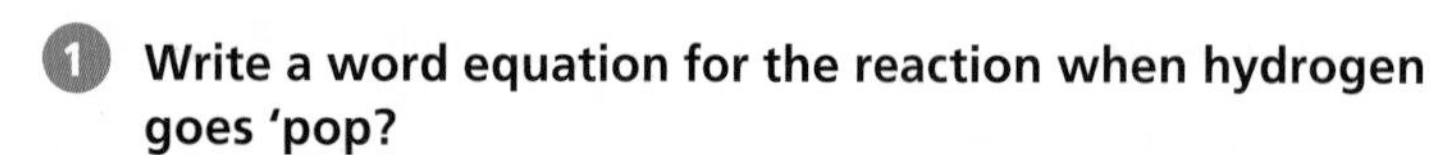
1 Write a word equation for the reaction when hydrogen goes 'pop?

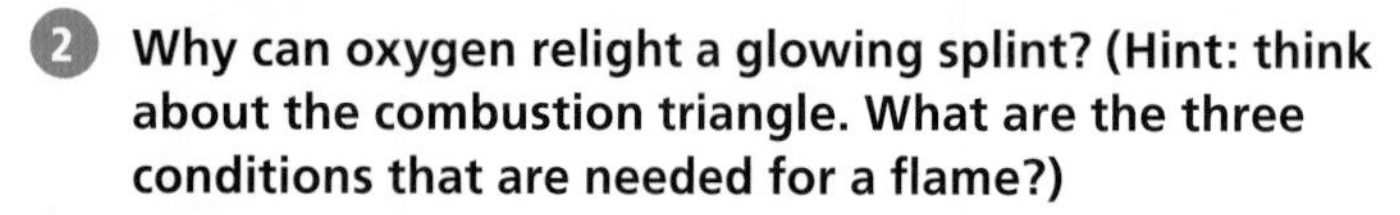
2 Why can oxygen relight a glowing splint? (Hint: think about the combustion triangle. What are the three conditions that are needed for a flame?)

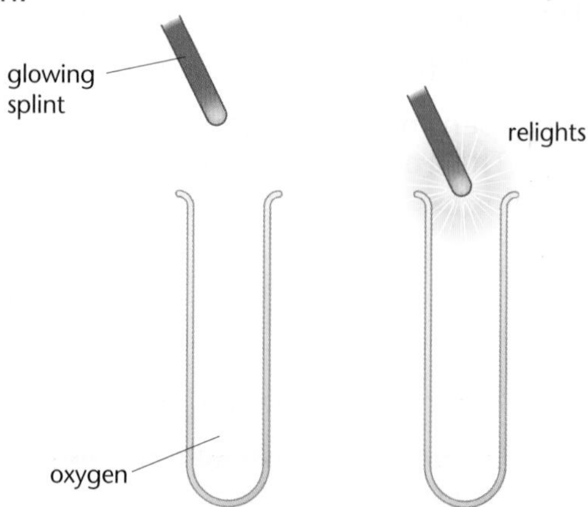

Figure 4.27 Testing for oxygen

Testing for carbon dioxide

The test for carbon dioxide uses an aqueous solution of calcium hydroxide (which is often called 'limewater'). When carbon dioxide is shaken with or bubbled through limewater the limewater turns milky (it looks cloudy).

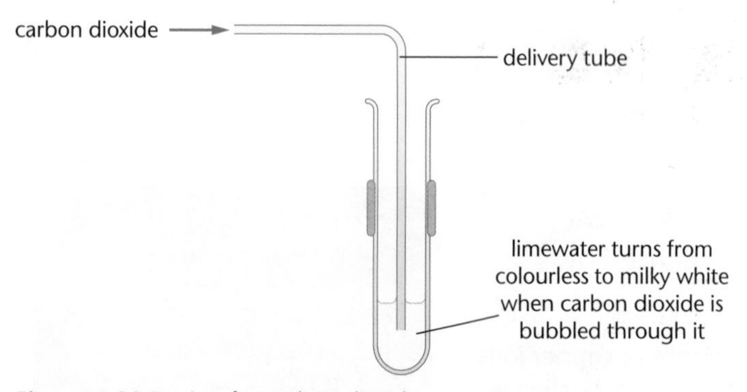

Figure 4.28 Testing for carbon dioxide

DID YOU KNOW?

Some airships used to be filled with hydrogen. The Hindenburg airship caught fire in May 1937, with disastrous consequences, as it tried to moor in New York after a voyage across the Atlantic.

What is happening in this test? Limewater is calcium hydroxide. When carbon dioxide is passed through limewater, the product formed is calcium carbonate. This white chalky substance gives a milky appearance to the solution as it is gradually deposited.

REMEMBER!

After a while the limewater will become colourless again if carbon dioxide continues to pass into it.

3. **Write the word equation for the reaction between limewater and carbon dioxide.**
4. **Find out why the limewater turns colourless again if CO_2 continues to be passed through the limewater.**

Testing for chlorine

The test for chlorine uses litmus paper. When damp litmus paper is put into chlorine gas the litmus paper is bleached and turns white. Chlorine can also be bubbled into water and tested with litmus paper, which bleaches.

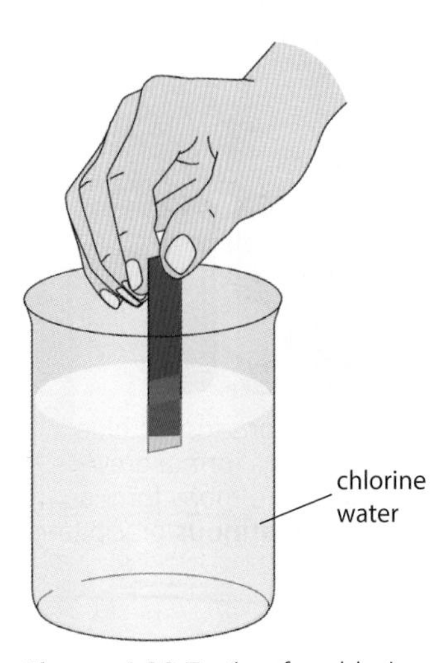

Figure 4.29 Testing for chlorine gas dissolved in water

5. **What colour is chlorine gas?**
6. **Which halogen gives a brown-red gas when warmed?**
7. **Which halogen gives a purple gas when warmed?**
8. **Two students tested some gases to find their identity. Fill in the table with their results or the reagents they used.**

Sample	Test	Result	Gas	Formula
D	Glowing splint		Oxygen	O_2
E	Lighted splint	Burned with 'pop'		
F	Passing though limewater	Went milk and then colourless		
G	Testing with damp litmus paper		Chlorine	

Metal hydroxides

Learning objectives:

- recognise the precipitate colour of metal hydroxides
- explain how to use sodium hydroxide to test for metal ions
- write balanced equations to produce insoluble hydroxides.

KEY WORDS

hydroxide
precipitate
gelatinous
ion

Some metal ions can be detected by a precipitation reaction. What is a precipitation reaction? It is a reaction when two solutions react together to make a solid and another solution. Many metals make hydroxide precipitates that are jelly-like. Can these gelatinous precipitates be used in detection?

Precipitate

Figure 4.30 A precipitate forming from two solutions

Coloured precipitates

A compound that contains a transition element is often coloured.

- Copper compounds are often blue.
- Iron(II) compounds are often pale green.
- Iron(III) compounds are often orange–brown.

If sodium **hydroxide** solution is added to a solution of these metal compounds they form **precipitates** of different colours.

Metal in compound solution	copper	iron(II)	iron(III)
Colour	blue	light green	brown

Cu^{2+} Fe^{2+} Fe^{3+}

Figure 4.31 Cu^{2+} ions form a blue precipitate. Fe^{2+} ions form a grey/green precipitate. Fe^{3+} ions form an orange/brown **gelatinous** precipitate.

This precipitation reaction can be used to identify these metals in solution.

When these solutions react together with sodium hydroxide, they make a solid. The reaction is called a *precipitation reaction*.

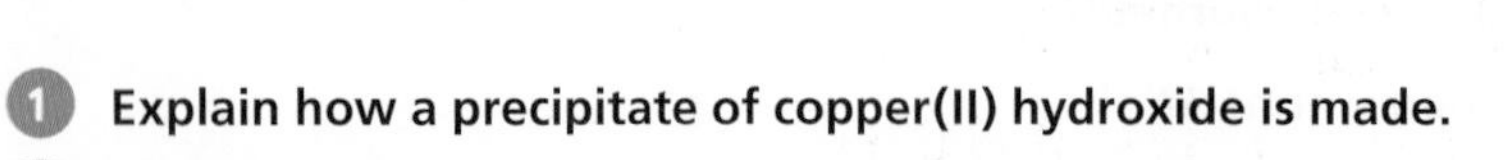

1 Explain how a precipitate of copper(II) hydroxide is made.

2 Sodium hydroxide was added to a metal sulfate solution. A grey/green gelatinous precipitate formed. Identify the metal ion.

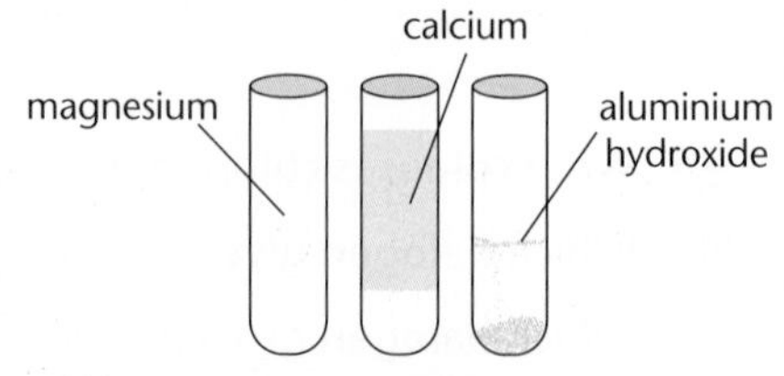

Figure 4.32 Magnesium ions and calcium ions form hydroxides that are white precipitates. Aluminium hydroxide redissolves

White precipitates

Sodium hydroxide also forms precipitates with other metal **ions**, such as magnesium, calcium and aluminium.

All three metal ions form white precipitates with sodium hydroxide. So other tests need to be used to distinguish between them.

Aluminium can be distinguished from calcium and magnesium using sodium hydroxide.

KEY INFORMATION

Look up how to distinguish magnesium ions from calcium ions using a flame test.

It forms a white precipitate at first just like the other two metal ions. But aluminium hydroxide precipitate then dissolves if excess sodium hydroxide solution is added.

3 **An ionic compound formed a white precipitate with sodium hydroxide. The precipitate did not dissolve in excess sodium hydroxide and gave a red flame test. Identify the metal ion.**

4 **Labels have fallen off two bottles. The labels say copper sulfate and magnesium sulfate. One substance, A, makes a blue solution and a blue precipitate with sodium hydroxide. The other, B, makes colourless solution and a white precipitate with sodium hydroxide. Which label identifies A and which identifies B?**

Balancing equations

We can write a balanced symbol equation to describe a precipitation reaction using these steps:

Step 1: Write a word equation for the reaction. For example:

copper sulfate + sodium hydroxide → copper hydroxide + sodium sulfate

Step 2: Write in the formulae for the reactants and products:

$CuSO_4 + NaOH \rightarrow Cu(OH)_2 + Na_2SO_4$ unbalanced

Step 3: Balance the equation:

$CuSO_4 + 2NaOH \rightarrow Cu(OH)_2 + Na_2SO_4$ balanced

HIGHER TIER ONLY

This can also be done by using ionic equations:

For example: $Cu^{2+} + 2OH^- \rightarrow Cu(OH)_2$

Neither the Na^+ nor the SO_4^{2-} ions are needed in the equation so they are not written down – they are called 'spectator ions'. Only the ions that make the precipitate are included.

Two OH^- ions needed to react with one Cu^{2+} ion because the cation forms a 2^+ ion.

A model of some ions forming a precipitate while the spectator ions remain in solution is shown in Figure 4.33.

5 **Write a balanced symbol equation for the precipitation of magnesium hydroxide from magnesium chloride.**

6 **Explain why three OH^- ions are needed to react with the metal ion that makes a brown precipitate.**

7 **A solution of an ionic compound was known to contain chloride ions. Aqueous sodium hydroxide was added to the solution and a brown precipitate formed. Write a full and an ionic equation for the reaction. Include state symbols.**

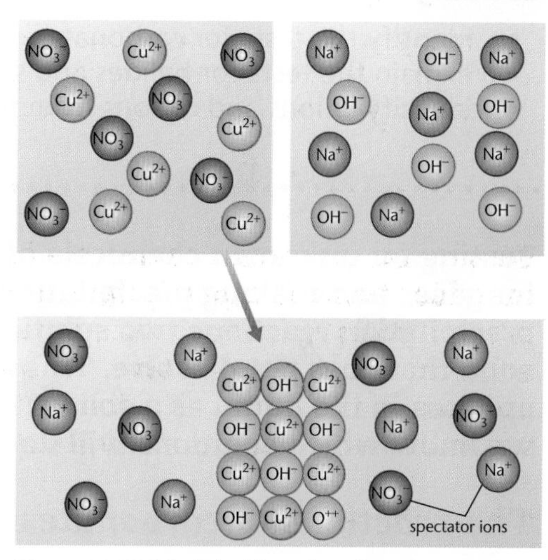

Figure 4.33 Model of ions forming a precipitate

DID YOU KNOW?

Aluminium hydroxide redissolves in excess NaOH because it is *amphoteric* and makes sodium aluminate. You are not expected to write equations for this reaction.

Tests for anions

Learning objectives:

- identify the tests for carbonates
- explain the tests for halides and sulfates
- identify anions and cations from the results of tests.

KEY WORDS

carbonates
halides
sulfates
limewater

Testing for unknown chemicals often involves testing for gases and making precipitation reactions. In precipitation reactions two solutions react to form a solid that does not dissolve. This chemical suddenly appears in the liquid as a solid. This is a precipitate. Do we know which solutions will do this?

The reactions of carbonates

Most **carbonates** are insoluble. They form white powders if they are carbonates of metals of Group 1 or 2. They form distinctive coloured powders if they are carbonates of many of the transition metals. Examples are green copper carbonate and brown iron(III) carbonate.

Carbonates react with dilute acids to form carbon dioxide gas. The test for carbon dioxide is that it turns with **limewater** milky.

The carbon dioxide turns the limewater milky at first but after more carbon dioxide is introduced the solution clears again.

Some carbonates, including sodium carbonate and potassium carbonate, are soluble in water and produce solutions containing carbonate ions. They too react with acids to release carbon dioxide.

sodium carbonate + nitric acid → sodium nitrate + water + carbon dioxide

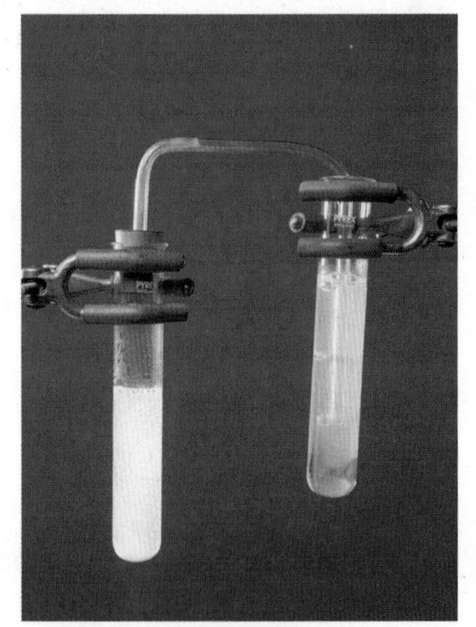

Figure 4.34 The test for carbon dioxide is that it turns limewater milky

1. **Describe how you would test for carbon dioxide.**
2. **Explain how you would test whether a copper compound was a carbonate.**

Testing for halides and sulfates

Halide ions in solution can be tested using *silver nitrate*.

Sulfate ions in solution can be tested using *barium chloride*.

The halides are chlorides, bromides and iodides. If they are tested with silver nitrate solution in the presence of dilute nitric acid, they produce precipitates.

metal halide + silver nitrate → metal nitrate + silver halide

Silver chloride is white, silver bromide is cream and silver iodide is yellow.

Precipitate	silver chloride	silver bromide	silver iodide
Colour	white	cream	yellow

If sulfate ions in solution are tested with barium chloride solution in the presence of dilute hydrochloric acid they produce a white precipitate.

metal sulfate + barium chloride → metal chloride + barium sulfate

3. **Describe what you would see if you added silver nitrate solution to potassium iodide solution.**
4. **Tap water samples give a positive test for chloride ions. Suggest why.**
5. **Name the reagent you would use if you were testing a solution for sulfates.**

DID YOU KNOW?

These kinds of tests are used to test for ions in tap water.

Balanced equation for precipitation

The balanced symbol equations for precipitation reactions are written with state symbols following the formulae. It is important to know which is the solid formed in the reaction.

silver nitrate + sodium chloride → silver chloride + sodium nitrate (white precipitate)

$AgNO_3(aq) + NaCl(aq) \rightarrow AgCl(s) + NaNO_3(aq)$

silver + sodium → silver bromide + sodium
nitrate bromide (cream precipitate) nitrate

$AgNO_3(aq) + NaBr(aq) \rightarrow AgBr(s) + NaNO_3(aq)$

silver + sodium → silver iodide + sodium
nitrate iodide (yellow precipitate) nitrate

$AgNO_3(aq) + NaI(aq) \rightarrow AgI(s) + NaNO_3(aq)$

magnesium sulfate + barium chloride → magnesium chloride + barium sulfate

$BaCl_2(aq) + MgSO_4(aq) \rightarrow BaSO_4(s) + MgCl_2(aq)$

6. **Write a balanced symbol equation for the reaction between silver nitrate solution and potassium iodide solution.**
7. **Write a balanced symbol equation for the reaction between silver nitrate solution and calcium chloride, $CaCl_2$**

REMEMBER!

Remember to add up the number atoms on each side of the equation to make sure it is balanced.

Flame tests

Learning objectives:

- carry out flame test procedures
- identify the colours of flames of ions
- identify substances from the results of the tests.

KEY WORDS

flame test
lilac
crimson
yellow

Fireworks are spectacular sights with colours sprayed out against the night sky. These colours originate from chemicals added to the firework that burn different colours in flames. You can make these flame colours in the lab and use them to identify unknown substances in detective work.

Figure 4.35 Potassium ions give this spectacular display its distinctive colour

Producing flame colours

The procedure for carrying out **flame tests** includes eye safety:

- Put on safety glasses.
- Clean the wire with hydrochloric acid and hold in a blue flame.
- Moisten a flame-test wire with dilute hydrochloric acid.
- Dip the flame-test wire into the sample of solid chemical.
- Hold the flame-test wire in a blue Bunsen burner flame.
- Record the new colour of the flame.

1. **State the colour of the Bunsen burner flame that the test wire and sample is held in.**
2. **Explain why it is necessary to clean and moisten the test wire.**

KEY INFORMATION

If a sample containing a mixture of ions is used some flame colours can be masked.

Flame colours

The flame colour is matched with the metal known to make that colour when it is in a compound. For example:

Metal in compound	lithium	sodium	potassium	calcium	copper
Flame colour	**crimson**	**yellow**	**lilac**	orange-red	green
Flame colour					

Figure 4.36 Common flame test colours

3. **An ionic chloride produces a green colour in a flame. Name the ionic compound.**
4. **An ionic compound produces an orange colour in the flame. Suggest a reason why.**

5 **Some students want to test the colour chemicals that the Whizzbang firework company is going to add to its fireworks. They want to produce three fireworks, a red, a green and a lilac one. Name three metal compounds they should add.**

6 **Suggest why flame tests are not always reliable.**

Identifying substances

Flame tests can be used to identify the metal ions (cations) in compounds. Lithium, sodium, potassium, calcium and copper ions produce distinctive colours in flame tests as identified in the table above.

Flame tests for metal ions (cations) can be combined with others tests for cations and anions to identify unknown compounds. These will be covered next.

Colour of flame	Name of cation	Result for anion	Compound identified
green		nitrate	
lilac		sulfate	
orange-red		chloride	
crimson		carbonate	
yellow		nitrate	

7 **Explain how a flame test could be used to distinguish between a sample of potassium chloride and calcium chloride.**

8 **Dilute hydrochloric acid was added to solid compound R. Fizzing was observed. Compound R gave a red flame when a flame test was carried out.**

a **Identify compound R.**

b **Write a balanced equation for the reaction with the acid. Include state symbols.**

9 **A small mass of substance V was reacted with chlorine gas. A white solid, W, was formed. A flame test on W gave a yellow flame.**

a **Identify substance V.**

b **Identify substance W.**

c **Write an equation for the reaction of V with chlorine. Include state symbols.**

Figure 4.37 Carrying out a flame test

DID YOU KNOW?

When a flame test is used outer electrons is 'excited' out of their shell and when they return energy is given out as light and a specific colour is produced. The colours are different because the outer electrons of different atoms are excited by different amounts.

Instrumental methods

Learning objectives:

- identify advantages of instrumental methods compared with the chemical tests
- describe some instrumental techniques
- explain the data provided by instrumental techniques.

KEY WORDS

spectroscopy
separation technique
gas-liquid chromatography

Classical techniques for analysing elements and compounds include flame tests, precipitation tests, testing for gases and other chemical tests. Titrations can also be used to measure quantities. Nowadays, we use many more instrumental techniques. You may have seen some of them if you watch the forensic scientists on TV detective programmes.

The advantages of instrumental techniques

Elements and compounds have long been identified by using 'wet chemistry' techniques. You have already practised some of these. Over the decades instrumental methods have been developed to perform these tests and to introduce new techniques that can't be performed by hand. The advantages of these new instrumental methods are that they are:

- more rapid
- more accurate
- more sensitive.

Figure 4.38 Analysing using gas chromatography

There are many types of instrumental methods. These can be divided into groups.

Spectroscopy	Mass spectrometry
Electrochemical analysis	Thermal analysis
Separation techniques	Microscopy

We have already looked at **separation techniques** such as paper and thin layer chromatography. An instrumental technique using similar principles is **gas-liquid chromatography**. It too has a mobile phase and a stationary phase and separates components in a mixture.

1. **Look at the chromatogram in Figure 4.39. Identify the gas that comes through in 12 minutes and identify which gas is most abundant.**
2. **Explain why an instrumental method of chromatography may have advantages over paper chromatography.**

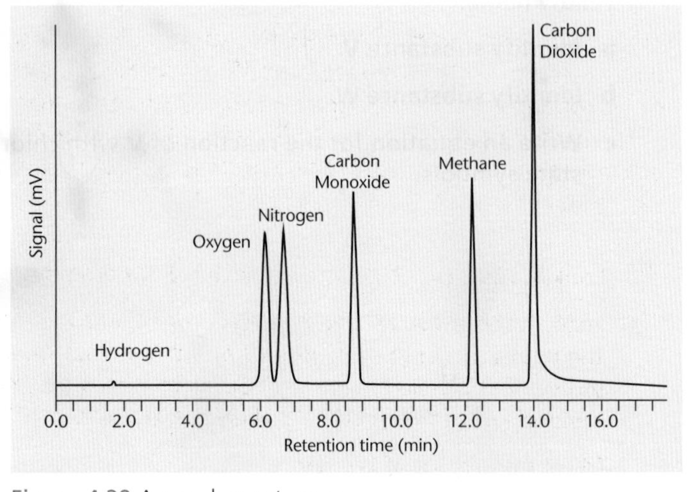

Figure 4.39 A gas chromatogram.

Some examples of instrumental techniques

Another example of a separation technique is electrophoresis. This can be used to separate fragments of DNA and used to identify proteins.

Successful crime detection and conviction for crimes has increased considerably since the detection and matching of DNA has been possible, even on very small sample.

You will be familiar with using light microscopes in the laboratory, which magnify samples up to 400 times. There are more powerful microscopes that can magnify up to 10 000 000 times called electron microscopes. Another type is a scanning tunnelling microscope that images surfaces at the atomic level.

Another example of instrumentation methods is **spectroscopy**. You will see this type of technique described later in this topic, using flame emission spectroscopy. This can be used as an instrumental method instead of flame tests.

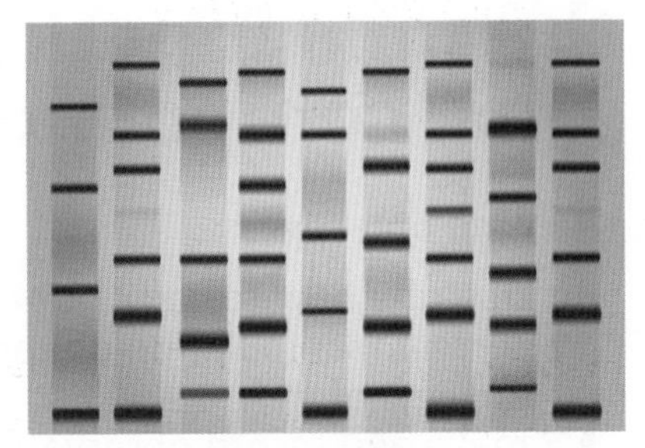

Figure 4.40 DNA fragments separated and identified by electrophoresis

3 **Suggest why the analysis of DNA is useful in crime detection.**

4 **Work out how many times more powerful an electron microscope is compared to a light microscope.**

REMEMBER!

You do not need to remember the names of these techniques or how they work, just the advantages.

How do instrumental methods work?

Each method works in a different way but essentially they all have the same features.

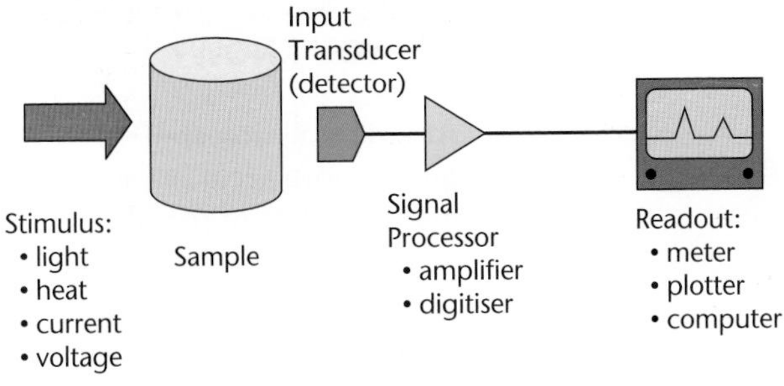

Figure 4.41 A schematic diagram of an analytical instrument

There is a stimulus that enters the sample. This is followed by a detector, then a signal amplifier and finally a recorder or output reader. The output needs to be interpreted and is often in the form of a graph, lines or blocks along a scale. The advantage of instrumental methods is that very small samples can give results.

KEY INFORMATION

You do not need to remember the details of how to interpret a mass spectrum.

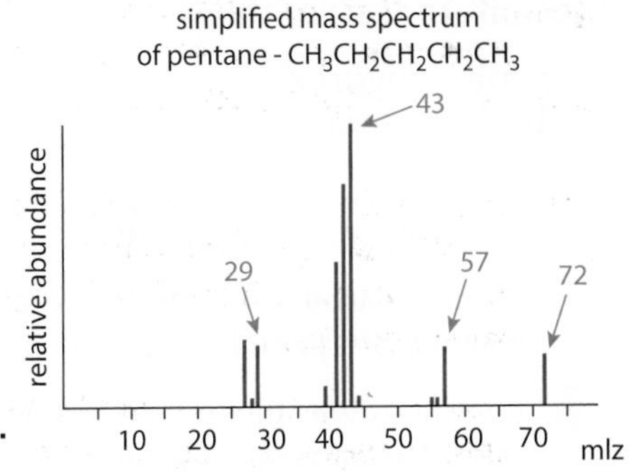

Figure 4.42

5 **Look at Figure 4.39. Explain what the chromatogram shows.**

6 **Look at the mass spectrum in Figure 4.42. In simple terms, the peaks represent the molecular mass of fragments of the molecule if it is whole or split. Suggest which fragment of the molecule of pentane $CH_3CH_2CH_2CH_2CH_3$ the peak at 72 represents.**

[A_r of C is 12 A_r of H is 1]

PRACTICAL

Use chemical tests to identify the ions in unknown single ionic compounds

KEY WORDS

precipitate
flame test
hydrochloric acid
barium chloride
silver nitrate

Learning objectives:

- describe how to carry out experiments safely using the correct manipulation of apparatus for the qualitative analysis of ions
- make and record observations using flame tests and precipitation methods
- identify unknown ions in chemical compounds.

The skill of observation is needed for qualitative analysis. Scientists note colour changes, precipitation and the evolution of gases as part of a toolkit for identifying ions. Can you use these techniques for finding the identity of some unknown compounds?

These pages are designed to help you think about aspects of the investigation rather than to guide you through it step by step.

A number of different techniques are needed to carry out an identification. This topic looks at a few of the specific techniques needed to identify ions safely.

DID YOU KNOW?

Atomic emission spectra give a more accurate identification of metal ions.

Carrying out analysis

Here are some of the techniques that might be needed to identify single ions:

Testing for the gases:

- carbon dioxide
- hydrogen
- oxygen
- chlorine

Flame tests

Precipitate tests with:

- sodium hydroxide
- **barium chloride**
- **silver nitrate**

take one solution

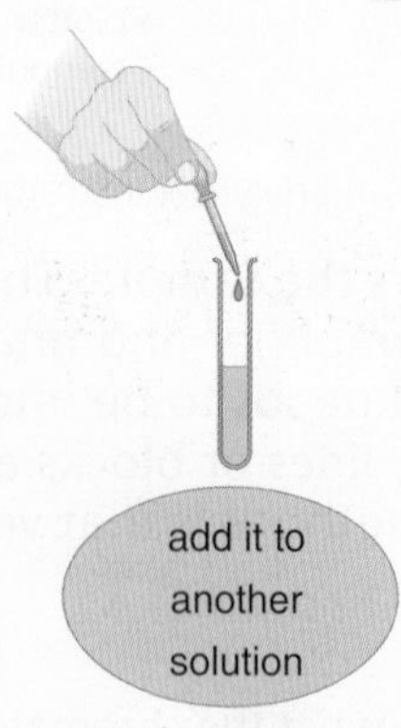

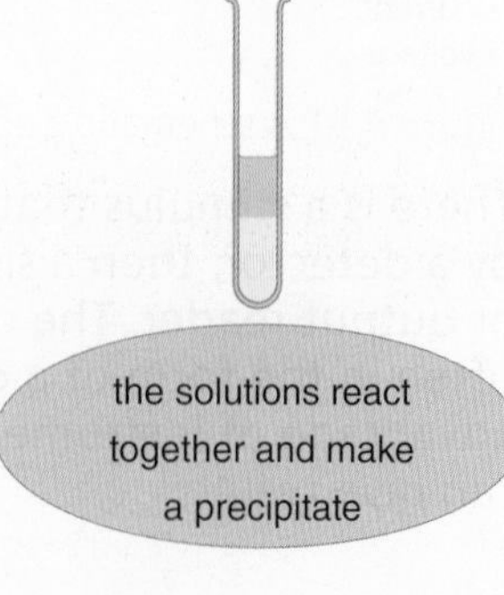

Figure 4.43

Think about these questions:

1. **When a carbonate is added to acid, carbon dioxide gas is given off. Explain how you would test for the gas.**
2. **A student was given blue, green and brown transition metal salts. The salts were soluble. Describe how the student could use sodium hydroxide to identify the metal ions. State the safety precautions that would be needed.**

3 A sample of white crystals was thought to be a salt of a Group 1 metal. Suggest what may be observed in a flame test.

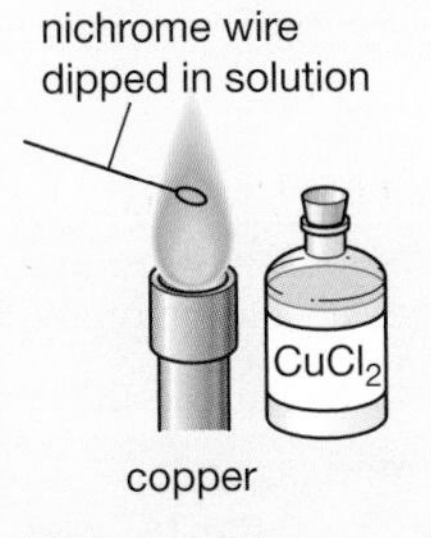

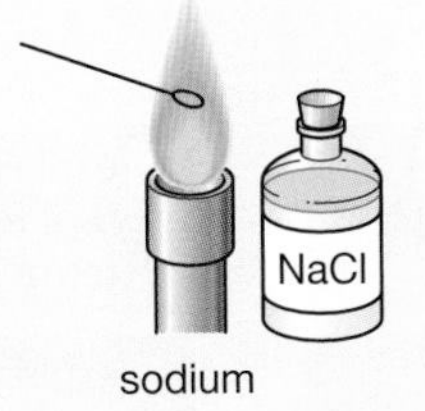

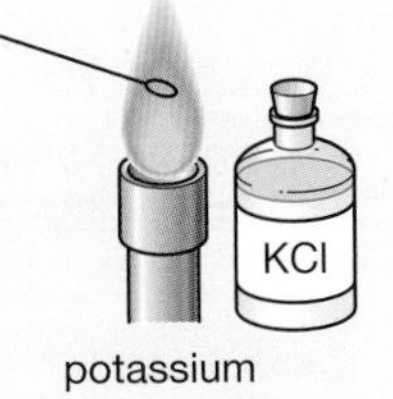

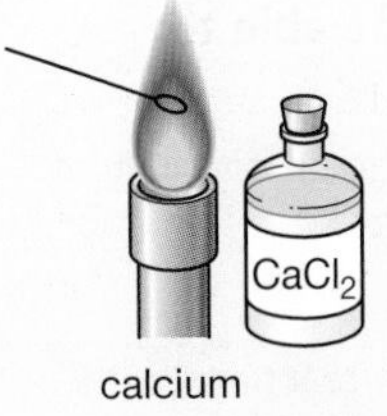

Figure 4.44

Making and recording observations

In order to do an investigation to analyse unknown compounds it is essential that the tests have been tried on a known ion and the observation made. This experience is then used to identify the results of tests on an unknown compound to try to identify the ions in the compound.

The order of the tests is:

a Flame tests
b Precipitation with sodium hydroxide
c Testing for carbon dioxide
d Testing with silver nitrate
e Testing with barium chloride

Use your knowledge of these tests to think about these questions:

4 Describe what happens when sodium hydroxide is added to magnesium sulfate and calcium sulfate.

5 Explain how another test can distinguish between the magnesium ion and the calcium ion.

6 Describe what happens when sodium hydroxide is added to aluminium sulfate.

> **KEY INFORMATION**
>
> Safety glasses need to be worn all of the time during these tests.

Analysing unknown compounds

Jo and Sam had observed these results:

Compound	Flame test	Test with NaOH	Test for CO_2	Test with $AgNO_3$	Test with $BaCl_2$	Cation	Anion
A	Orange-red	White ppte	Turns milky	None	None		
B	None	Brown ppte	None	White ppte	None		
C	Yellow	None	None	Cream ppte	None		
D	Lilac	None	None	None	White ppte		
E	Green	Blue ppte	None	Yellow ppte	None		

7 Identify compound A

8 Identify compound B

9 Identify compound C

10 Identify compound D

11 Identify compound E

Check your progress

You should be able to:

describe the unreactivity of the noble gases →	explain the trend down Group 0 of increasing boiling point →	explain the trend down Group 0 of increasing boiling point in terms of atomic mass
predict the reactions with water of Group 1 elements lower than potassium →	predict and explain the relative reactivity down the groups →	explain the trend down the group of increasing reactivity by electron structure
recall the colours of the halogens and the order of reactivity of chlorine, bromine and iodine →	describe the order of reactivity and explain the displacement of halogens →	predict displacement reaction outcomes of halogens other than chlorine, bromine and iodine.
explain that a stable outer shell of electrons makes noble gases unreactive →	predict the properties of 'unknown' elements from their position in the group →	explain the trend of increasing reactivity in terms of electron structure
explain that transition metals have higher melting points and are stronger and harder than Group 1 metals →	explain that transition metals are less reactive than Group 1 metals and form coloured solutions →	explain that transition metals form ions of different charges and are useful as catalysts
describe the reactions, if any, of metals with water or dilute acids to place these metals in order of reactivity →	explain how the reactivity is related to the tendency of the metal to form its positive ion →	deduce an order of reactivity of metals based on experimental results
use experimental results of displacement reactions to confirm the reactivity series →	use the reactivity series to predict displacement reactions →	write ionic equations for displacement reactions
carry out flame test procedures →	identify the colours of flames of ions →	identify substances from the results of the tests
recognise the precipitate colour of metal hydroxides →	explain how to use sodium hydroxide to test for metal ions →	write balanced equations of producing insoluble hydroxides
identify the tests for carbonates →	explain the tests for halides and sulfates →	identify anions and cations from the results of tests
identify advantages of instrumental methods compared with the chemical tests →	describe some instrumental techniques →	explain the data provided by instrumental techniques
describe flame emission spectroscopy →	identify the advantages of instrumental methods compared with the chemical tests →	interpret an instrumental result given appropriate data in chart or tabular form, using a reference set

Worked example

Sam and Alex are researching some properties of Group 1 metals.

1 Shade the section of the periodic table where the Group 1 metals are found.

1 **H** hydrogen 1.0			
3 **Li** lithium 6.9	4 **Be** beryllium 9.0		
11 **Na** sodium 23.0	12 **Mg** magnesium 24.3		
19 **K** potassium 39.1	20 **Ca** calcium 40.1	21 **Sc** scandium 45.0	22 **Ti** titanium 47.9
37 **Rb** rubidium 85.5	38 **Sr** strontium 87.6	39 **Y** yttrium 88.9	40 **Zr** zirconium 91.2
55 **Cs** caesium 132.9	56 **Ba** barium 137.3	57-71 lanthanides	72 **Hf** hafnium 178.5
87 **Fr** francium	88 **Ra** radium	89-103 actinides	104 **Rf** rutherfordium

This is incorrect. The first column needs to be shaded.

The two metals they are researching are sodium and potassium.

2 Write down two properties that these metals have.

They are shiny when cut

They have a very high density

The first property is correct. However, sodium and potassium float on water so have a density less than water. The student may be confusing Group 1 metals with transition metals, which have high density.

Sam and Alex find out that sodium and potassium react with water. They find that sodium reacts with water to make sodium hydroxide and that hydrogen is given off.

3 Write a word equation for the reaction

sodium + water → sodium hydroxide

The reactants are correct but hydrogen needs to be written as a product on the right hand side.

Sam says that potassium reacts more vigorously than sodium but Alex says that they are in the same group so they react the same.

4 a Explain why Sam is correct about the trend.

The lower down the group the better they react

This answer could be expressed more clearly by substituting the word 'better' with 'more vigorously'.

b Explain why Sam is correct using ideas about the structure of atoms.

The bigger the atom the quicker the reaction

This answer needs more detail. The further away the outer electron is from the nucleus the more easily it is 'lost', as the pull by the positive nucleus on the negative electron is less.

End of chapter questions

Getting started

1. What gas is given off when sodium reacts with water? **1 Mark**

2. Which of the following is a noble gas? **1 Mark**

   a oxygen

   b helium

   c chlorine

   d nitrogen

3. Jo and Gita test compounds of Group 1 metals with a flame test.

   Give two differences between a metal and a non-metal.

   Match the flame colour that they see with the correct metal of the compound. **2 Marks**

yellow	lithium
crimson	potassium
lilac	sodium

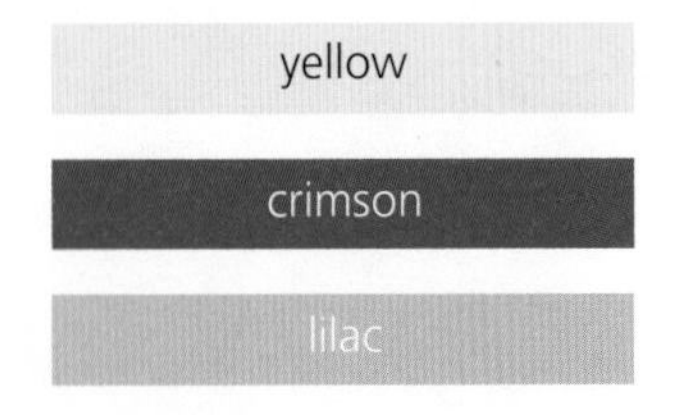

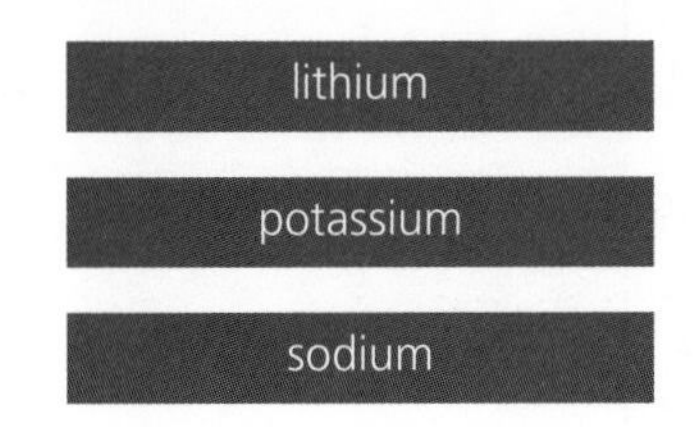

4. Which metal is most reactive?

   a sodium

   b iron

   c zinc

   d magnesium **1 Mark**

5. The diagram below shows four different metals reacting with sulfuric acid solution.

   a Determine the order of **decreasing** reactivity for metals **D** to **G**.

   b Explain what makes one metal more reactive than another. **2 Marks**

D
E
F
G
mild fizzing
no fizzing
Copious quantities of fizzing
In between D and F for fizzing

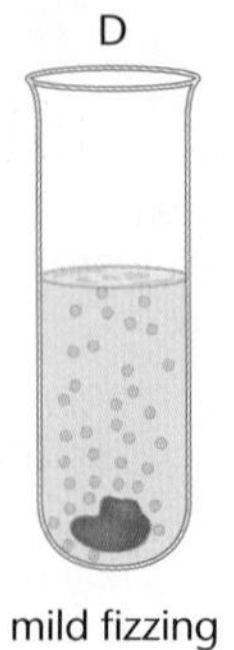

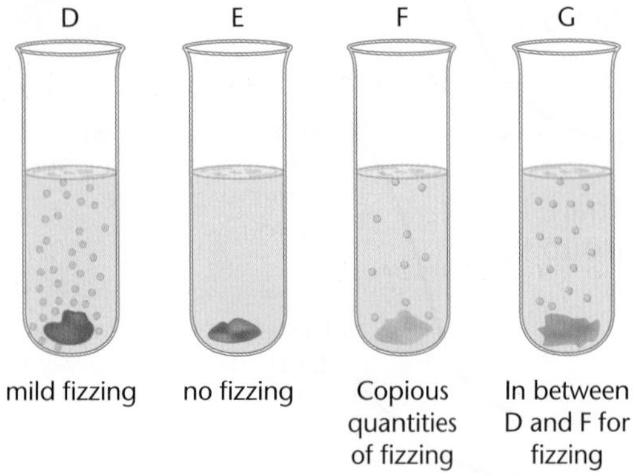

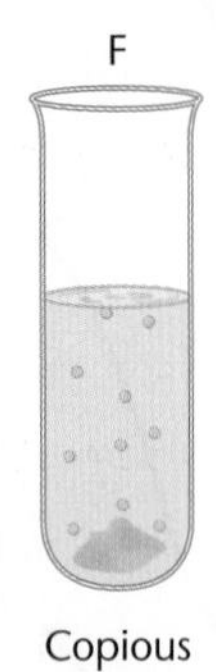

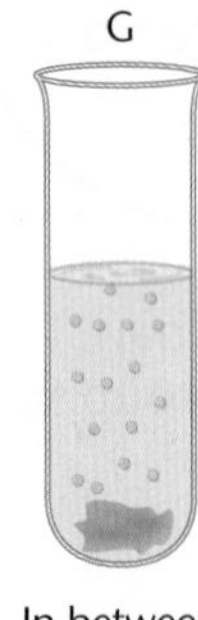

6. What would you see when aqueous sodium hydroxide is added to copper sulfate solution? **1 Mark**

   a brown precipitate

   b brown solution, no precipitate

   c blue precipitate

   d blue solution, no precipitatete

7. What solution is used to test for aqueous sulfate ions and what would be observed? **1 Mark**

   a silver nitrate, cream precipitate

   b sodium hydroxide, green precipitate

   c barium chloride, white precipitate

   d sulfuric acid, fizzing

8. Match the result to the observation of adding silver nitrate solution **2 Marks**

cream precipitate	chloride ion present
yellow precipitate	iodide ion present
white precipitate	bromide ion present

Going further

9. When sodium reacts with water hydrogen is given off. Identify the other product. **1 Mark**

10. To which group does chlorine belong? **1 Mark**

    a 0

    b 1

    c 6

    d 7

11. Describe the trend in reactivity of the Group 1 metals. **2 Marks**

12. Halogen X has a boiling point of 59°C and halogen Y a boiling point of 184°C. **2 Marks**

    a Justify which halogen has the lower atomic number.

    b Explain which halogen is more reactive.

13. Which metal oxide can be reacted with carbon to produce the metal? **1 Mark**

    a aluminium oxide

    b sodium oxide

    c iron(III) oxide

    d magnesium oxide

14 **Some experiments were carried out on metals and solutions of metal salts.** 2 Marks

Zinc was added to copper(II) sulfate solution. Copper was deposited.
Metal X was added to zinc sulfate solution. Zinc was not deposited.
Metal X was added to copper(II) sulfate solution. Copper was deposited.

Place the metals in order of reactivity. Justify your answer.

15 **Describe two tests that can identify calcium ions.** 2 Marks

More challenging

16 **Francium is a highly radioactive and rare alkali metal.** 2 Marks

	Melting point / °C	Density / gcm^{-3}
Rb	39.3	1.53
Cs	28.4	1.93
Fr	D	E

Predict the melting point, D, and density, E, of francium using the data in the table.

17 **Sodium and chlorine are in the same row of the periodic table, yet have different chemical and physical properties. Compare the two elements.** 6 Marks

18 **In an experiment a student reacted metals with solutions of sulfates of the other metals. One metal, Z, was unknown. These are the results:**

	Mg	Fe	Cu	Z
$MgSO_4$		✗	✗	✗
$FeSO_4$	✓		✗	✓
$CuSO_4$	✓	✓		✓
XSO_4	✓	✗	✗	

Describe where Z lies in the reactivity series, giving reasons. 2 Marks

19 **Sodium hydroxide can be used to test for transition metal ions. Suggest why this method is unlikely to be effective for testing mixtures of transition metal ions.** 1 Mark

20 **Ionic substance D gave a white precipitate with sodium hydroxide and silver nitrate. Explain why you could not fully identify the ions in D without further tests.** 2 Marks

Most demanding

21 **The table shows some data and properties for an element.**

Deduce which type of element it is, the group it belongs to and why it reacts so violently with water.

4 Marks

Atomic number	Electron pattern	Reaction with water
37	2,8,8,18,1	Violently, giving off hydrogen and forming an alkali

22 **An ionic compound is thought to contain iron(II) ions and bromide ions.**

2 Marks

Describe the tests that could confirm this.

23 **When a flame test is carried out on substance R, a red flame is seen. Substance R reacts with dilute hydrochloric acid. The gas given off turns limewater milky. When heated strongly, R decomposes into two substances. The gas produced also turns limewater milky.**

a **Identify the gas given off when R reacts with acid and decomposes.**
b **Write a balanced chemical equation for the reaction with hydrochloric acid and for the decomposition.**
c **It was suspected that substance R was contaminated with lithium. Suggest why another method should be used to confirm the identity of the metallic element in substance R.**

4 Marks

Total: 45 Marks

MONITORING AND CONTROLLING CHEMICAL REACTIONS

IDEAS YOU HAVE MET BEFORE:

MEASURING QUANTITIES

- Balances measure mass in grams.
- Measuring cylinders measure volume in cm^3.
- Stopwatches measure time in minutes and seconds.

MAKING SALTS AND CRYSTALS

- When solids are filtered off some solid stays on the paper.
- When solutions are crystallised some crystals stay in the dish.
- Waste of chemicals can be reduced by using careful techniques.

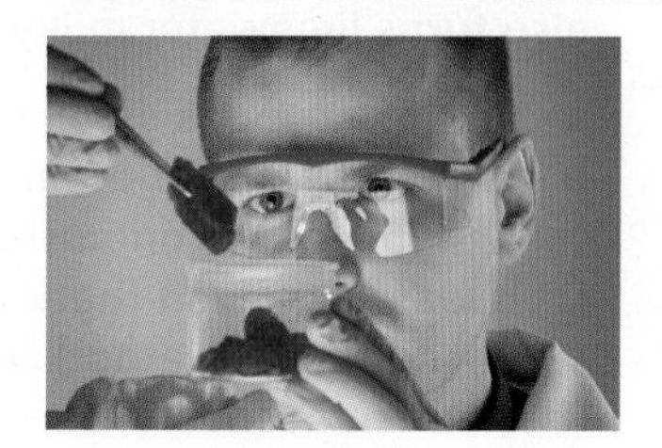

MEASURING REACTIONS

- Mass is measured in grams using a digital balance.
- Some chemical reactions show a colour change.
- If a gas is given off in a reaction, its volume can be measured.

CONTROLLING RATES OF REACTIONS

- Small pieces of sugar dissolve quicker than lumps.
- Boiling water will cook eggs quicker.
- Concentrated acid is more hazardous than dilute acid.

REVERSIBLE REACTIONS AND EQUILIBRIUM

- Copper sulfate can change from blue to white.
- Ammonia is a gas.
- When making chemicals we want to make as much as possible.

5

IN THIS CHAPTER YOU WILL FIND OUT ABOUT:

HOW CAN WE MEASURE AMOUNTS OF SUBSTANCES?

- Moles are the chemical measure of amounts of substances.
- Grams and moles interconvert using the molar mass.
- Amounts to be made can be predicted using equations.

MAXIMISING CHEMICAL YIELDS

- Theoretical yields in chemical reactions are rarely achieved.
- Losses must be minimised so that percentage yields are high.
- Atom economy is maximised so that waste products are low.

HOW CAN WE MEASURE REACTION RATES?

- Amount reacted can be measured by collecting gas volumes.
- Amount reacted can be measured by loss of mass.
- Amount reacted can be measured by a colour change.

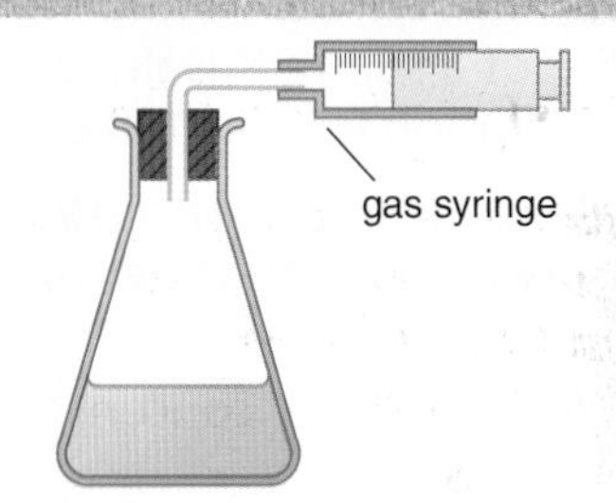

WHAT FACTORS AFFECT RATES OF REACTIONS?

- Surface area, pressure and concentration affect reaction rate.
- Increasing temperature causes more successful collisions.
- Catalysts lower activation energy, making reactions faster.

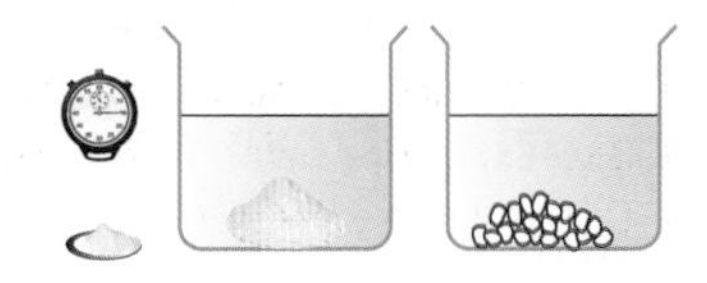

HOW CAN REACTIONS BE IN EQUILIBRIUM?

- A reversible reaction has a forward reaction and a backward reaction.
- Increasing pressure can change the position of equilibrium.
- Le Chatelier's principle can be applied to systems in equilibrium.

Concentration of solutions

Learning objectives:

- relate mass, volume and concentration
- calculate the mass of solute in solution
- relate concentration in mol/dm^3 to mass and volume.

KEY WORDS

solute
solvent
solution
concentration

We have learned before that a solution forms when a solute is dissolved in a solvent and that solutions can be dilute or concentrated. Dilutions are often critical, such as making the correct formulations of medicines or baby milk formula. How do we make sure we have the correct concentration?

HIGHER TIER ONLY

Checking the correct units

Many chemical reactions take place in **solutions** and often the **concentration** needs to be known. If a volume of **solvent** is used, say 100 cm^3, the number of particles of **solute** in the solvent is lower if the solution is more dilute and is higher if the solution is more concentrated.

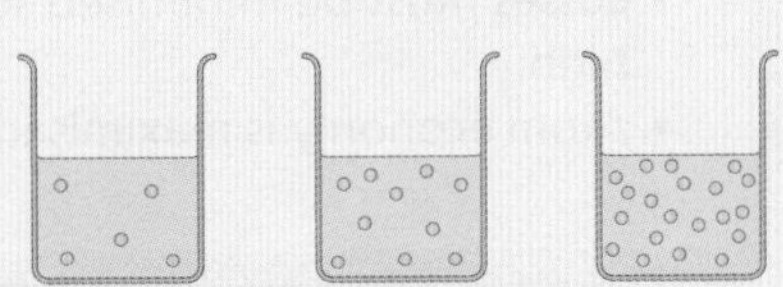

Figure 5.1 The concentration increases as the number of solute particles in a fixed volume increases

If 2.5 g of a solute X is dissolved in 100 cm^3 solvent then the concentration is 2.5 g/100 cm^3

This is because the *concentration* of a solution can be measured as the *mass per specified volume of solution*

or $$\text{concentration} = \frac{\text{mass of solute}}{\text{volume}}$$

The standard unit of concentration used in the laboratory is grams per dm^3 so the units are then g/dm^3.

1 dm^3 is 1000 cm^3 so the concentration of X is 25 g/dm^3

1 **Put the following solutions into order with the most dilute first:**

a **20 g/100 cm^3**
b **20 g/1000cm^3**
c **8 g / 50cm^3**

2 **Calculate the concentration of the following solutions in g/dm^3.**

a **3.2 g in 100 cm^3**
b **3.2 g in 250 cm^3**
c **6.4 g in 500 cm^3**

Calculating from concentrations

A solution has a concentration of 6 g/dm^3. What is the mass of solute Z that was added to 100 cm^3 to make this solution?

$$\text{Concentration} = \frac{\text{mass}}{\text{volume}}$$

$$\text{Concentration} = \frac{6}{1000}\ g/dm^3 \quad \text{and}$$

$$\text{Concentration} = \frac{x}{100}\ g/100\ cm^3$$

$$\text{Therefore: } \frac{x}{100} = \frac{6}{1000} \qquad \text{so } x = \frac{6 \times 100}{1000} = 0.6\ g$$

KEY INFORMATION

1 dm^3 is commonly called 1 litre but we use 1 dm^3 when using units in scientific contexts.

3 **A solution has a concentration of 4.2 g/dm^3. Calculate the mass of solute dissolved in 250 cm^3 of solution.**

4 **A solution has a concentration of 5.4 g/100 cm^3. Calculate the mass of solute dissolved in 35 cm^3 of solution.**

Using moles in concentrations

Many solutions in used in chemical reactions have solutes expressed as amounts of substances, measured in moles.

A relationship exists between the amount in moles, concentration in mol/ dm^3 and volume in dm^3:

concentration = amount in moles ÷ volume

amount in moles = concentration × volume

volume = amount in moles ÷ concentration.

A way to remember this is by using a formula triangle, as in Figure 5.2. Don't forget these formula triangles are only an aid to help you check your answer in a test.

You should work from the first principles of the equation and carry out the same operations on both sides of the equation in order to calculate a new value in the relationship.

For example:

A solution has a concentration of 0.2 mol/dm^3

To find the amount of moles in 500 cm^3 of a 0.2 mol/dm^3 solution we need first to make sure that the units match.

The volume in cm^3 converts into dm^3 so: 500 cm^3 = 0.5 dm^3.

Then we use the formula:

amount in moles = concentration × volume
= 0.2 mol/dm^3 × 0.5 dm^3 = 0.1 mol

DID YOU KNOW?

Remember Avogadro's constant? Concentrations are expressed in mol/dm^3 because chemical equations are used to express the ratios of reactants and products in moles. The number of particles in 1 mole is always the same.

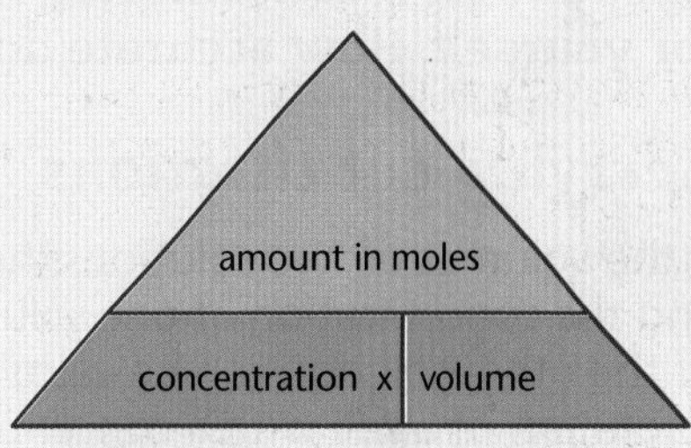

Figure 5.2 Cover the quantity needed to find the formula to use

5 **A solution with a concentration of 2 mol/dm^3 contains 1 mole of solute. Calculate the volume, in dm^3, of the solution.**

6 **A solution of 0.18 moles has a volume of 0.6 dm^3. Calculate the concentration in mol/dm^3**

7 **Calculate the amount of solute, in moles in 500 cm^3 of a solution with a concentration of 3 mol/dm^3**

8 **4.90 g of H_2SO_4 was dissolved in 200 cm^3 of water. Calculate the resulting concentration of the acid solution in mol/dm^3**

9 **0.0250 moles of HNO_3 was dissolved in 125 cm^3 of water. Calculate:**

a the concentration of HNO_3 in mol/dm^3

b the concentration of HNO_3 in g/dm^3

10 **The concentration of some HCl is 18.25 g/dm^3**

a Calculate the concentration of HCl in mol/dm^3

b Work out the number of molecules of HCl per dm^3. The Avogadro constant is 6.02×10^{23} mol^{-1}

Using concentrations of solutions

Learning objectives:

- describe how to carry out titrations
- calculate the concentrations in titrations in mol/dm^3 and in g/dm^3
- explain how the concentration of a solution in mol/dm^3 is related to the mass of the solute and the volume of the solution.

KEY WORDS

titration
burette
pipette
indicator

Do you remember finding out about neutralisation of acid and alkalis using indicators? We looked at titration curves and saw a sudden change of pH when neutralisation happened. A titration can also be used to calculate how concentrated a solution is. You can do this for yourself. How accurate do you think you could be?

Carrying out titrations

If the volumes of two solutions that react completely are known and the concentration of one solution is known, the concentration of the other solution can be calculated. In this example we will use a solution of alkali with a concentration of 0.12 mol/dm^3.

The volumes of acid and alkali solutions that react with each other can be measured by **titration** using an **indicator**.

The procedure for a titration is:

1. Wear safety glasses. Alkali is particularly harmful for eyes.
2. Fill a **burette** with acid. Record the start volume of acid.
3. **Pipette** 25 cm^3 of the alkali into a conical flask using a pipette filler.
4. Put a few drops of a suitable indicator into the flask with the alkali.
5. Add the acid slowly from the burette into the alkali, swirling the solution in the conical flask gently.
6. When the indicator suddenly changes colour record the volume reading – this is the end point. This is a trial run.
7. Repeat steps 2–6 but this time stop just before the volume of acid used in the trial run is added. Now add drop by drop to get the sudden colour change. Record the second set of readings.
8. Repeat steps 2–6 until three volumes of acid used are the same (within 0.1 cm^3).

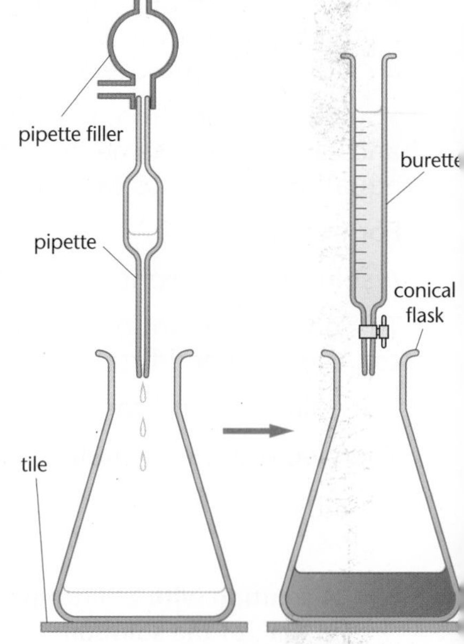

Figure 5.3 Acid in the burette and alkali in the flask using a pipette

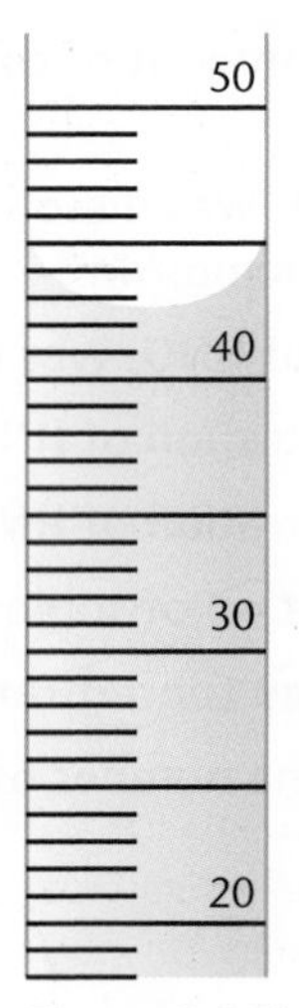

Figure 5.4 The final volume of acid

1 Describe how you must read the volume of acid in the burette in Figure 5.4 and write down the reading.

2 Which piece of titration apparatus delivers a fixed volume and which piece delivers a changeable volume?

Calculating with concentrations

These are some example readings in a titration.

Volume of acid reacting with 25.0 cm^3 of alkali				
	Trial	First	Second	Third
Start cm^3	1.5	4.1	2.7	18.6
End cm^3	28.8	30.9	29.6	45.3
Titre cm^3	27.3			

The mean titre can be calculated from these results. We have already seen that the concentration of a solution can be measured in g/dm^3 and in mol/dm^3.

So knowing the concentration of a solution and the volume used we can calculate the number of moles in a solution using the equation:

Amount of moles = concentration × volume

In this example titration we used 25.0 cm^3 of alkali. That is 0.025 dm^3. At the beginning we identified that its concentration was 0.12 mol/dm^3.

So the amount of moles of alkali we always used in this titration was:

Amount of moles = concentration × volume = 0.025 × 0.12
= 0.003 moles

KEY INFORMATION

A titre is the total volume of solution used.

Volume of acid at end – volume of acid at start.

DID YOU KNOW?

Titration is an accurate technique, but there is a chance of experimental error. If the readings are close, the technique is reliable. Small differences result from normal experimental error, so an average reading is calculated.

3 Explain why the trial run will not be included in the calculation of the mean titre.

4 Complete the results table and calculate the mean titre.

Calculating concentrations of unknown solutions

The acid 'HA' exactly neutralised the alkali 'BOH' as:

$$HA + BOH \rightarrow BA + H_2O$$

So now that we know the number of moles of the alkali in the flask we also know the number of moles in the acid solution that exactly neutralised the alkali. 0.003 moles of alkali was neutralised by 0.003 moles of acid.

But 0.003 moles of acid were in the mean titre volume of 26.8 cm^3

$$\text{So the concentration of acid} = \frac{\text{amount of moles}}{\text{volume in dm}^3} = 0.11 \text{ mol/dm}^3$$

If the molar mass of the acid was 63 then the concentration in grams of this solution would be:

Concentration = molar mass × amount of moles = 63 × 0.11
= 6.93 g/dm^3

5 23.8 cm^3 of a solution of 0.11 mol/dm^3 of hydrochloric acid reacts with 25.0 cm^3 of sodium hydroxide solution. What is the concentration of the sodium hydroxide solution in mol/dm^3? The equation for this reaction is:
$HCl + NaOH \rightarrow NaCl + H_2O$

PRACTICAL

Finding the reacting volumes of solutions of acid and alkali by titration

Learning objectives:

- describe how safety, the correct manipulation of apparatus and the accuracy of measurements are managed when titrations are carried out
- make and record observations and accurate measurements using burettes
- calculate the concentration of a solution from the concentration and volume of another.

KEY WORDS

titration
concentration
acid
alkali

These pages are designed to help you think about aspects of the investigation rather than to guide you through it step by step.

Using instruments for measuring volumes accurately is an important skill for many scientists. The ability to calculate the concentration of solutions in both mol/dm³ and g/dm³ from the reacting volumes is a skill necessary to achieve the higher grades. Can you recognise the neutralising volume from a titration curve when measuring pH?

Managing titrations safely

A number of different skills are needed to carry out a **titration** with safety and accuracy. The volumes of the solutions used are either measured as a fixed volume or a variable volume. The point of neutralisation is determined by use of an indicator.

Think about these questions:

1. **Name the type of reaction that takes place between the acid and the alkali.**
2. **Name the piece of apparatus that is needed to measure a fixed volume of the alkali.**
3. **Identify the piece of apparatus that is needed to measure the variable volume of the acid.**
4. **Suggest two safety measures that should be used during a titration.**

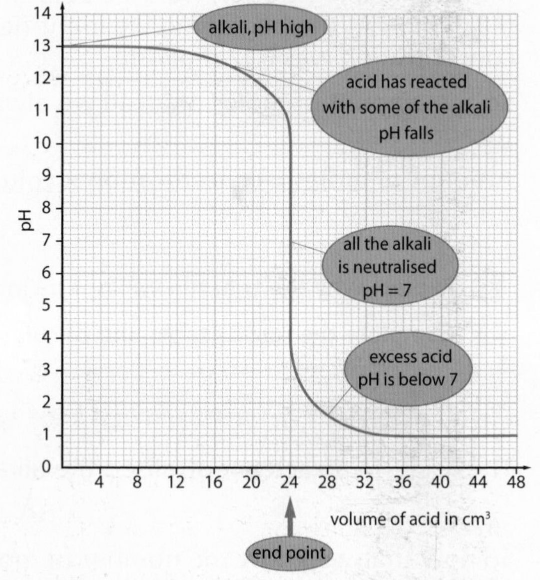

Figure 5.5

DID YOU KNOW?

Acid is added from the burette to neutralise the alkali, seen by the indicator colour change.

Obtaining accurate measurements in a titration

When acid is added to the burette it does not need to be filled right up to the 0.0 cm³ mark as the volume used can be calculated, but the section below the tap needs to be filled.

The readings on the burette are taken by looking directly in line with the meniscus of the liquid. They are noted down in a recording table, measured in cm^3.

The order of recordings are:

- The final reading at neutralisation is taken and the 'rough' volume of acid used is calculated
- The procedure is repeated again but towards the final expected volume, the acid is added drop by drop until the colour just changes
- The 'titre' is noted and three more titres are obtained with volume reading within 0.1 cm^3

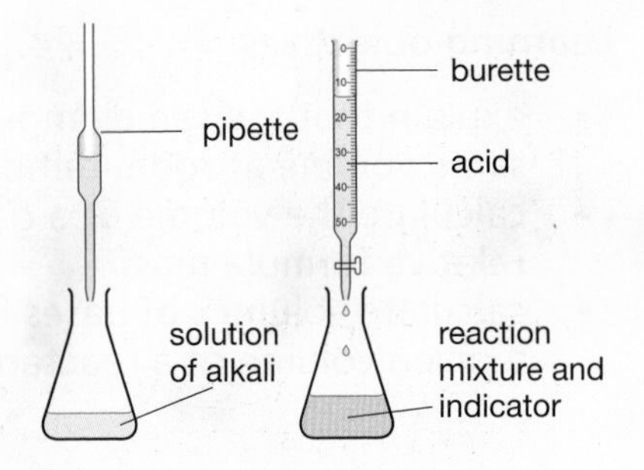

Figure 5.6

Jo had recorded these results and calculated the mean titre:

Volume of nitric acid	Rough volume	More accurate trials		
		1st trial	2nd trial	3rd trial
1st reading cm^3	1.5	1.8	2.4	15.2
2nd reading cm^3	28.5	28.4	29.1	41.8
Volume used cm^3	27.0	26.6	26.7	26.6

5 **Calculate the mean titre.**

6 **Jo calculated the mean titre as 26.725 cm^3. Explain why this is incorrect.**

7 **The mean titre is 26.6 cm^3 and not 26.63 cm^3. Explain why.**

HIGHER TIER ONLY

Determining the concentration of one of the solutions in mol/dm³ and g/dm³

- First write the equation of acid + alkali.
- Check that it is an equation of the type:
  $HA + BOH \rightarrow BA + H_2O$.
  This means that 1 mole of acid reacts with 1 mole of alkali.

Jo had used acid and alkali that reacted 1 mole:1 mole ratio.

8 **The concentration of Jo's alkali in mol/dm³ is 0.12. Calculate the number of moles in 25 cm^3.**

This number of moles exactly neutralised the same number of moles of Jo's acid.

9 **This same number of moles was contained in Jo's mean titre of acid. Calculate how many moles this would be in 1000 cm^3.**

This is the concentration of the acid in mol/dm³.

10 **Jo used HNO_3. Work out the molar mass of this acid. This will be the mass that is dissolved in 1 mol/dm³.**

11 **Calculate the concentration of Jo's nitric acid in g/dm³.**

12 **In a titration, 10.0 cm^3 of 0.250 mol/dm³ NaOH was titrated with HCl. The mean titre of HCl was 11.2 cm^3. Calculate the concentration of HCl in mol/dm³.**

DID YOU KNOW?

If the reaction is of the type:
$H_2A + 2BOH \rightarrow B_2A + 2H_2O$
then 1 mole of acid reacts with 2 moles of alkali and you will need to take that into account in your calculation.

KEY SKILL

The answer to question 9 was the number of moles in 1 dm^3 of Jo's acid. The molar mass of Jo's acid was calculated in question 10. You can work out the mass of acid by rearranging the equation:
moles = mass/molar mass.

Amounts of substance in volumes of gases

Learning objectives:

- explain that the same amount of any gas occupies the same volume at room temperature and pressure (rtp)
- calculate the volume of a gas at rtp from its mass and relative formula mass
- calculate volumes of gases from a balanced equation and a given volume of a reactant or product.

KEY WORDS

gas volume
room temperature and pressure
decimetre cubed
atmospheric pressure

Over 200 years ago it was found that 'gases combined in whole number ratios' so two volumes of hydrogen reacted with one volume of oxygen to make two volumes of water vapour. So, hypothesised Avogadro, wouldn't each volume of gas have the same number of molecules?

HIGHER TIER ONLY

Comparing volumes of gases

Let's think of this 'box' of gas particles as a volume of helium gas at room temperature and 1 atmosphere pressure. The model is represented by a few atoms when really there would be vast numbers.

The three gases below occupy exactly the **same volume** as the 'box' of He.

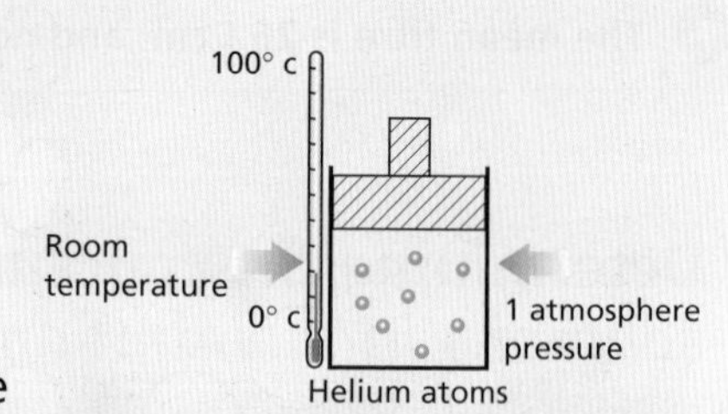

Figure 5.7 Model of helium gas at room temperature and 1 atmosphere pressure

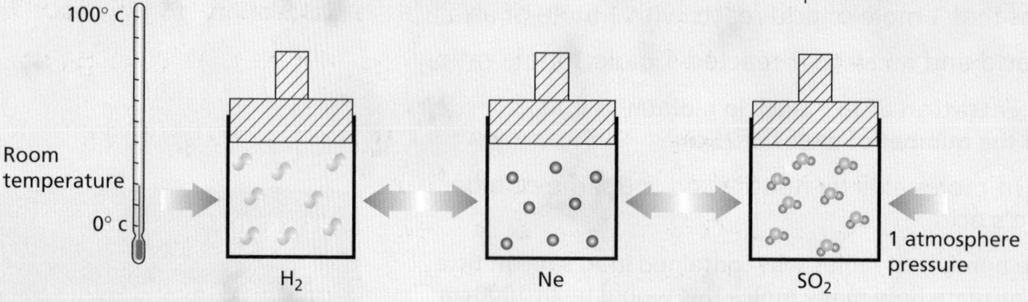

Figure 5.8 These three gases occupy exactly the same volume as the He box in Figure 5.7. There are exactly the same number of particles in each. Will the mass of each box be the same?

There is exactly the same number of particles in each. Will the mass of each box be the same?

The answer is no, as each individual particle has a different mass so the collections of particles will have different collective masses. We have to take account of this mass difference.

Let's l compare them.

He	
4	

The v lerably more mass than the same volume with H

We fi red in *grams* and the *volume* it occupies is measu n the gas would always occupy *24 dm³*.

1 **W t would occupy 24 dm³ at rtp.**
A

2 **C dioxide would occupy.**

3 **E ies the sa**

Mole

We ha substance' is a *mo*

One m stance in gram and is 6.02 × 10^{23})

So foll

'equ e volume under the .'

If those sure then:

'the erature and pres dm³'

DID YOU KNOW?

It is the number of particles that matters. It does not matter how large the atoms or molecules are. This is because the space between molecules in a gas is enormous.

KEY INFORMATION

The number of moles of gas can be found if its volume is known:

number of moles = volume of gas in dm³ ÷ 24

4 **Work out the volume (at rtp) of 1.5 moles of chlorine gas (Cl_2)**

5 **Calculate the volume (at rtp) of 7g of N_2 (A_r N is 14)**

Volumes from balanced equations

The volumes of gaseous reactants and products can be calculated from the balanced equation for the reaction.

For example, if propane reacts with oxygen the equation is: $C_3H_8 + 5O_2 \rightarrow 3CO_2 + 4H_2O$

If 24 dm³ of C_3H_8 is used then 5 × 24 dm³ of O_2 is needed (120 dm³)

If 2.4 dm³ of C_3H_8 is used then 3 × 2.4 dm³ of CO_2 is made (7.2 dm³)

In other words if *v* is volume and *n* is the number of moles then:

$$\frac{v_1}{n_1} = \frac{v_2}{n_2} \quad \text{or} \quad \frac{v_1 \times n_2}{n_1} = v_2$$

6 **Calculate the volume of water vapour made at rtp when 1.5 dm³ of C_3H_8 reacts with oxygen as in the equation above.**

7 **Explain why 10 dm³ of CO_2 is formed at rtp when 2 dm³ of pentane C_5H_{12} reacts with O_2**

8 **Work out the volume of hydrogen in dm³ needed to react with 42 g of nitrogen using the equation below.**
N_2 (g) + $3H_2$ (g) $\rightleftharpoons$ $2NH_3$ (g)

KEY CONCEPT

Percentage yield

Learning objectives:

- calculate the percentage yield from the actual yield
- identify the balanced equation for calculating yields
- calculate theoretical product amounts from reactant amounts.

KEY WORDS

yield
percentage yield
theoretical yield
actual yield

Making chemicals for worldwide use is one of the biggest industries in the UK. Over two million scientists are employed directly by the industry. How do they make sure all the chemicals used are turned into products without wasting them?

Product yield

The more *reactant* that is used, the more *product* is made.

To get the best amount of product,all reactants need to be *converted* to products during the reaction.

Even if the exact amount of reactants to use is *calculated* and reacted, there still may not be 100% product.

Some of the product may be lost when it is separated from the reaction mixture. It can be lost during these stages:

- loss in filtration – small amounts stay on the filters
- loss in evaporation – some chemicals evaporate
- loss in transferring liquids – it sticks to glass vessels

Figure 5.9 Making sure all of the product is retrieved

The amount of a product obtained is known as the *yield.*

The **percentage yield** of a reaction is a way of comparing the mass of product made (the **actual yield)** to the mass we expect to make (the *maximum* **theoretical** mass).

$$\%\text{Yield} = \frac{\text{Mass of product actually made}}{\text{Maximum theoretical mass of product}} \times 100$$

- 100% yield means that no product has been lost.
- 0% yield means that no product has been collected.

KEY INFORMATION

Remember that the law of conservation means that matter cannot be destroyed.

1. **Jo makes a substance by pouring two solutions into a beaker and then filtering off the solid. Suggest two reasons why the yield is less than 100%.**

2. **A reaction produces less than half the expected product. Choose the correct percentage yield from the list.**

   30% 50% 70% 100%

Calculating the percentage yield

To *calculate* percentage yield two things must be known:

- the amount of product made, the 'actual yield'
- the amount of product that should have been made.

For example, a company making sulfuric acids gets an actual yield of 74 tonnes. They predicted a yield of 85 tonnes.

$$\text{Percentage yield} = \frac{\text{actual yield}}{\text{predicted yield}} \times 100 = \frac{74}{85} \times 100 = 87\%$$

3 **Alex makes some copper oxide. The reaction produces 48 g. The predicted yield was 80 g. Calculate the % yield.**

4 **Sam made some $MgCO_3$. The % yield was 75% and the predicted yield was 84 g. Calculate the actual yield.**

Calculating theoretical yields

You can make insoluble salts such as lead iodide by precipitation, filtering, washing the insoluble salt and leaving it to dry. What is the **theoretical yield**?

2 molecules KI need 1 molecule of PbI_2:

$$2KI + Pb(NO_3)_2 \rightarrow PbI_2 + 2KNO_3$$

If we used the twice the formula mass of KI we should get the formula mass of PbI_2 (A_r of K is 39, A_r of I is 127, A_r of Pb is 207).

$$M_r \text{ of } 2KI = 2\ (39 + 127) = 2 \times 166 = 332$$

$$M_r \text{ of } PbI_2 = 207 + (2 \times 127) = 207 + 254 = 461$$

So if 332 g of KI reacts with $Pb(NO_3)_2$ we should get 461 g of PbI_2. This is the theoretical maximum amount or predicted yield.

5 **Potassium iodide and lead nitrate solutions were mixed in a beaker. The insoluble lead bromide product was filtered off. The expected yield was 20 g but the actual yield was 18 g.**

a **Calculate the the percentage yield.**

b **Suggest one reason why some product was lost.**

6 **Sodium hydroxide was the limiting reactant in the reaction with sulfuric acid.**

$$H_2SO_4 + 2NaOH \rightarrow Na_2SO_4 + 2H_2O$$

a **Calculate the theoretical maximum mass of sodium sulfate that can be produced from 20 g of sodium hydroxide.**

b **In another experiment, 15.5 g of sodium sulfate was formed. Calculate the theoretical mass for a yield of 85%.**

DID YOU KNOW?

The other reactant should either be in exactly the same proportions or in excess. The reactant used to work out the theoretical maximum should also be the *limiting reactant*.

Figure 5.10 Yellow lead iodide is made by precipitation

Atom economy

Learning objectives:

- identify the balanced equation of a reaction
- calculate the atom economy of a reaction to form a product
- explain why a particular reaction pathway is chosen.

KEY WORDS

atom economy
sustainability
reaction pathway
by-product

When we are making chemical products it is important that we make useful product, so that we do not have waste. This is important for sustainable development. Is there a way of calculating how much product is useful?

Figure 5.11 Magnesium and oxygen react to make magnesium oxide and no other product

The atom economy

Atom economy is a measure of starting materials that end as *useful products.* We need *high atom economy.*

Atom economy is a way of measuring the *amount of atoms* that are wasted or lost. 100% atom economy means that all the atoms have changed into the desired product.

For example,

magnesium + oxygen → magnesium oxide

All the magnesium atoms and all the oxygen atoms are used to make magnesium oxide. The atom economy is 100%.

If less than 100% some reactant atoms will be wasted.

An example is making magnesium sulfate for bath salts from:

magnesium oxide + sulfuric acid → magnesium sulfate + water

As two products are made, there is a lower atom economy, so the reaction is more wasteful.

Figure 5.12 Making bath salts has an atom economy of less than 100%

1. **Ethene and water react to make ethanol only. Choose the correct atom economy for this reaction.**
   a 50% **b** 0 % **c** 100%
2. **Sugar and water react to make ethanol and carbon dioxide. The atom economy is:**
   a 100% **b** greater than 0 but less than 100% **c** 0%

Figure 5.13 One way of making ethanol

Calculating atom economy

The atom economy (or *atom utilisation*) can be found using a formula and the *balanced equation of the reaction.* It is found as a percentage.

$$\text{Percentage atom economy of reaction} = \frac{\text{Relative formula mass of desired product}}{\text{Sum of relative formula masses of all reactants}} \times 100$$

M_r is used as for relative *formula mass* (A_r is relative *atomic* mass).

$$\%\ \text{atom economy} = \frac{M_r \text{ desired product}}{\text{Sum of } M_r \text{ of all reactants}} \times 100$$

Look at this example:

A company makes magnesium sulfate $MgSO_4$ for use as bath salts. They need to find the best method.

A_r: Mg = 24, O = 16, H = 1, S = 32

M_r of $MgSO_4$ (desired) = 24 + 32 + (16 × 4) = 24 + 32 + 64 = **120**

Method 1:	Method 2:
$MgO + H_2SO_4 \rightarrow MgSO_4 + H_2O$	$MgCO_3 + H_2SO_4 \rightarrow MgSO_4 + H_2O + CO_2$
M_r reactants:	M_r reactants:
24 + 16 + 2 + 32 + 64 = 138	24 + 12 + 48 + 2 + 32 + 64 = 182
% atom economy = $\frac{120}{138}$ × 100 = 86.9%	% atom economy = $\frac{120}{182}$ × 100 = 65.9%

Which method is best? The higher the atom economy, the fewer atoms are in the wasted product, so the first method is a less wasteful process.

3 **A company has two methods for making a drug of M_r = 568. Method 1 uses reactants with an M_r of 676 and Method 2 with M_r of 624. Calculate the atom economy for each method.**

4 **Suggest how a reaction that has 64% atom economy for the desired product could also have 100% atom economy.**

REMEMBER!

You need to be able to recall and use this formula:

$$\%\ \text{atom economy} = \frac{M_r \text{ desired product}}{\text{Sum of } M_r \text{ of all reactants}} \times 100$$

Choosing reaction pathways

Every industrial process wants to go along a **reaction pathway** that has as high an atom economy as possible to reduce the formation of unwanted products. We saw this in making ethanol in Q1 and Q2. However, as well as atom economy, the *yield*, the *rate of reaction*, the *equilibrium position* in a *reversible reaction* and the usefulness of **by-products** also need to be considered. We meet these ideas in later chapters.

Looking at by-products, some reactions can give a low atom economy, e.g. hydrogen for vehicles is made from water: $2H_2O \rightarrow O_2 + 2H_2$

Using the atom economy formula we find this atom economy is 12.5%. However, if oxygen were the desired product, this reaction would have an atom economy of 87.5%.

If *both* products are useful, the overall atom economy is then at 100%.

Figure 5.14 When hydrogen used for cars comes from water this is a low atom economy reaction

5 **$S + O_2 \rightarrow SO_2$ is a reaction that happens when coal burns. Evaluate whether this is a 'green' reaction, taking atom economy into account.**

6 **$2NH_4Cl(s) + Ca(OH)_2(s) \rightarrow CaCl_2(s) + 2NH_3(g) + 2H_2O(g)$ makes NH_3, ammonia. It can also be made from N_2 and H_2. Compare the reactions for sustainability.**

DID YOU KNOW?

In the reaction to make ammonia we need to consider the yield, rate and equilibrium position even though the atom economy is 100%

MATHS SKILLS

Change the subject of an equation

Learning objectives:

- to use an equation to demonstrate conservation
- to rearrange the subject of an equation
- to carry out a multi-step calculation.

KEY WORDS

equation
rearrange
multi-step calculation
multiply

If we are doing a titration or a rate of reaction investigation we will need to measure quantities. Sometimes this requires a multi-step calculation. In this process you will often need to change the subject of an equation. There are several ways to do this. Do you have a way of remembering how to change the equations?

KEY INFORMATION

You will need to know how to change the subject of an equation by using all four operations: add, subtract, multiply and divide.

Conservation of mass

If we use a total mass of reactants in a chemical reaction the products made will be of the same total mass. This is known as the conservation of mass. In this reaction:

magnesium oxide	+ sulfuric acid	→	magnesium sulfate	+ water
MgO	+ H_2SO_4	→	$MgSO_4$	+ H_2O
40 g of MgO		will make	120 g of $MgSO_4$	+ 18 g of H_2O

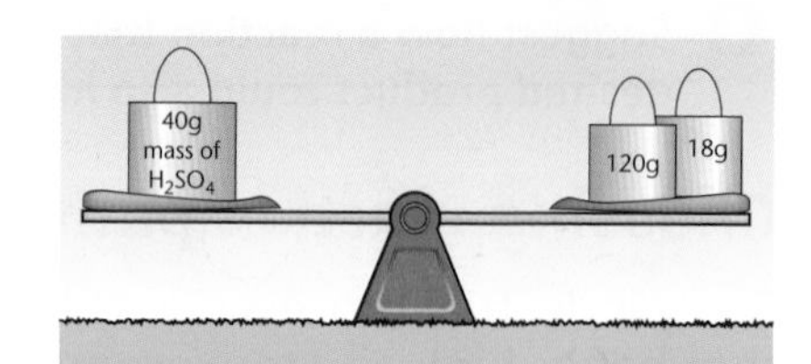

Figure 5.15

Why do we get the bigger mass of product? It is because the mass of sulfuric acid has to be added in.

This is because:

mass of MgO + mass of H_2SO_4 = mass of $MgSO_4$ + mass of H_2O

What mass of sulfuric acid will be used?

40 + mass of H_2SO_4 = 120 + 18

The subject of the equation needs to be the mass of sulfuric acid. In order to isolate the mass of the sulfuric acid, we need to subtract the mass of the MgO. In order to keep the equation balanced, we must also subtract 40 g from the right hand side.

As the equation is balanced we need to do the *same to both sides* to move the 40 g.

Therefore we need to *subtract 40 g from both sides.*

(40 – 40) + mass of H_2SO_4 = 120 + 18 – 40

mass of H_2SO_4 = 98

1. **Use the equation of the reaction to work out how much $MgSO_4$ is made from 24.5 g of H_2SO_4 and 10 g of MgO, when 4.5 g water is made. Write out your working to show how you change the subject of the equation.**

2. **Use the equation: $MgCO_3$ + H_2SO_4 → $MgSO_4$ + H_2O + CO_2 to find out how much $MgCO_3$ was needed to react with 49 g of H_2SO_4 to make 91 g of the total product. Write out your working to show how you change the subject of the equation.**

Calculating the mass needed

If we are trying to make a product like Epsom bath salts we need to know how much reactant to start with. These salts are magnesium sulfate, which we can make in two ways:

$MgO + H_2SO_4 \rightarrow MgSO_4 + H_2O$ *or* $MgCO_3 + H_2SO_4 \rightarrow MgSO_4 + H_2O + CO_2$

We can use the relative atomic mass (A_r) to calculate the relative formula mass (M_r) of any of the substances in this equation and so find a ratio of how much reactant gives a quantity of product.

For example: A_r: Mg is 24, O is 16, S is 32, H is 1
so: M_r: MgO is 40 M_r: $MgSO_4$ is 120

We can therefore say that 40 g of MgO will give 120 g of $MgSO_4$. So the ratio is expressed as:

$$\frac{\text{Mass of MgO}}{\text{Mass of MgSO}_4} = \frac{40}{120}$$

If we use 2.0 g MgO to find out how much $MgSO_4$ we can make we will need to change the subject of the ratio equation by keeping it balanced at all times:

$$\frac{2}{\text{mass of MgSO}_4} = \frac{40}{120}$$ Step 1 multiply through by the mass of $MgSO_4$

$$\frac{2 \times \text{mass of MgSO}_4}{\text{mass of MgSO}_4} = \frac{40 \times \text{mass of MgSO}_4}{120}$$ Step 2 cancel the fractions

$$2 = \frac{40 \times \text{mass of MgSO}_4}{120}$$ Step 3 multiply both sides by 120

$$2 \times 120 = \frac{40 \times \text{mass of MgSO}_4 \times 120}{120}$$ Step 4 cancel the fraction

$$2 \times 120 = 40 \times \text{mass of MgSO}_4$$ Step 5 divide both sides by 40

$$\frac{2 \times 120}{40} = \frac{40 \times \text{mass of MgSO}_4}{40}$$ Step 6 cancel the fractions to isolate the mass of the $MgSO_4$

$$\frac{2 \times 120}{40} = \text{mass of MgSO}_4$$ Step 7 calculate the left hand side

$$6 = \text{mass of MgSO}_4$$ Step 8 Is this a reasonable answer? Yes

DID YOU KNOW?

You can do steps 1–5 all at the same time.

3 **What is the mass of $MgSO_4$ produced by 62.5 g of MgO?**

4 **What is the mass of $MgSO_4$ produced by 25.2 g of $MgCO_3$?**

Finding the concentration of an acid

A titration can be used to find the concentration of an acid using a fixed volume of alkali of known concentration. (Assuming 1 mole alkali reacts with 1 mole acid) the equation used for calculation is:

concentration of alkali × volume of alkali = concentration of acid × volume of acid

concentration of alkali / mol / dm³	volume of alkali / cm³	concentration of acid / mol / dm³	volume of acid / cm³
0.12	25	*z*	27.2
0.15	25	0.13	*y*

Use the data in the table for Q5 and Q6. Show how you change the subject of the equation in both.

5 **Calculate the concentration of acid (*z*).**

6 **Calculate the volume of acid (*y*).**

Measuring rates

Learning objectives:

- explain how to measure the amount of gas given off in a reaction
- explain how to measure the rate of a reaction
- read data from graphs to interpret stages of a reaction.

KEY WORDS

balance
gas syringe
measuring cylinder
volume

It is difficult to measure the rate of very fast or very slow reactions. However, the rate of some reactions can be measured easily. Here are some of the experimental methods to measure gas volume.

Ways to measure reaction times

There are many ways to measure the time taken for a chemical reaction. Three of these ways are measuring:

- the loss in mass of reactants over time
- the volume of gas produced over time
- the time for a solution to become opaque or coloured.

During the reaction between calcium carbonate and dilute acid a gas is given off.

calcium carbonate + hydrochloric acid
→ calcium chloride + water + carbon dioxide

If this is done in a flask, the carbon dioxide gas will escape. The mass of the flask's contents will go down. The decrease in mass can be measured on a digital **balance**.

The decrease in mass can be measured each minute and plotted on a graph.

The time for the reaction between magnesium and dilute hydrochloric acid can also be measured by this method.

magnesium + hydrochloric acid → magnesium chloride + hydrogen

Look at Figure 5.16 to answer questions 1 and 2.

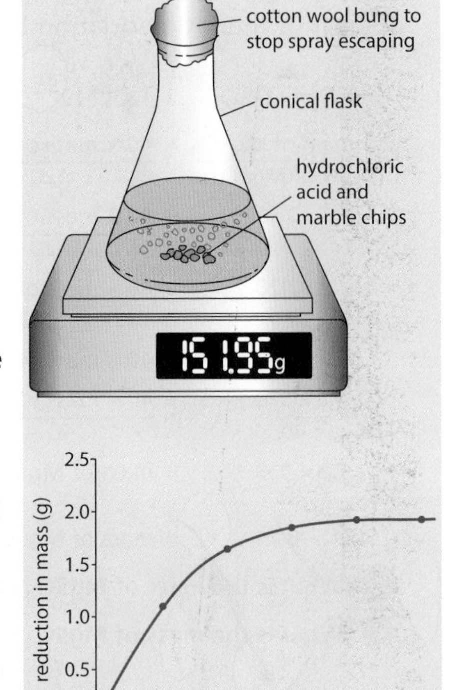

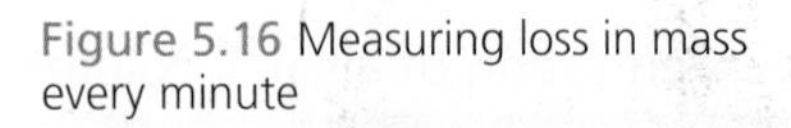
Figure 5.16 Measuring loss in mass every minute

1. **Estimate the decrease in mass after two minutes.**
2. **During which minutes was the graph steepest? Suggest why this is.**

KEY SKILLS

Remember to make sure you have a cotton wool plug in the flask to prevent damage to the balance from the fizzing contents.

Volumes of gas

It is often easier to measure the volume of a gas made rather than very small changes in mass.

Figure 5.17 shows three ways of collecting gas given off in a reaction using an upturned **measuring cylinder**, an upturned burette or using a **gas syringe**.

The **volume** of hydrogen collected in the gas syringe can be measured every 10 seconds.

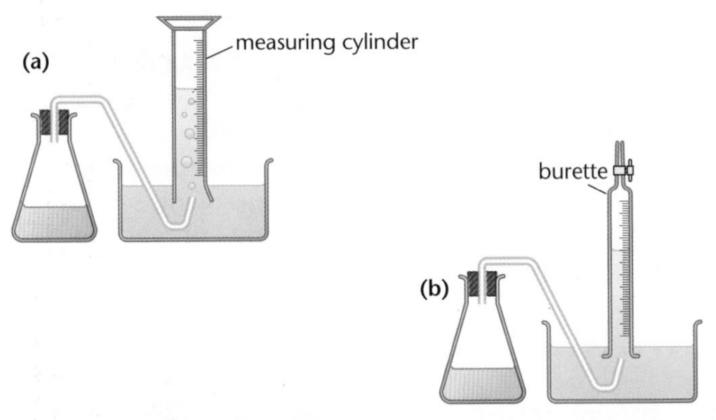

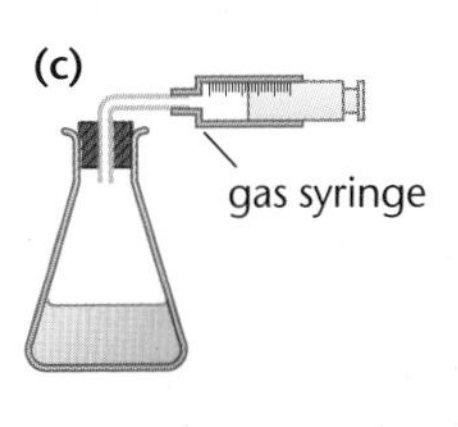

Figure 5.17 Three ways of collecting and measuring volumes of a gas

The results of the experiment can be plotted on a graph, as shown in Figure 5.18. The slope of the line shows how fast the reaction is. The steeper the slope is the faster the reaction.

Use Figure 5.18 to answer questions 3 and 4.

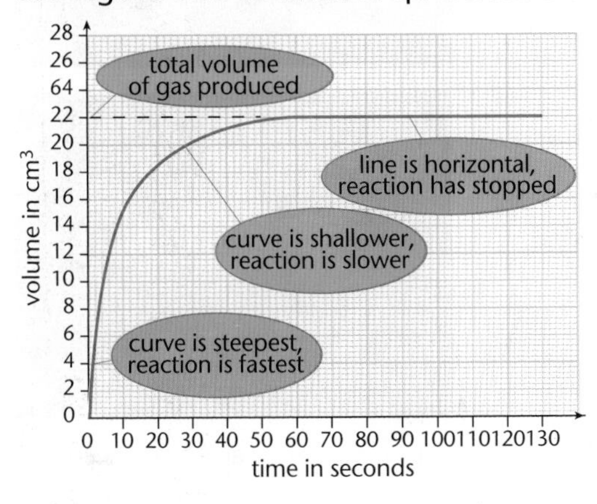

Figure 5.18 Gas collected and measured every 10 seconds

3 **Determine the volume of gas produced in the first 20 seconds.**

4 **State the time taken for the reaction to stop.**

DID YOU KNOW?

If an upturned burette or measuring cylinder is used it must be filled with water before it is turned upside down. The volume is read off the scale.

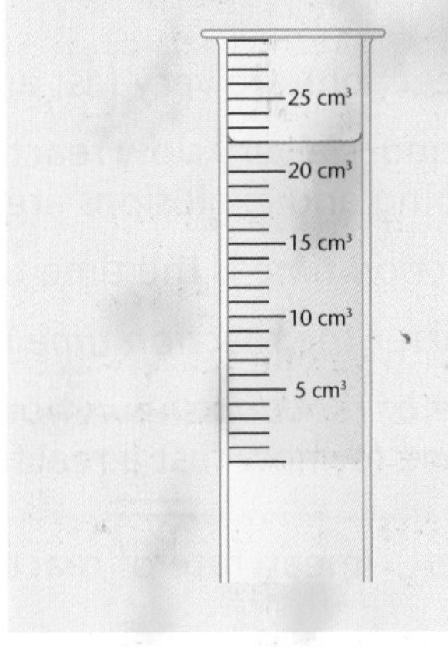

Measuring degree of opaqueness

In the reaction between sodium thiosulfate and hydrochloric acid a precipitate of sulfur is slowly produced.

$$Na_2S_2O_3(aq) + 2HCl(aq) \rightarrow 2NaCl(aq) + H_2O(l) + SO_2(g) + S(s)$$

The flask is put over a paper with a large X marked on it. When the reactants are mixed the time taken for the cross to 'disappear' is measured with a timer.

5 **The concentration of sodium thiosulfate can be varied. Explain how you could adapt the experiment to measure the effect of this on rate.**

6 **Give two reasons why the method may lack accuracy and lead to errors. Suggest one modification to improve the accuracy for each of these.**

Figure 5.19 The precipitate of sulfur forms slowly to obscure the X.
The time is taken when the cross just 'disappears'.

Calculating rates

Learning objectives:

- calculate the mean rate of a reaction
- draw and interpret graphs of reaction times
- draw tangents to the curves as a measure of the rate of reaction.

KEY WORDS

axes
construction lines
gradient
rate of reaction

We can measure the progress of a reaction by collecting gas, measuring the loss of mass or the degree of opaqueness. If we do this in set time intervals we can then calculate the rate of a reaction.

Reaction time and reaction rate

In a chemical reaction reactants are made into products.

reactants → products

The **rate of reaction** measures how much product is made each second.

Some reactions are very fast and others are very slow:

- Rusting is a very slow reaction.
- Burning and explosions are very fast reactions.

The *reaction time* is the time taken for the reaction to finish.

The shorter the reaction time is, the faster the reaction.

The *rate of reaction* shows how much product is formed in a fixed time (or how fast a reactant disappears).

So either: $\text{mean rate of reaction} = \dfrac{\text{quantity of reactant used}}{\text{time taken}}$

or $\text{mean rate of reaction} = \dfrac{\text{quantity of product formed}}{\text{time taken}}$

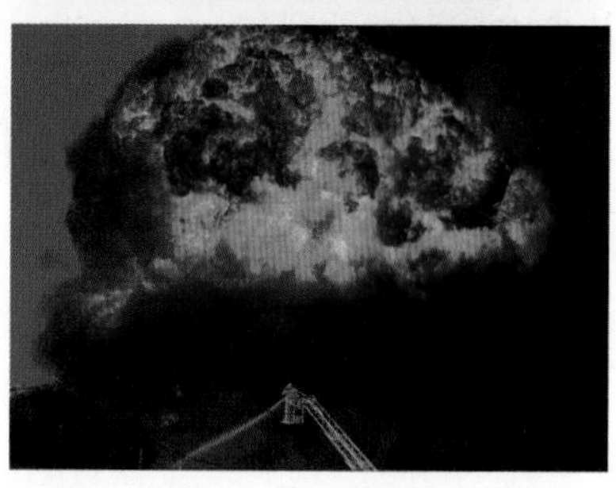

Figure 5.20 Slow and fast reactions

DID YOU KNOW?

The rate of a reaction is an important factor in running industrial processes.

1 Describe the following reactions in terms of rate and time taken.

a Glue setting

b Methane combusting

c Diamond and oxygen reacting at room temperature

2 Hydrochloric acid was added to lumps of calcium carbonate. In the first 50 seconds, 30 cm³ of gas was collected. Calculate the mean rate of reaction.

Changes of rate

The rate changes during a reaction.

Reactions are usually faster at the start, and then they slow down as the reactants are used up.

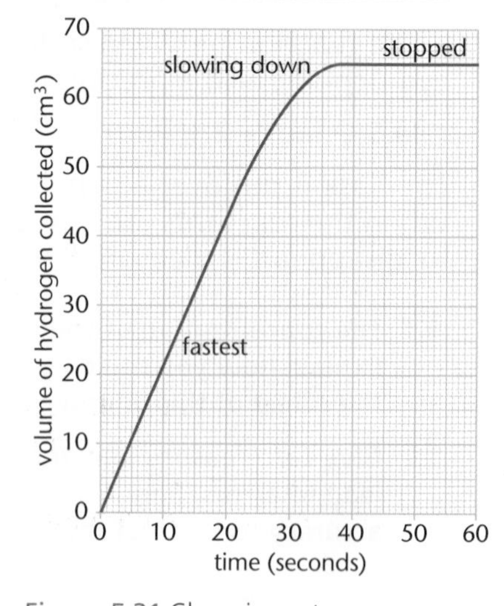

Figure 5.21 Changing rates

An example is the reaction between magnesium ribbon and hydrochloric acid.

$$Mg + 2HCl \rightarrow MgCl_2 + H_2$$

Hydrogen is a gas so it is easy to measure how much is produced.

Look at Figure 5.21. The steeper the gradient (slope) on the graph is, the faster the reaction. As the **gradient** decreases, the reaction slows down. When the line is horizontal the reaction has finished, so no more product is being made. This is because one of the reactants has been used up.

The rate of reaction can be worked out from the gradient of a graph.

The gradient is found by drawing **construction lines**. Choose a part of the graph where there is a straight line (not a curve). Measure the value of y and x and remember to use the scale of the graph carefully. Then divide y by x. Determine the units from those given on the **axes**.

In Figure 5.22 the gradient = $\frac{y}{x} = \frac{21.5}{10} = 2.15\ cm^3/s$

It is very important to check the units you are using.

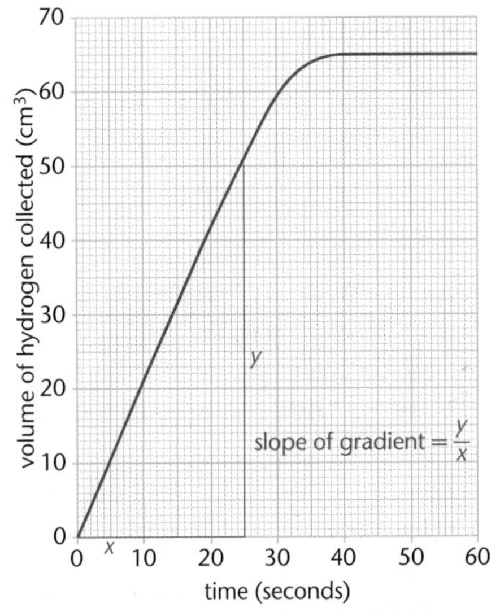

Figure 5.22 Calculating gradients

Quantity of reactant or product	Unit of quantity	Unit of rate
Mass	g	g/s
Volume	cm^3	cm^3/s
Moles	mol	mol/s

REMEMBER!

Sometimes it is easier to measure the mass of a product formed. The rate of reaction is then measured in g/s or g/min.

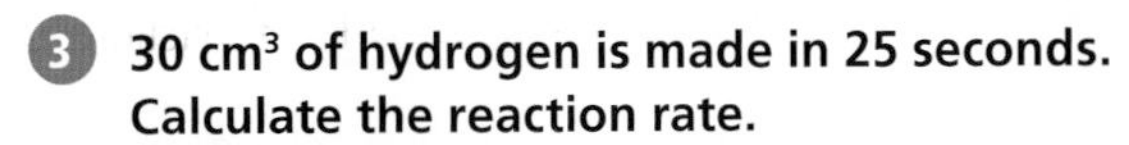

3 **30 cm^3 of hydrogen is made in 25 seconds. Calculate the reaction rate.**

4 **Look at Figure 5.23. Draw construction lines to calculate the gradient of the red line.**

5 **Explain the difference in gradients between the red and blue lines in Figure 5.23.**

HIGHER TIER ONLY

Rates at specific times

The rate changes during a reaction. The tangent to a curve can be chosen at a specific point on a graph and the gradient calculated as a measure of rate of reaction at a specific time.

6 **Use Figure 5.23 to calculate the gradients to determine the rates of reaction at**
a 25 seconds and
b 40 seconds on both the red and blue lines.

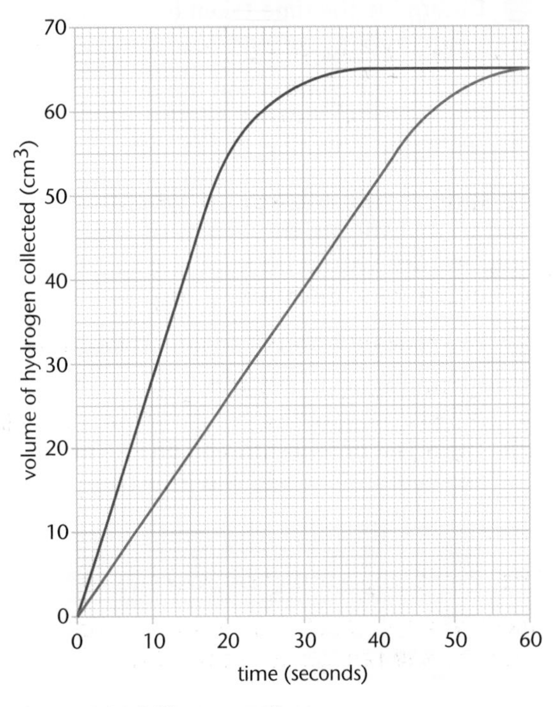

Figure 5.23 Different gradients

Factors affecting rates

Learning objectives:

- identify which factors affect the rate of reactions
- explain how changes of surface area affect rates
- explain how rates are affected by different factors.

KEY WORDS

catalyst
concentration
pressure
temperature

There are many ways to alter the rate of a reaction. Gases reacting together can be put under pressure to push the particles nearer to each other. Catalysts can be used to help particles have a surface on which to bind. Some of these ways are easy to investigate.

Factors affecting the rate of reaction

Factors that affect the rates of chemical reactions include the:

- **temperature**
- **concentrations** of reactants in solution
- **pressure** of reacting gases
- surface area of solid reactants
- presence of **catalysts**

In the reaction between sodium thiosulfate and hydrochloric acid a precipitate of sulfur is made. This happens slowly enough to measure the time that it takes to make a large X disappear under the reaction flask. (See topic 5.13).

This reaction can be carried out five times at *different temperatures*.

A graph can be drawn of the results.

1. **Determine the time taken for the reaction to reach the end point at 35 °C.**
2. **What is the trend when the temperature increases?**

KEY SKILLS

Remember that if you change the temperature (one factor) you must keep everything else the same.

Figure 5.24 The time is taken when the cross just disappears. This is repeated at different temperatures.

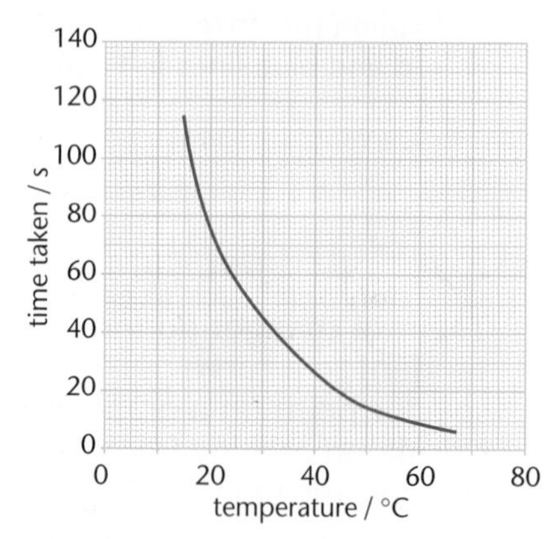

Figure 5.25 Temperature against time taken for X to disappear

Changing concentrations

The sodium thiosulfate and hydrochloric acid reaction can be used to investigate the effect of changing concentrations. The temperature would be kept the same each time, but different *concentrations* of sodium thiosulfate would be used.

3. **The concentration of sodium thiosulfate was increased without changing the temperature. Describe what would happen to the time taken for the cross to disappear and the reaction rate.**
4. **The effect of concentration on rate of reaction between calcium and hydrochloric acid was investigated, the results are shown in the table.**

	Temperature / °C	Concentration of HCl / mol dm^3
Experiment 1	20	0.1
Experiment 2	25	0.2

Suggest whether the investigation was valid.

Surface to volume ratio

The calcium carbonate and dilute hydrochloric acid reaction can be used to investigate the effect of changing the surface area of solid reactants. The acid concentration would be kept the same each time but the size of the calcium carbonate pieces would change, while keeping the mass the same.

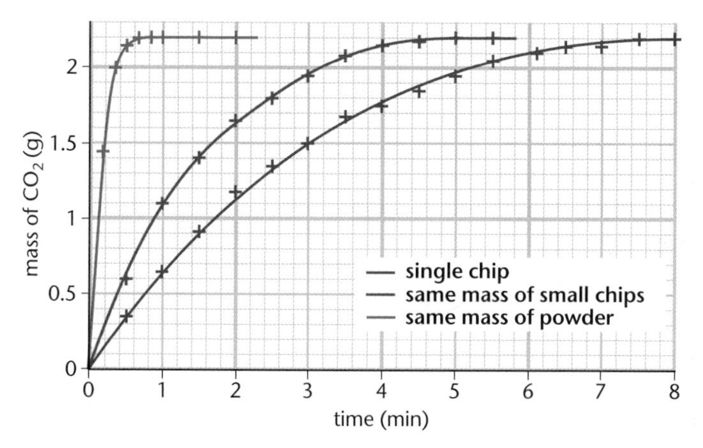

Figure 5.26 Graph of mass of CO_2 lost using three sizes of marble chips

The reason that the rate of reaction changes with the size of pieces of a reacting solid is because there is a bigger surface area that the volume of acid can attack if the chip is cut into smaller pieces. Look at Figure 5.27. The large cube has six large faces for attack. If the cube is divided, those six faces remain but there are other surfaces that were 'locked inside' that are now available for attack so the reaction of the acid can produce more CO_2 per minute.

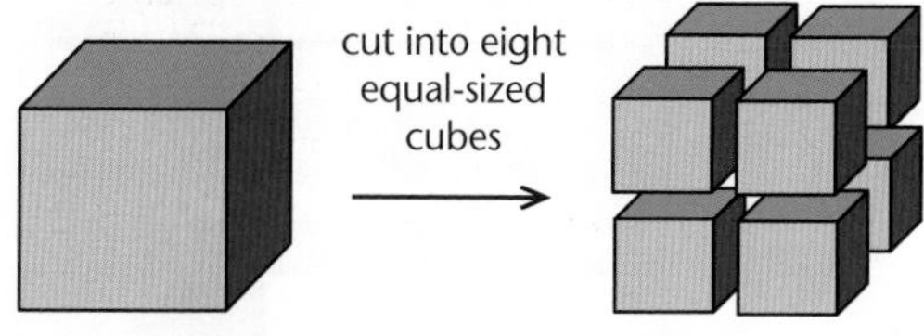

Figure 5.27 Increasing surface area

5 **Determine the mass of CO_2 lost using a single chip after three minutes.**

6 **Describe the pattern linking the size of the chip and the time taken to produce carbon dioxide.**

7 **Work out the mean rate of reaction, including units, for each of the three reactions in Figure 5.26.**

8 **Explain why the same mass of CO_2 was produced in each of the three experiments.**

9 **Describe three ways that the rate of the Mg/HCl reaction could be altered. Would these increase or decrease the rate?**

KEY INFORMATION

Remember that if you change the concentration of sodium thiosulfate you must keep the volume the same. You must also keep the concentration and volume of hydrochloric acid the same and the temperature the same too.

DID YOU KNOW?

You can either measure the volume of carbon dioxide given off using a gas syringe or the loss in mass using a digital balance (see topic 5.9).

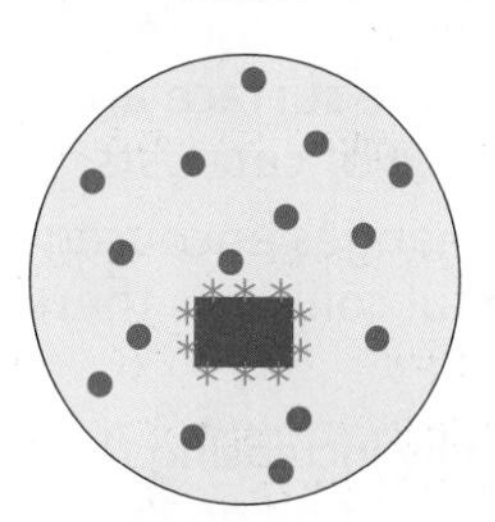

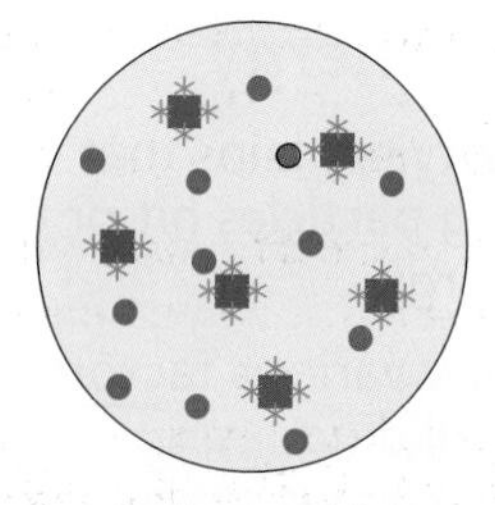

Figure 5.28 Increased surface area can be attacked by acid

Collision theory

Learning objectives:

- describe a reaction by particles colliding
- explain the effects of changes of factors on rates of reaction using collision theory
- describe activation energy.

KEY WORDS

activation energy
collision
effective
frequency

We know that we can change the rate of reaction by altering the concentration, temperature, pressure and other factors. We know how we can measure the changes in reactions. Now we need to be able to *explain* what is happening by developing a theory – collision theory.

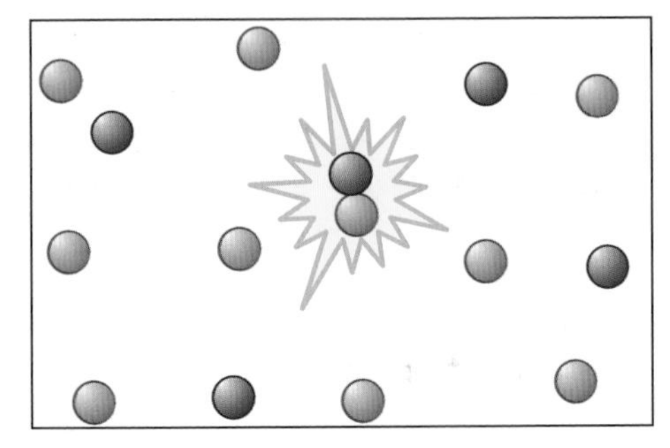

Figure 5.29 Collisions and one successful collision between reacting particles

Collisions

Chemical reactions take place when reactant particles collide with each other and form new products.

Not all collisions are successful, a reaction only takes place when a **collision** is successful.

To make the reaction slower, the number of successful collisions needs to decrease.	To make the reaction faster, the number of successful collisions needs to increase.

A reaction can be made to go faster by:

- increasing the temperature
- increasing the concentration (of reactants in solution)
- increasing the pressure (if the reactants are gases)
- increasing the surface area of solid reactants
- the presence of catalysts

All of these changes encourage more successful collisions. The more successful collisions there are, the faster the reaction.

KEY INFORMATION

Remember that increasing the surface area of solid marble chips by breaking them down into powder will allow more surface area for the particles of acid to collide with and react.

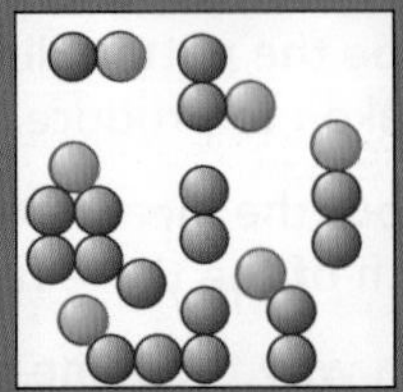

Figure 5.30 Increasing surface area increases the chances of successful collisions

1 **Explain why increasing the pressure on two gases reacting will make the reaction go faster.**

Activation energy

Collision theory explains that chemical reactions can occur only when reacting particles hit or collide with each other with sufficient energy.

It explains how various factors affect rates of reactions in terms of particles colliding.

As the concentration increases the particles become more crowded.

For example, at the low concentration in Figure 5.31 there are four particles of A. At the higher concentration there are ten particles of A in the same volume.

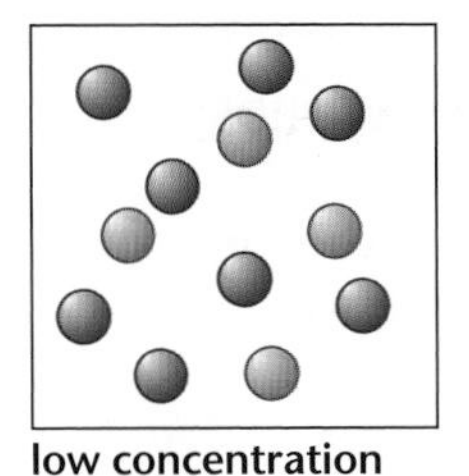

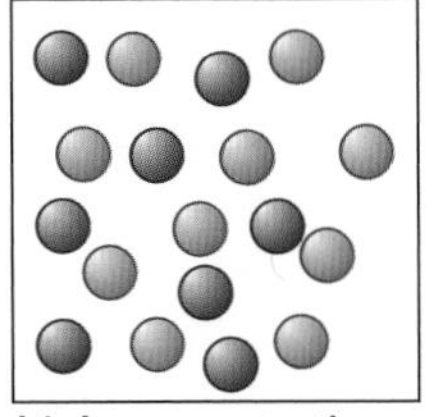

reacting particle of substance **A**
reacting particle of substance **B**

Figure 5.31 Increasing concentration increases chances of successful collisions

The particles are more crowded in the same volume, there are more collisions, and so there are more chances of successful collisions and the rate of reaction increases.

Increasing the temperature also increases the rate of reaction.

Particles move faster as the temperature increases. The reacting particles have more kinetic energy and so the number of successful collisions increases and the rate of reaction increases.

For a successful collision to occur each particle must have sufficient energy to react. This minimum amount of energy is the **activation energy**.

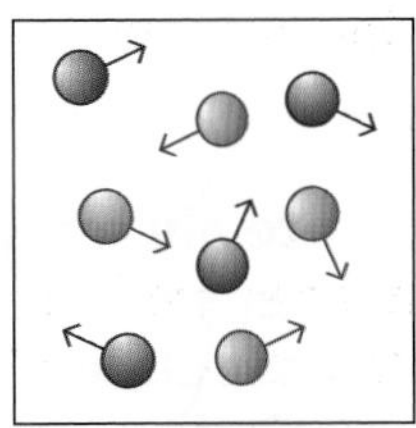

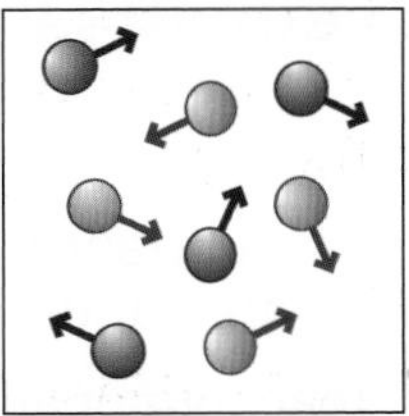

reacting particle of substance **A**
reacting particle of substance **B**

Figure 5.32 Increasing temperature increases chances of successful collisions

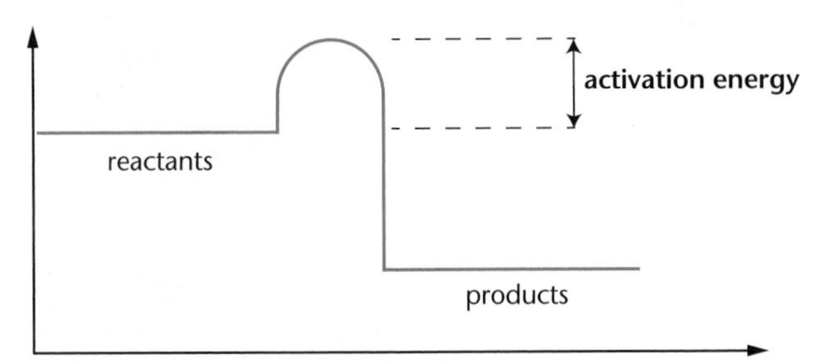

Figure 5.33 There is a minimum energy that has to be overcome for colliding particles to react. This is the activation energy.

2 Explain, using collision theory, how increasing the temperature of the reaction between magnesium ribbon and dilute hydrochloric acid increases the rate of reaction.

Collision frequency

It is not just the number of collisions that determines the rate of a reaction, it is the **collision frequency**.

Collision frequency describes the number of successful collisions between reactant particles that happen each second. The more successful collisions there are per second, the faster the reaction.

3 Explain how increasing the concentration of the acid increases the rate of reaction of:

$$CaCO_3(s) + 2HCl(aq) \rightarrow CaCl_2(aq) + H_2O(l) + CO_2(g)$$

4 A typical collision frequency for gas particles is 5×10^{10} Calculate the time taken for 1×10^6 collisions to occur.

DID YOU KNOW?

There are ways to reduce activation energy using catalysts.

Catalysts

Learning objectives:

- identify catalysts in reactions
- explain catalytic action
- explain activation energy.

KEY WORDS

activation energy
catalysis
catalyst
catalytic action

Scientists are always searching for new catalysts that can make a specific reaction faster, more energy efficient and more cost effective. Catalysts are now always used in catalytic converters in car exhausts to change polluting exhaust gases into less harmful gases.

Catalysts

Catalysts are substances that change the speed of a chemical reaction but are not used up during that reaction.

Different reactions need different catalysts.

A catalyst:

- is a substance added to a chemical reaction to make the reaction go faster (or to start) a reaction
- is only needed in small amounts
- remains unchanged at the end of a reaction.

When a catalyst is added to a reaction the *same amount* of product is produced but in a shorter time.

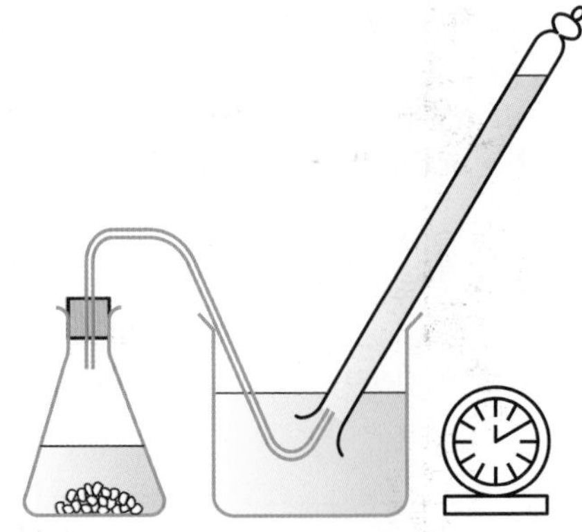

Figure 5.34 Measuring the time taken to collect 100 cm^3 of hydrogen

For example:

Sam and Alex investigate the reaction between zinc and sulfuric acid. Zinc sulfate solution and hydrogen gas are formed.

They want to find a catalyst for this reaction.

They measure the time it takes to collect 100 cm^3 of hydrogen in a gas syringe.

They add a different substance each time.

Their results are shown in Table 5.1.

KEY INFORMATION

Remember that when zinc is added to a solution of copper chloride the reaction displaces *copper* from the solution.

Experiment number	Substance	Time to collect 100 cm^3 of gas (seconds)	Colour of substance at the start	Colour of substance at the end
1	no substance	150	—	—
2	magnesium chloride	150	white	white
3	copper chloride	15	green	pink
4	copper powder	25	pink	pink

Table 5.1

1. Explain whether each substance is a catalyst or not.

Specific catalysts

Catalysts change the rate of chemical reactions but are not used up during the reaction.

Some other properties of a catalyst are that:

- only a small mass of catalyst is needed to catalyse a large mass of reactants
- they are specific to one single reaction.

Most catalysts only make a specific reaction faster. They do not make all reactions faster.

Examples of catalysts that are widely used are:

- vanadium pentoxide in the Contact process
- iron in the Haber process
- zeolites in the cracking of long-chain hydrocarbons
- rhodium-based catalysts in a **catalytic** converter.

2 **Write a balanced symbol equation for the reaction between zinc and sulfuric acid with a copper catalyst.**

KEY INFORMATION

You do not need to know the names of catalysts other than those in the specific processes you are learning about. Remember that the catalyst is not written in the equation as a reactant. You may sometimes see it written over the arrow in the equation.

Activation energy pathways

A catalyst does not increase the number of collisions per second. Instead, it works by making the collisions that take place more *successful.*

For reactions to take place the successful collisions have to have enough energy to overcome the **activation energy**.

Catalysts create a new pathway for the reaction. This new pathway has a lower *activation energy.* Catalysts therefore increase the rate of reaction by providing a different activation energy pathway.

A reaction profile for a catalysed reaction is shown in Figure 5.35.

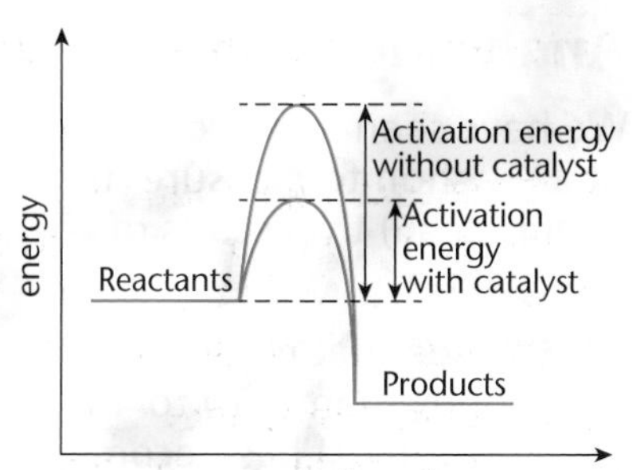

Figure 5.35 Catalysts lower the activation energy of a reaction

Enzymes are very important in maintaining the life and health of organisms and need special conditions to work effectively.

DID YOU KNOW?

In a biological system it is an *enzyme* that acts as a catalyst. They also work by lowering the activation energy pathway.

3 **Use an energy profile diagram to explain how vanadium pentoxide allows sulfur dioxide particles and oxygen particles to react more quickly in the Contact process.**

4 **Hydrogen peroxide is a strong oxidising agent and is used as bleach and disinfectant. When it decomposes, it produces oxygen and water. The activation energy for the decomposition has been measured.**

a **Write a balanced equation for the decomposition of hydrogen peroxide.**

b **Identify the most effective catalyst in the table and explain how it works.**

c **Suggest a value for the activation energy without a catalyst. Justify your answer.**

d **Hydrogen peroxide is produced in cells in the body. Explain why catalase is also present in cells.**

Catalyst	Activation energy in kJ/mol
None	?
Catalase enzyme	23
Platinum, Pt	49
Iodide ions, I^-	56
Manganese(IV) oxide, MnO_2	58

Factors increasing the rate

Learning objectives:

- analyse experimental data on rates of reaction
- predict the effects of changing conditions on rates of reactions
- use ideas about proportionality to explain the effect of a factor.

KEY WORDS

excess
analyse
graph
steeper

We have already seen how to measure the change of the rate of reactions and which factors can be altered to increase the rate. We have also developed a theory to explain these changes. We now need to ensure that we present the results so that evidence can be analysed.

Analysing data

We have seen how we can collect the results of a reaction to measure its rate. Look at (Figure 5.36) the results of collecting a gas in a gas syringe.

We can see that more gas is given off in 0 to 20 seconds than in 20 to 40 seconds. The rate of gas given off has become less. The rate has decreased.

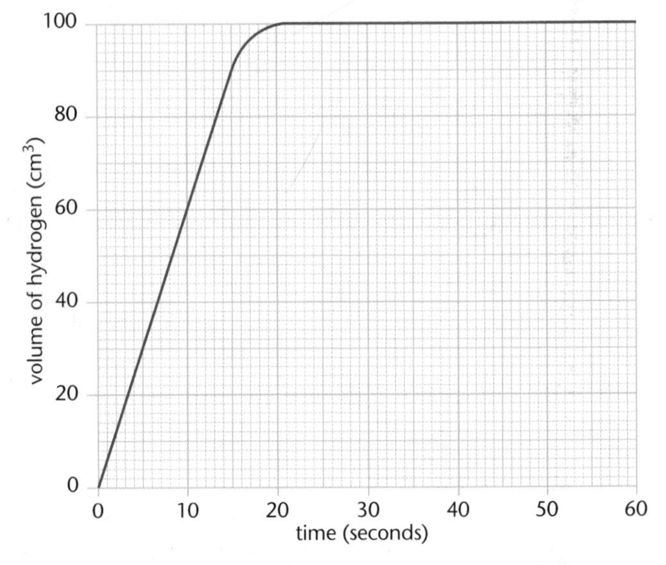

Figure 5.36 Collecting gas between 0 and 60 seconds

1. **When does the reaction stop?**
2. **Identify two substances that produce hydrogen when they react.**
3. **Explain what happens to the rate of gas given off each second between 21 and 60 seconds.**
4. **Calculate the average rate during the first 10 seconds.**

Increasing the rate of reaction

We have seen that we can make the reaction faster by:

- increasing the concentration
- increasing the temperature
- increasing the pressure (of gases)
- increasing the surface area (of a solid)

Figures 5.37 and 5.38 show the rate of reaction changes with increasing (a) concentration (b) temperature.

We can see that there was 30 cm^3 of gas made in the first 10 seconds when the high concentration acid is used and that increasing the concentration gives a **steeper** curve.

DID YOU KNOW?

In Figure 5.37 the red line shows the results of the experiment with dilute hydrochloric acid and the blue line shows results with acid that is more concentrated.

The gradient of the blue line is greater than that of the red line. This shows that the rate of reaction is faster when more concentrated acid is used.

In Figure 5.38 the blue line shows the results with the acid solution at a temperature of 30 °C. Its gradient is greater than that for the red line, which describes the experiment in which the acid solution was at a temperature of 20 °C.

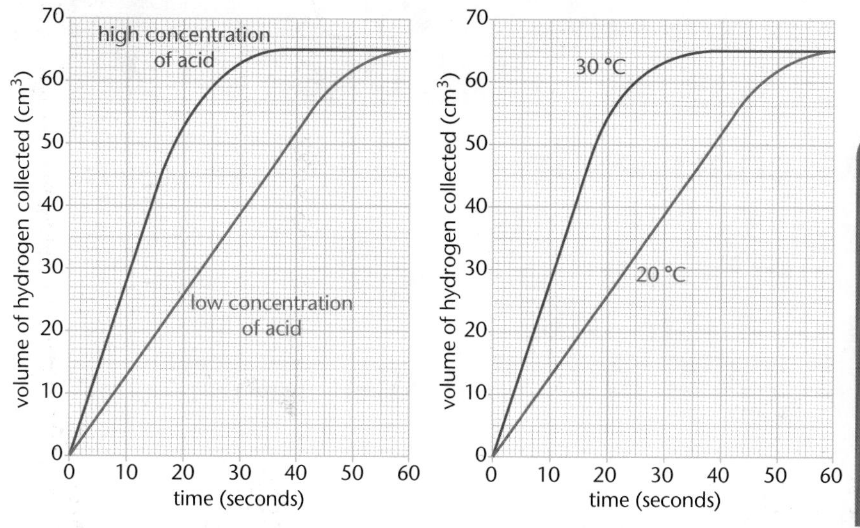

Figure 5.37 Increasing the concentration Figure 5.38 Increasing the temperature

KEY INFORMATION

Figures 5.37 and 5.38 show the results from four different experiments. The total volume of hydrogen produced during all experiments is the same. This is because excess acid and the same mass of magnesium are used.

5 **Fill in tables 5.2 and 5.3 and explain what the data shows.**

6 **Determine the total volume of gas made. Explain why it is the same for both the reactions at 20 °C and 30 °C.**

7 **Predict and sketch what the curved line will look like if the reaction is carried out at 40 °C.**

Volume of gas made after 10 seconds	
Low acid concentration	High acid concentration
	30 cm^3

Table 5.2

Volume of gas made after 10 seconds	
Low temperature	High temperature
	28 cm^3

Table 5.3

Changing the mass

We can alter the rate of reaction by changing the *factors* but the amount of product will remain the same.

However, if we reduce the amount of one reactant used, then less product will be formed.

8 **Look at Figure 5.39. Explain what the data shows about:**

a **the amount of product formed in both reactions**

b **the rate of reaction in both reactions**

9 **Predict and sketch what the curved line will look like if the reaction is carried out with 0.0495 g of magnesium.**

10 **An investigation was carried out into the rate of reaction between solid calcium carbonate, $CaCO_3$, and nitric acid. The mass of calcium carbonate was kept the same as was the temperature. The results are shown in the table below.**

	Time taken to produce 10 cm^3 of CO_2/seconds	Total volume of CO_2 produced/cm^3
Experiment 1	27	40
Experiment 2	14	80

a **Write a balanced equation for the reaction. Include state symbols.**

b **Explain the results of experiments 1 and 2.**

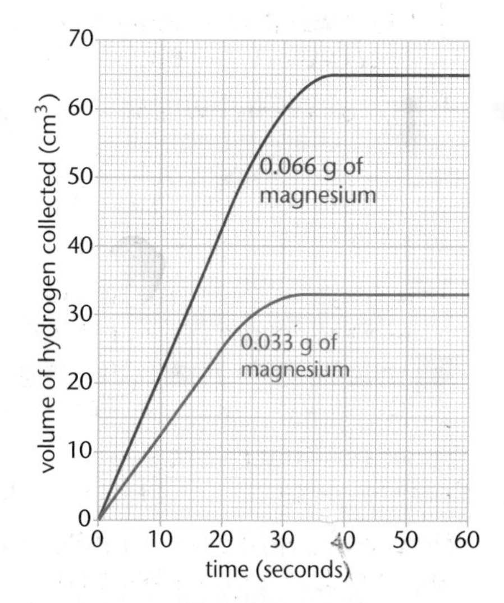

Figure 5.39 Changing the mass of reactant

PRACTICAL

Investigate how changes in concentration affect the rates of reactions by a method involving the production of a gas and a method involving a colour change

KEY WORDS

concentration
volume
rate
tangent

Learning objectives:

- use scientific theories and explanations to develop a hypothesis
- plan experiments to test the hypothesis and check data
- make and record measurements using gas syringes
- evaluate methods and suggest improvements and further investigations.

Rates of reaction are very important for industrial processes and many chemical engineers are involved in controlling rates on large scale processes. Several factors affect the rate of a reaction. Are you able to spot the trends if one variable is changed?

These pages are designed to help you think about aspects of the investigation rather than to guide you through it step by step.

Developing a hypothesis

A number of different skills are needed to carry out a *full* investigation. First a hypothesis needs to be developed from previous knowledge or observations. Consider an investigation that shows the **rate** of reaction of acid with magnesium increasing when the **concentration** of the acid increases. Your idea, from this conclusion, may be 'the more particles of acid there are, the more chances of successful particle collisions taking place'.

After this evaluation another question could be: what other variable change might increase the number of successful particle collisions? Could increasing temperature do this?

DID YOU KNOW?

You can make an experiment so that a colour change happens at an exact time, by adjusting the concentration of the solutions.

1. **Suggest a hypothesis.**
2. **The reaction you would follow is acid on carbonate. Identify the variable that should be changed.**
3. **Suggest which variables need to be kept the same.**

Planning the investigation: Figure 5.40 shows a diagram of the apparatus that might be used to follow the reaction of acid on magnesium.

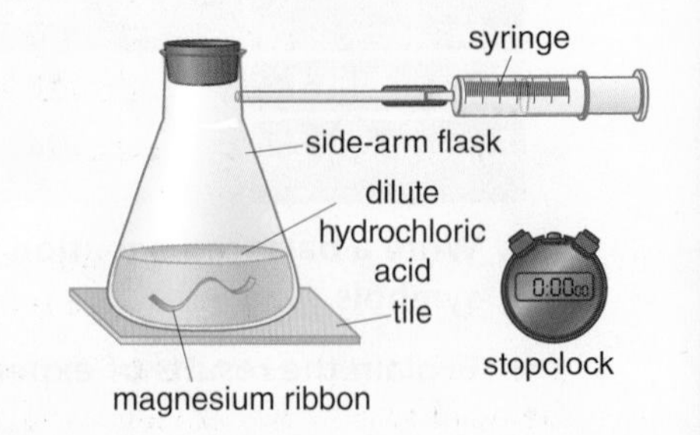

Figure 5.40

Think about these questions:

4. **Explain why a gas syringe is used.**
5. **What would need to happen quickly when the magnesium is put into the flask?**

Measurements need to be accurately made, and the data collected, recorded and processed. Graphs of the results need to be drawn and, if possible, **tangents** to the curves drawn to calculate rates at different points of the reaction.

Analysing the results

Sam and Alex had investigated the effect of temperature on the reaction between acid and calcium carbonate (using the same apparatus as above) and had gathered these results:

Time in s	Volume of gas collected from acid at specific temperatures in cm^3						Rate of reaction in cm^3/s at 20 °C
	20 °C	20 °C	30 °C	30 °C	40 °C	40 °C	
0	0	0	0	0	0	0	
10	12	11	24	45	49	50	
20	23	21	45	47	64	64	
30	34	35	62	61	64	64	
40	52	51	64	64	64	64	
50	63	62	64	64	64	64	
60	64	64	64	64	64	64	

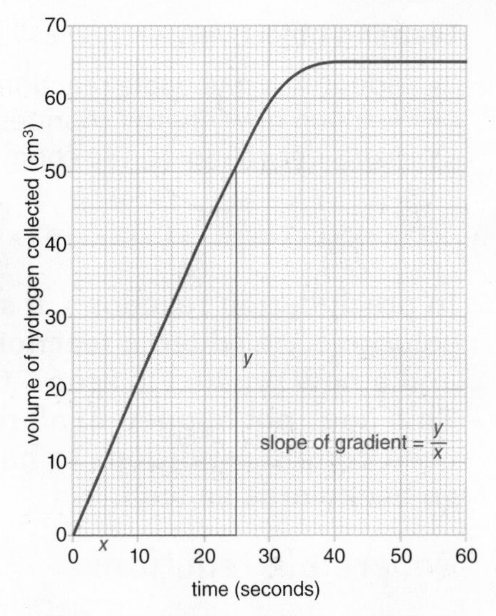

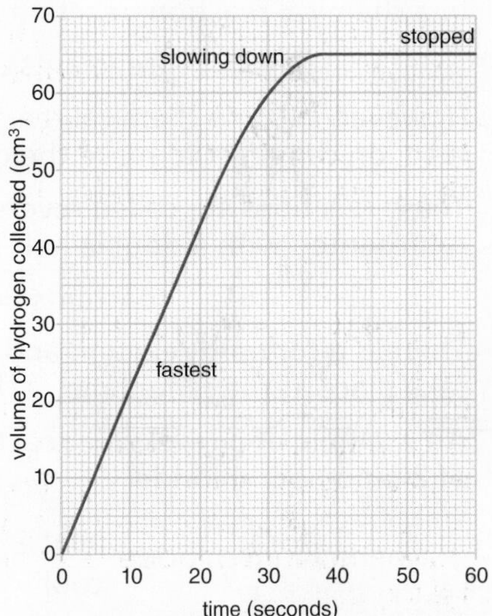

6. **Identify the relationship that Sam and Alex were exploring.**
7. **Identify the anomaly in the data. Suggest what Sam and Alex should do with this data point.**
8. **Plot three lines on a graph of time against volume of gas collected, one line for each temperature.**
9. **Suggest what their conclusion might be.**
10. **Work out the average rate of reaction over 40 seconds for the 20°C reaction.**

Evaluating the experiment

Sam and Alex were testing the hypothesis: 'does an increase in temperature increase the rate of reaction between acid and calcium carbonate?' Using their conclusion in question 9, discuss how well the data supports their conclusion.

HIGHER TIER ONLY

11. **Explain whether the evidence supports their conclusion or whether could someone use the same data to support a different conclusion.**
12. **Explain how they made sure that their data was *reliable*.**
13. **Explain how the apparatus for their investigation and their method of use ensured that their findings were *valid*.**
14. **Suggest one way the apparatus or method could have been improved.**
15. **Suggest one way they could find out if their specific hypothesis could be extended more generally.**
16. **Explain, using the particle model, why their conclusion may be applied to this reaction and other reactions.**

KEY SKILL

It is important to be very methodical in recording results. Tables need to be clearly labelled.

Reversible reactions and energy changes

Learning objectives:

- identify a reversible reaction
- explain how energy changes occur in reversible reactions
- consider changing the conditions of a reversible reaction.

KEY WORDS

backwards
exothermic
forwards
reversible

In precipitation reactions, it seems that when the two solutions are added a reaction happens immediately and a new product is seen – the precipitate is made. This does not happen in all reactions. The reaction may need certain conditions to happen and the reaction may go backwards as well.

Reversible reactions

In a chemical reaction, reactants react to make products.

$$\text{reactants} \rightarrow \text{products}$$

However, in some chemical reactions, the products of the reaction can react to make the original reactants again.

These reactions are called **reversible** reactions and are represented with a double half-headed arrow.

$$\text{reactants} \rightleftharpoons \text{products}$$

There is a **forward** reaction and a **backward** reaction. These can take place at the same time.

An example of a reversible reaction is the heating and cooling of ammonium chloride.

$$\text{ammonium chloride} \underset{\text{cool}}{\overset{\text{heat}}{\rightleftharpoons}} \text{ammonia + hydrogen chloride}$$

The white ammonium chloride decomposes to two colourless gases. When the gases are cooled they re-form into white ammonium chloride.

Figure 5.41 Heating ammonium chloride is a reversible reaction

1. **Suggest whether the following are reversible or irreversible.**
   - **a** **Wood burning**
   - **b** **Ice melting**
   - **c** **Formation of ammonia from hydrogen and nitrogen**
2. **Explain whether dissolving sodium chloride in water is reversible or not.**

Energy changes in reversible reactions

Another example of a reversible reaction is the heating of blue copper sulfate.

At first it is a hydrated blue crystal. When heated it loses the water of crystallisation to become white anhydrous copper sulfate. The reaction is represented by:

hydrated copper sulfate(blue) $\underset{\text{exothermic}}{\overset{\text{endothermic}}{\rightleftharpoons}}$ anhydrous copper sulfate (white) + water

The forward reaction takes place when the crystals are heated. Energy is transferred in so it is an endothermic reaction.

The backward reaction takes place when the water is added to the anhydrous copper sulfate. Energy is transferred out so it is an exothermic reaction.

The same amount of energy is transferred out as the amount of energy transferred in.

3 Explain what the term 'endothermic' means.

Figure 5.42 Blue copper sulfate turns white on heating and returns to blue if water is added

DID YOU KNOW?

Anhydrous means 'without water'.

HIGHER TIER ONLY

Changing direction

A reversible reaction has a forward reaction and a backward reaction that take place at the same time.

$$A + B \rightleftharpoons C + D$$

What do you think happens if the forward and backward reactions take place at the same rate? We will meet this idea in the next pages.

However, if a reversible reaction is **exothermic** in one direction, it is endothermic in the opposite direction. The same amount of energy is transferred in each case.

The direction of reversible reactions can be changed. This can be done by changing the conditions.

For example, nitrogen and hydrogen react to form ammonia:

$$\text{nitrogen} + \text{hydrogen} \rightleftharpoons \text{ammonia}$$

$$N_2 + 3H_2 \rightleftharpoons 2NH_3$$

The ammonia made in the forward reaction breaks down to form nitrogen and hydrogen again.

The pressure can be increased and the reaction will move to make more product. If the pressure is decreased there will be more backward reaction instead.

4 *Suggest* why increasing the pressure in the reversible reaction

$$N_2 + 3H_2 \rightleftharpoons 2NH_3$$

may favour the forward reaction. Think about particle theory.

5 An iron catalyst is used in the Haber process. Describe how it affects the energy change for the forward and reverse reactions.

Equilibrium

Learning objectives:

- describe how equilibrium is reached
- explain what happens to the forward and reverse reactions
- predict the effects of changes on systems at equilibrium.

KEY WORDS

equilibrium
counteract
reverse reaction
Le Chatelier's principle

In a reversible reaction, there is a forward reaction and a backward reaction. Sometimes these take place at the same time. The symbol $\rightleftharpoons$ is used to show that a reaction is reversible. When a balance is reached between the rates of these competing reactions, the reaction has reached equilibrium.

Reaching equilibrium

In a reversible reaction there is a forward reaction and a **reverse reaction**. What would be the outcome if these were happening at the same time?

Equilibrium is reached when the forward and reverse reactions occur at exactly the *same rate* but the apparatus must prevent the escape of the reactants and the products. This is called a closed system.

1 Explain why an open system with a gas escaping will not reach equilibrium.

2 Explain how a saturated solution of salt is at equilibrium in terms of rate.

DID YOU KNOW?

Nitrogen and hydrogen are used to make ammonia in the Haber process. Ammonia is needed for another process to make fertilisers for growing better crops.

DID YOU KNOW?

In the Haber process an equilibrium is established between the reactants nitrogen and hydrogen and the product ammonia.

Equilibrium position

At equilibrium:

- the rate of the forward reaction equals the rate of the backward reaction
- the concentrations of reactants and products do not *change*.

However, when the backward and forward rates balance and are equal, the *concentrations* of the reactants and products do not need to be equal. There is usually more of one than the other.

Reactants	$\rightleftharpoons$	Products
If the concentration of reactants is greater than the concentration of products, we say that the position of equilibrium is on the *left*.		If the concentration of the reactants is less than the concentration of the products, the position of equilibrium is on the *right*.

DID YOU KNOW?

In the Haber process the position of equilibrium changes depending on the conditions of temperature and pressure.

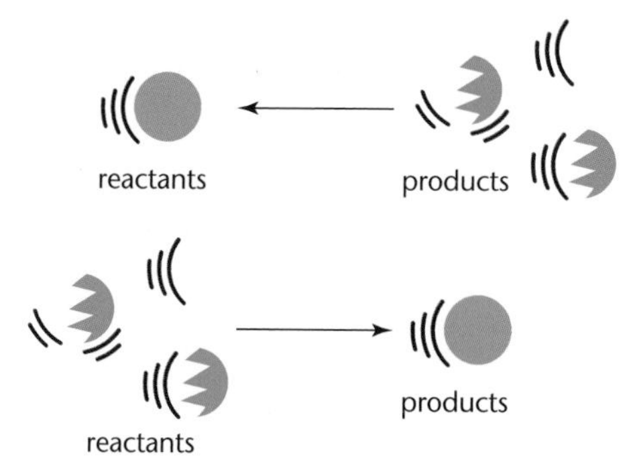

Figure 5.43 Position of equilibrium

Remember: this only works if the chemicals cannot escape.

Initially, the forward reaction rate is fast, but then it slows as the reactants are used up.

At the same time, the rate of the backward reaction increases as more products are available to react. Eventually the backward reaction has the same rate as the forward reaction. Equilibrium has been reached.

3 The reaction: A+ B $\rightleftharpoons$ C + D is at equilibrium. The concentration of A + B is higher than the concentration of C + D. Suggest where the *position* of equilibrium lies.

HIGHER TIER ONLY

Le Chatelier's principle

The relative amounts of all the reactants and products at equilibrium depend on the conditions of the reaction.

If a system is at equilibrium and a change is made to any of the conditions, then the system responds to **counteract** the change. This is known as **Le Chatelier's principle**.

The effects of changing conditions on a system at equilibrium can be predicted using Le Chatelier's principle.

The reaction: A + B $\rightleftharpoons$ C + D is at equilibrium.

If the concentration of A and/or B is increased then the system will respond by making more C and D and so reducing the amount of A and B. The system will respond to the change by counteracting it.

4 Explain what will happen to the reaction

$$A + B \rightleftharpoons C + D$$

if it is in equilibrium and then the amount of C and D is increased.

5 Hydrochloric acid was added to bismuth(III) oxychloride and a clear solution formed. Describe and explain what you would see if water was added followed by hydrochloric acid.

$$BiOCl(s) + 2\ HCl(aq) \rightleftharpoons BiCl_3(aq) + H_2O(l)$$

DID YOU KNOW?

The Contact process is a reversible reaction to produce sulfur trioxide, needed to make sulfuric acid, from sulfur dioxide and oxygen. Changing conditions can alter the amount of product formed.

KEY INFORMATION

You will find out about how conditions change the position of equilibrium in the following topics.

Changing concentration and equilibrium

Learning objectives:

- identify reactants and products in a reversible reaction
- explain how changing concentrations changes the position of equilibrium
- interpret data to predict the effect of a change in concentration.

KEY WORDS

equilibrium position
concentration
dinitrogen tetroxide
reversible reaction

Reversible reactions are used in making the propellant for space rockets and for making fertilisers. These reactions can make more products by changing the position of equilibrium. By removing the product as it forms, more can be made. This means that space rockets can be powered and more food can be grown.

HIGHER TIER ONLY

Removing products

Some rocket propellants use **dinitrogen tetroxide** N_2O_4. Dinitrogen tetroxide is made from nitrogen dioxide (NO_2) in a **reversible reaction**.

$$2NO_2(g) \rightleftharpoons N_2O_4(g)$$

NO_2 + NO_2 $\rightleftharpoons$ N_2O_4

Figure 5.45 Two molecules rearrange to form one larger molecule

Figure 5.44 Dinitrogen tetroxide is made from nitrogen dioxide and used as a propellant

An equilibrium is set up in this reaction.

According to Le Chatelier's principle, if the **concentration** of one component is altered the equilibrium will respond to counteract the change.

If $N_2O_4(g)$ is removed as it is formed, the concentration of it decreases so the equilibrium moves to the *right*, as more is made to replace it.

If more NO_2 (the reactant) is introduced, the concentration of it increases so again the equilibrium moves to the right, making more product N_2O_4

When a product of a reaction is a gas and it is removed this in effect reduces the concentration of the product. So the position of the equilibrium moves to the right.

KEY INFORMATION

Remember that the (g) following the formula is a state symbol, and stands for gas.

1. **In the reaction**

   $2SO_2(g) + O_2(g) \rightleftharpoons 2SO_3(g)$

   Predict which gas would need to be removed to make more product.

Changing concentrations

If the concentration of one of the reactants or products is changed, the system is no longer at equilibrium and the concentrations of all the substances will change until equilibrium is reached again.

- If the concentration of a reactant is increased, more products will be formed until equilibrium is reached again.
- If the concentration of a product is decreased, more reactants will react until equilibrium is reached again.

An increase in the amount of the product is the goal of manufacturers of bulk chemicals and so methods to increase the amount of product are important. Many of these reactions involve equilibrium reactions – one example is the Haber process.

Making ammonia and sulfuric acid

The Haber process is used to make ammonia (NH_3), a starting chemical for the production of fertilisers.

The reaction is:

$$N_2(g) + 3H_2(g) \rightleftharpoons 2NH_3(g)$$

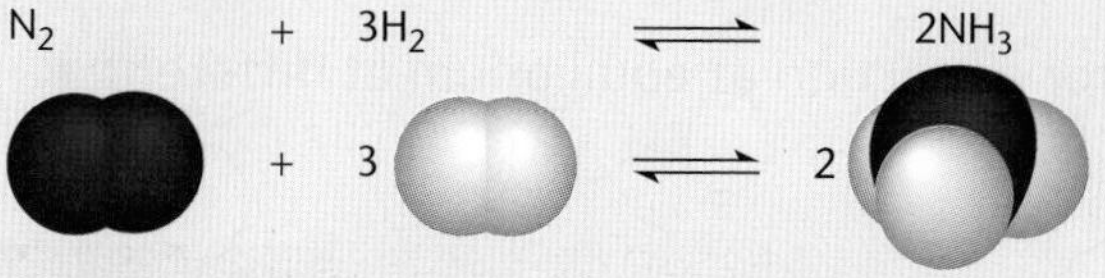

N_2 + $3H_2$ $\rightleftharpoons$ $2NH_3$

+ 3 $\rightleftharpoons$ 2

Figure 5.46 Four molecules rearrange to form two molecules

DID YOU KNOW?

The yield of ammonia from this process is very low so the gases are recycled to start the process again.

If ammonia, (NH_3), is liquefied and removed during the process more nitrogen and hydrogen (the reactants) will react to make more ammonia (the product) and attempt to restore equilibrium.

The Contact process is used to make sulfur trioxide (SO_3), a starting chemical for the production of sulfuric acid to make fertilisers, paints and plastics.

The reaction at equilibrium is:

$$2SO_2(g) + O_2(g) \rightleftharpoons 2SO_3(g)$$

If sulfur trioxide, (SO_3), is removed during the process more sulfur dioxide and oxygen will react to produce more sulfur trioxide by restoring equilibrium.

KEY INFORMATION

You will need to remember details of the Haber process but the Contact process is an extra example.

2 Methanol is produced in industry by the reaction of hydrogen with carbon monoxide. Suggest how more methanol could be produced.

$$CO(g) + 2H_2(g) \rightleftharpoons CH_3OH(g)$$

3 Lead(II) chloride is a white solid. It is sparingly soluble in water and the following equilibrium is set up. Explain what you would observe if sodium chloride was added.

$$PbCl_2(s) \rightleftharpoons Pb^{2+}(aq) + 2Cl^-(aq)$$

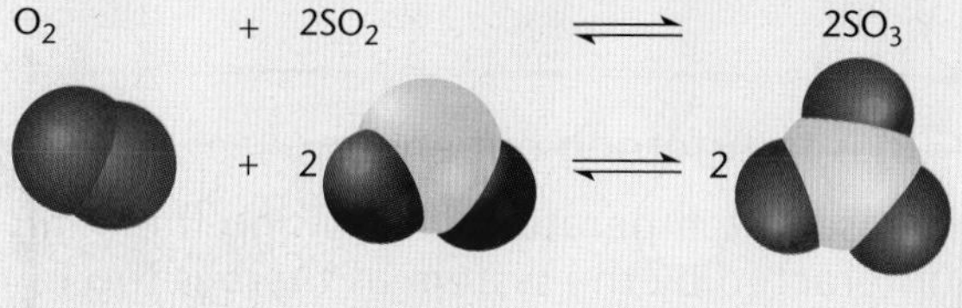

O_2 + $2SO_2$ $\rightleftharpoons$ $2SO_3$

+ 2 $\rightleftharpoons$ 2

Figure 5.47 Three molecules rearrange to form two molecules

Changing temperature and equilibrium

Learning objectives:

- explain how exothermic reactions behave
- explain how endothermic reactions behave
- apply Le Chatelier's principle to reactions in equilibrium.

KEY WORDS

endothermic
equilibrium position
exothermic
Le Chatelier's principle

Exothermic and endothermic reversible reactions behave in different ways when the temperature is increased or decreased. Applying Le Chatelier's principle to each case will explain how the yield of a product can be maximised.

HIGHER TIER ONLY

Changes with increasing temperature

The percentage change in reactants and products for a reaction at equilibrium at different temperatures is shown in Table 5.4.

This reaction is: $A \rightleftharpoons B$

Temperature in °C	Percentage of ***reactants*** at equilibrium	Percentage of ***products*** at equilibrium
20	38	62
30	45	55
40	52	48
50	56	44

Table 5.4

You can see that as the temperature increases the percentage of products decreases.

Is this an **exothermic** reaction or an **endothermic** reaction?

Le Chatelier's principle states: if a system is at equilibrium and a change is made to any of the conditions, then the system responds to *counteract* the change.

If this principle is applied to this reaction, then the temperature increase is causing the reaction to go *backwards.*

KEY SKILLS

Remember to apply Le Chatelier's principle every time there is a change in the system of a reversible reaction.

To make the reaction go forwards (to produce more product) we can predict that the temperature must be lowered.

Why is this?

This will be because the forward reaction ($A \rightarrow B$) is transferring out energy.

As it is transferring out energy any external increase in temperature (i.e. raising the temperature of the system) will need to be counteracted by the system as there is too much energy in the system. So the reaction goes backwards.

We can deduce then, that this forward reaction is an exothermic reaction.

The opposite is true for a forward reaction that is endothermic. The external temperature will need to be raised.

So, in summary for formation of products:

If the temperature of a system at equilibrium is *increased*:	If the temperature of a system at equilibrium is decreased:
the relative amount of products at equilibrium *increases* for an *endothermic* reaction	the relative amount of products at equilibrium *decreases* for an *endothermic* reaction
the relative amount of products at equilibrium *decreases* for an *exothermic reaction.*	the relative amount of products at equilibrium *increases* for an *exothermic* reaction.

1 Use Table 5.4 to predict the percentages of A and B in the equilibrium mixture at (a) 45 °C and (b) 55 °C.

2 Explain why an increase in temperature would favour the making of products in an endothermic reaction.

Example in industrial processes

The Haber process, used to make ammonia, is an exothermic reaction.

$$N_2(g) + 3H_2(g) \rightleftharpoons 2NH_3(g)$$

This means that to increase the *yield* of products the temperature of the reaction needs to be lowered.

For the Contact process that converts sulfur dioxide to sulfur trioxide:

$$2SO_2(g) + O_2(g) \rightleftharpoons 2SO_3(g)$$

Figure 5.48 shows that *lowering* the temperature increases the % conversion by altering the position of the equilibrium.

3 Use Figure 5.48 to explain whether the Contact process is endothermic or exothermic in the forward direction.

4 A substance F is in equilibrium with substance E.

$$E(g) \rightleftharpoons 2F(g)$$

Some data is given in the table below.

	% Conversion to F at different temperatures and pressures		
	1 atmosphere	5 atmospheres	10 atmospheres
20°C	15%	10%	2%
100°C	35%	24%	4%

a Explain the variation in % conversion with temperature.

b Justify whether the reaction is exothermic or endothermic.

c Using Le Chatelier's principle, suggest why % conversion changes with pressure.

DID YOU KNOW?

In the Haber process there is more than one factor to consider so an optimum set of conditions is employed. The example here is a theoretical response.

KEY INFORMATION

You do not need to remember details of the Contact process.

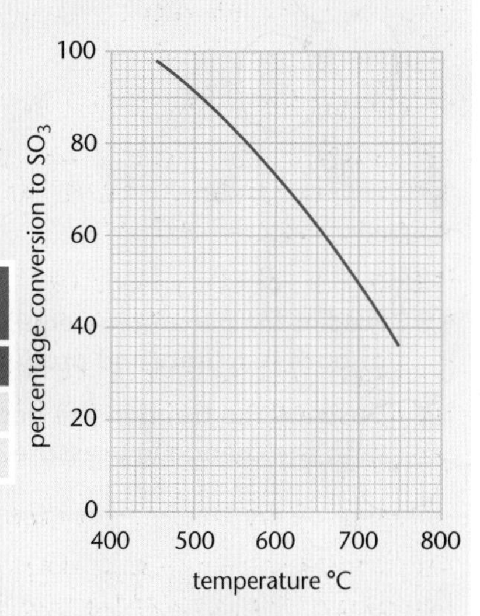

Figure 5.48 Graph showing % sulfur trioxide conversion from SO_2

Changing pressure and equilibrium

Learning objectives:

- predict the effects of changes in pressure
- explain why these effects occur
- interpret data to predict the effect of a change in pressure.

KEY WORDS

compromise
equilibrium position
molecule
pressure

The yield in industrial processes involving gases can be increased by changing the pressure, according to Le Chatelier's principle. The yield of product in all the reactions between gases can be maximised by making changes in all three conditions – concentration, temperature and pressure – but sometimes compromises have to be made.

HIGHER TIER ONLY

Pressure change and yield

When a reversible reaction takes place and reaches equilibrium, the position of the equilibrium shifts if the conditions are changed.

The data in Table 5.5 shows the percentage change in reactants and products for a reaction through a range of **pressures**.

Pressure in atmospheres	Percentage of reactants at equilibrium	Percentage of products at equilibrium
1	33	67
5	41	59
10	55	45
15	62	38

Table 5.5

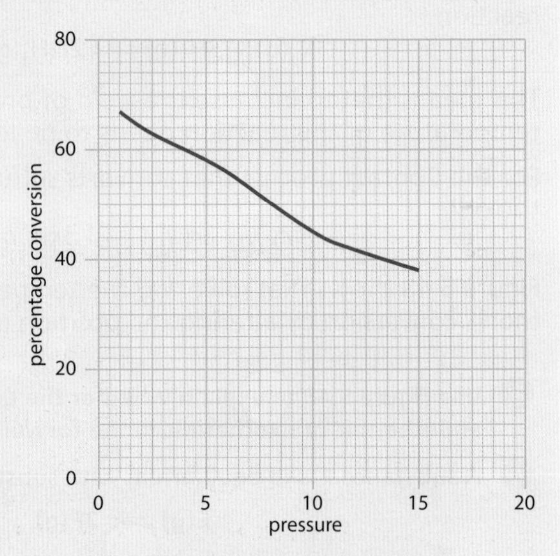

Figure 5.49 Graph of pressure against percentage of product

1. **Predict the percentage equilibrium composition of both reactants and products at 7.5 atmospheres.**
2. **Describe the trend in the percentage of products with the change in pressure.**

Examples of pressure change in processes

In the Haber process, the reaction between nitrogen and hydrogen to form ammonia is reversible:

$$N_2(g) + 3H_2(g) \rightleftharpoons 2NH_3(g)$$

The number of molecules as reactants is 4 (3 + 1)	The number of molecules as products is 2

In summary for gaseous reactions at equilibrium:

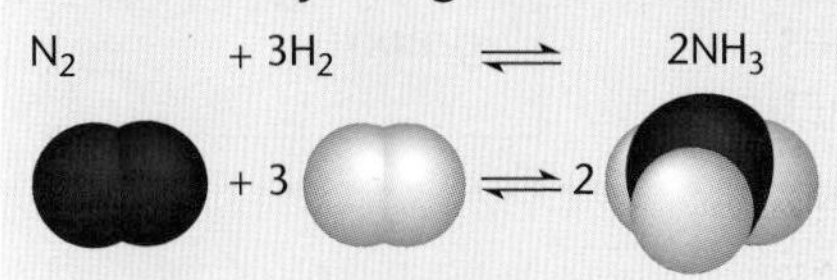

Figure 5.50 Four molecules react and become two molecules of product

an *increase* in pressure causes the **equilibrium position** to shift towards the side with the *smaller* number of molecules as shown by the symbol equation for that reaction

a decrease in pressure causes the equilibrium position to shift towards the side with the *larger* number of **molecules** as shown by the symbol equation for that reaction.

3 Explain the change in pressure needed to increase the percentage product in these gaseous reactions.

a $2A + 3B \rightleftharpoons 2C$

b $A + B \rightleftharpoons 3C$

c $A + B \rightleftharpoons 2C + D$

Reaching a compromise

For each process the conditions that can be changed are concentration, temperature and pressure. In practice all three have to be considered together and a **compromise** may be needed.

To obtain the most economic yield for the Haber process, the reaction is carried out at high pressure, as there are four gas molecules on the left of the equation and two on the right, so high pressure forces the equilibrium further to the right which increases the yield.

The temperature used, however, is a compromise. The forward reaction is exothermic, so high temperatures reduce the yield and drive the equilibrium to the left. However, high temperatures increase the rate of reaction, so the product is made faster.

4 Compare and explain the similarities and differences between the equilibrium conditions used in the Haber and Contact processes. The equilibria are both exothermic in the forward direction.

	Haber Process	Contact Process
Temperature / °C	400	400
Pressure / atmospheres	200	1–2

KEY INFORMATION

The changes that happen follow Le Chatelier's principle: if a system is at equilibrium and a change is made to any of the conditions, then the system responds to counteract or reverse the change.

DID YOU KNOW?

A catalyst is a small amount of a chemical that speeds up a particular reaction, without being used up. It does not alter the equilibrium position or the yield of product but makes it feasible or produces more every second.

MATHS SKILLS

Use the slope of a tangent as a measure of rate of change

Learning objectives:

- draw graphs from numeric data
- draw tangents to the curve to observe how the slope changes
- calculate the slope of the tangent to identify the rate of reaction.

KEY WORDS

tangent
rate
slope
graph

We have found that reactions go faster or slower by changing the temperature, concentration of a solution or the size of solid pieces. Can you read graphs well to see the effects by looking at the slopes of the lines? Can you calculate the slopes of those lines? Can you draw tangents at any point of the reaction lines to calculate the rate?

Drawing graphs from data

When you have investigated the reaction between magnesium and acid you will have collected hydrogen. You will have measured the volume of hydrogen given off every 10 seconds for 60 seconds.

This is an example of the data you would collect:

Time in seconds	0	10	20	30	40	50	60
Volume of H_2 in cm^3	0	22	42	59	66	66	66

1. **Draw a graph of the data in the table. Mark where the reaction is: a fastest b slowing down c stopped**

2. **Explain in terms of collision theory when the rate of reaction is: a highest b zero**

Finding the gradient

To find the gradient, m, of a straight line, you need to find the coordinates of two points (x_0, y_0) and (x_1, y_1). The gradient of the line is the change in y (the vertical distance, compared to the change in x (the horizontal distance). This is calculated by dividing $(y_1 - y_0)$ by $(x_1 - x_0)$.

So $m = (y_1 - y_0) \div (x_1 - x_0)$.

Measure the value of y up the line (here it is volume of H_2) and the value is 50.0 cm^3.

Then measure the value of x along the axis (here it is time) and the value is 25 seconds.

To find the gradient divide y by x. So in this graph:

$m = (y_1 - y_0) \div (x_1 - x_0)$

$m = (50 - 0) \div (25 - 0)$

$m = \frac{50}{25}$

$m = 2$

Determine the units from those given on the axes so $\frac{y}{x} = \frac{50}{25} = 2$ cm³/second.

Quantity of reactant or product	Unit of quantity	Unit of rate
Mass	g	g/s
Volume	cm^3	cm^3/s
Moles	mol	mol/s

DID YOU KNOW?

Sometimes it is easier to measure the *mass* of a product formed. The rate of reaction is then measured in g/s or g/min

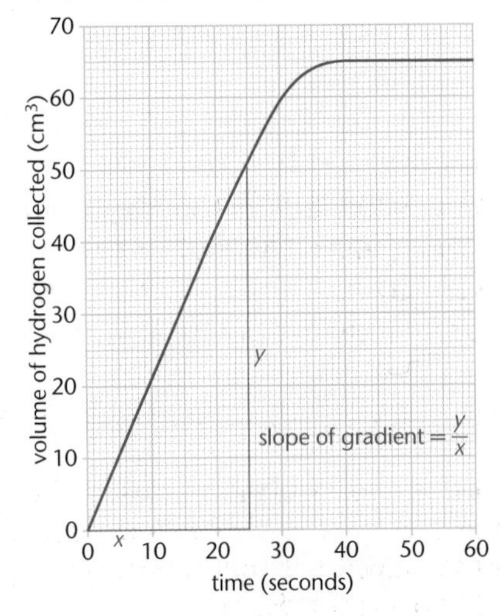

Figure 5.51 Calculating gradients

3 Look at Figure 5.51.

a Work out the volume of gas that was collected between 10 and 30 seconds?

b Calculate the rate of reaction during this 20 second interval.

c Calculate the gradient between 0 and 10 seconds.

d State why the gradient from part b) is different from that calculated in part c).

Drawing tangents to a curve

It is usually necessary to draw two graphs to compare the rate of a reaction when factors have changed. The results might look like Figure 5.52.

Gradients can be calculated for both lines and the rates can be compared.

Look at the blue line. There is some slowing down in the reaction. To measure the rate of reaction during this period a *tangent* to the curve needs to be drawn. Choose a point such as 25 secs and put your ruler so that it is at the best position to follow the curve direction *at that point*. Lengthen this tangent line so that it hits convenient major grid-lines. Draw a triangle using the tangent line as the hypotenuse. Measure the value of y and the value of x. Calculate the gradient $\frac{y}{x}$ at that point.

REMEMBER!

It is very important to check the units you are using.

4 Draw construction lines to calculate the gradients of the blue and red lines of Figure 5.52. What is the rate of reaction calculated from: a the blue line b the red line and comment on your findings.

5 On Figure 5.52 draw tangents to the curve at 22 seconds and 28 seconds. Calculate the rate of reaction at both points and comment on your findings.

6 Rate data from a reaction between a carbonate and an acid was plotted. Two tangents were taken from a volume–time graph. They were 2.7 cm³/minute and 8.5 cm³/minute.

a Explain which tangent was taken closer to the end of the reaction.

b Work out the volume of gas produced in 1 second using the tangent taken nearer the start of the reaction.

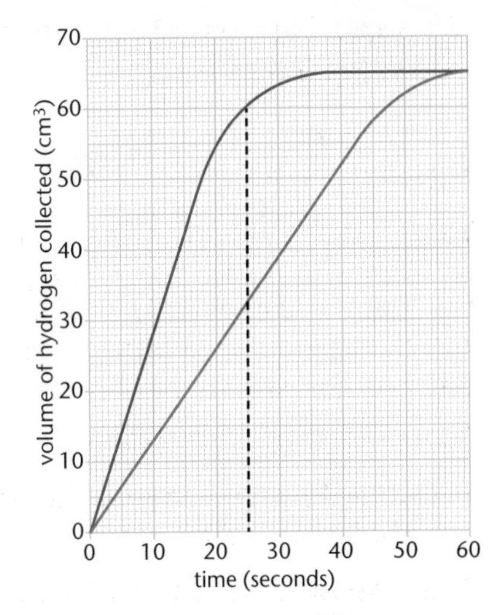

Figure 5.52 Different gradients

Check your progress

You should be able to:

recognise that when a reaction has stopped one of the reactants has been used up →	describe the reactant that is used up first in a reaction as the limiting reactant →	explain the effect of a limiting quantity of a reactant on the amount of products it is possible to obtain, using moles or grams
relate mass, volume and concentration →	calculate the mass of solute in a solution →	relate concentration in mol/dm^3 to mass and volume
describe how to carry out titrations →	calculate the concentrations in titrations in mol/dm^3 and in g/dm^3 →	explain how the concentration of a solution in mol/dm^3 is related to the mass of the solute and the volume of the solution
identify the balanced equation needed for calculating yields →	calculate the theoretical amount of products from the amounts of reactants →	calculate the percentage yield from the actual yield and the theoretical yield
identify the balanced equation for a reaction →	calculate the atom economy of a reaction to form a desired product →	explain why a particular reaction pathway is chosen to produce a product given the atom economy, yield, rate, equilibrium position and usefulness of by-products
explain that the same amount of any gas (in moles) occupies the same volume at room temperature and pressure (rtp) →	calculate the volume of a gas at rtp from its mass and relative formula mass →	calculate volumes of gases from a balanced equation and a given volume of a reactant or product
identify how to measure the amount of gas given off in a reaction →	calculate the mean rate of a reaction →	draw tangents to the curves as a measure of the rate of reaction
identify which factors affect the rate of reactions →	explain how the changes of surface area affect rates →	explain how rates are affected by different factors
analyse experimental data on rates of reaction →	predict the effects of changing conditions on rates of reactions →	explain the effects of changes of factor on rates of reaction using collision theory
identify catalysts in reactions →	explain catalytic action →	explain activation energy
identify a reversible reaction →	describe how equilibrium is reached →	predict the effects of changes on systems at equilibrium
identify reactants and products in a reversible reaction →	explain how changing concentrations changes equilibrium →	interpret data to predict the effect of a change in concentration
explain how exothermic reactions behave if the temperature of systems at equilibrium changes →	explain how endothermic reactions behave if the temperature changes →	interpret appropriate data to predict the effect of a change in temperature on reactions at equilibrium
predict the effects of changes in pressure →	explain why these effects of pressure change occur →	apply Le Chatelier's principle to reactions in equilibrium

Worked example

1 Explain how catalysts work.

They speed up a reaction because they lower the activation energy.

This answer is correct as the description of what catalysts do is linked to the explanation.

2 Draw the equation symbol used to show a reversible reaction.

$\rightleftharpoons$

The answer is correct

3 Draw an energy profile to show the activation energy needed for a reaction to take place.

Exothermic reaction

activation energy

reactants

products

Energy

Reaction

The profile is for a chemical reaction. The activation energy is the *difference between* the energy levels of the reactants and products not the initial energy level.

4 Explain how increasing temperature increases the rate of reaction.

The particles move faster so hit each other more likely.

This answer is worth one mark as a better word would be collide and an explanation is needed. A better answer is the particles move with more energy so there are more successful collisions

5 Look at the table. Fill in the missing number and describe the pattern shown by this reaction at equilibrium. Predict the percentage of product probably formed at 150°C and suggest what the owner of the factory making this product should do.

Temperature in °C	100	200	300	400
% reactants at equilibrium	22	37	52	59
% products at equilibrium	78	63	48	*41*

The higher the temperature the less products are made, so the owner should do this at a low temperature.

The data on the table is completed and the pattern is correctly described. The student has made a good suggestion to use a low temperature but needs to be careful to use the word lower not low. This reaction may not work at a low temperature and may not be fast enough. This time the 'benefit of doubt' is given. The answer is not complete though as they have missed the prediction. An answer of 60%–61% would give the fourth mark.

End of chapter questions

Getting started

1. Which one of the following statements about catalysts is true?
   - **a** They increase the amount of product formed.
   - **b** Large quantities of catalyst are needed.
   - **c** Catalysts are changed at the end of reaction.
   - **d** They increase the rate of reaction.

   1 Mark

2. How does rate change during a chemical reaction?
   - **a** Increases then decreases to zero.
   - **b** Decreases to zero.
   - **c** Stays the same.
   - **d** Increases from zero.

   1 Mark

3. Write down two factors that can be changed to make a reaction go faster. 2 Marks

4. The top is left off of a fizzy drink bottle. Explain why the carbon dioxide gas cannot reach equilibrium with dissolved carbon dioxide. 1 Mark

5. Magnesium reacts with hydrochloric acid solution. Explain why the rate increases when the concentration of hydrochloric acid increases. 1 Mark

6. Acid is added to sodium carbonate in a flask and left on a digital balance. Explain why the mass of flask and contents goes down. 2 Marks

7. A student was expecting to make 2.8 g of a chemical, but instead made 2.2g. Calculate the percentage yield. 2 Marks

Going further

8. One of the characteristics of an equilibrium is that it has to be a closed system. Give two other characteristics. 1 Mark

9. Nitrogen and oxygen gases react together to form nitrogen(II) oxide. Explain why increasing the pressure increases the rate of this reaction. 1 Mark

10. Hydrogen peroxide solution decomposes into oxygen and water. Suggest two ways that the rate of reaction can be followed experimentally. 2 Marks

11. Put these reactions in order of their atom economy by looking at their equations only, the desired product is in bold:
    - **a** i $Zn + 2HCl \rightarrow \mathbf{ZnCl_2} + H_2$ ii $Zn + 2HCl \rightarrow ZnCl_2 + \mathbf{H_2}$
      iii $2Mg + O_2 \rightarrow \mathbf{2MgO}$
    - **b** Check your answer by calculating the atom economy for each reaction.
      [Zn = 65, H = 1, Cl = 35.5, Mg = 24, 0 = 16]

    4 Marks

12. The activation energy for a reaction was measured with and without a catalyst. The values were 72 kJ/mol and 55 kJ/mol. Explain which value was the value for the catalysed reaction. 2 Marks

More challenging

13. Explain why the combustion of natural gas in a Bunsen burner cannot reach equilibrium. 1 Mark

14. In industry, there is a high risk of an explosion from flour, coal and other types of combustible dust. Suggest why. 1 Mark

15 Calculate the volume of gas occupied by 0.75 moles at rtp (Molar volume is 24 dm^3 at room temperature and pressure).

2 Marks

16 In an experiment a student reacted aqueous hydrochloric acid with solid potassium carbonate. The carbon dioxide was collected in a gas syringe.

Temperature °C	Volume of gas collected in 10 seconds cm^3		
18	30	32	31
23	40	45	44
28	59	60	58
33	73	76	72

a Explain which variable was being investigated.
b Suggest one variable which needs to be controlled.
c Identify any anomalous results.
d Describe the trend in the results.

4 Marks

Most demanding

17 Hydrogen and chlorine gases react violently to form hydrogen chloride. Explain the effect of decreasing temperature on the rate of this reaction.

2 Marks

18 The percentage yield of ammonia under different temperature and pressure conditions is given in the table.

Pressure/atm Temperature °C	200	300	400	500
350	43	56	64	71
400	32	36	49	58
450	21	25	35	40
500	15	18	22	28

Explain why a low temperature and a high pressure are chosen using the data in the table.

2 Marks

19 The Haber process is used in industry to make ammonia.

$$N_2 + 3H_2 \rightarrow 2NH_3$$

a Calculate the mass of nitrogen needed to make 1700 g of NH_3.
b Work out the mass of ammonia in 6dm^3 of ammonia gas. (1 mole of any gas occupies 24 dm^3 at rtp)
c Calculate the number of hydrogen molecules in 0.005 moles of hydrogen, H_2. (Avogadro's constant is 6.02×10^{23} mol^{-1}.).

6 Marks

Total: 38 Marks

GLOBAL CHALLENGES

IDEAS YOU HAVE MET BEFORE:

EXTRACTION OF METALS

- Gold is dug up as unreacted metal from the ground.
- Heating the green ore, malachite, turns it to copper oxide.
- Furnaces for melting iron were very important for making steel.

REUSING AND RECYCLING

- More household waste is recycled to reduce the landfill sites.
- Bottles and metal cans are collected separately for recycling.
- Houses are better insulated to cut down energy needs.

PROPERTIES OF HYDROCARBONS

- Bunsen burners provide either a blue flame or a yellow flame.
- We need fuels for heating and transport.
- Carbon atoms join with hydrogen atoms to make molecules.

VOLCANOES ERUPT PRODUCING GASES

- Volcanoes form on gaps between tectonic plates.
- Volcanoes produce lava and many gases.
- The Earth was very hot when it was formed and is cooling.

USING LESS FOSSIL FUEL CUTS DOWN POLLUTION

- Cars, ships and planes use fossil fuels to transport us.
- We use fossil fuels to heat buildings and drive machinery.
- Burning fossil fuels causes air pollution which is bad for health.

6

IN THIS CHAPTER YOU WILL FIND OUT ABOUT:

WHY ARE SOME METALS EXTRACTED BY REDUCTION WITH CARBON?

- Less reactive metals can be extracted by reduction with carbon.
- Reduction is the removal of oxygen, oxidation is gaining oxygen.
- Oxidation and reduction can be represented by ionic equations.

REDUCING RESOURCE WASTE

- Recycling reduces use of finite resources, saving for the future.
- Environmental impact is lessened by reducing energy consumption.
- Reducing corrosion saves valuable resources such as metals.

CRUDE OIL AND HYDROCARBONS

- How is crude oil separated?
- What are the hydrocarbons that make up crude oil?
- How does the size of hydrocarbons affect flammability?

WHAT WAS THE EARTH'S EARLY ATMOSPHERE?

- The early atmosphere arose from gases from volcanoes.
- Water vapour condensed to form the oceans.
- There was a high percentage of carbon dioxide and no oxygen.

HOW CAN WE REDUCE THE EFFECT OF HUMAN ACTIVITY?

- We can reduce the use of fossil fuels and use renewable energy.
- We can use sunlight more directly for energy needs.
- We can use resources more efficiently and fairly.

Extraction of metals

Learning objectives:

- identify substances reduced by loss of oxygen
- explain how extraction methods depend on metal reactivity
- interpret or evaluate information on specific metal extraction processes.

KEY WORDS

iron(III) oxide
reactivity
reduction
reduction with carbon

There has been a rapid rise in the need for iron to make steel due to urbanisation in the last decade. Which country is driving this need for more metal extraction?

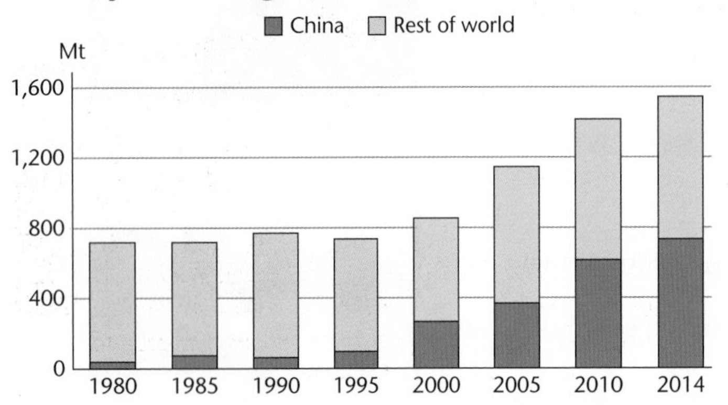

Figure 6.1 Graph of increase in steel consumption. What has happened to consumption by China?

Figure 6.2 Haematite, Fe_2O_3, is the common ore of iron

Reduction

Metals such as gold are found in the Earth as the metal itself because it is unreactive. Most metals, however, are found as compounds because they react with other elements. Chemical reactions are needed to extract the metal. The reactions needed depend on the **reactivity** of the metal.

Metals more reactive than carbon need to be extracted by electrolysis.	For example:	potassium sodium calcium magnesium aluminium
Metals less reactive than carbon can be extracted from their oxides by reduction with carbon.	For example:	zinc iron lead

Figure 6.3 Zinc blende, ZnS, is the common ore of zinc

Reduction involves the loss of oxygen.

In the reactions needed to extract zinc, two stages are needed.

a to convert the ore zinc blende (ZnS) to zinc oxide
b to convert the zinc oxide to zinc

The second stage is the **reduction** of zinc using carbon.

zinc oxide + carbon → zinc + carbon dioxide

1 **In the second stage of extracting, zinc oxide is changed to zinc as in the equation above. State which substance is oxidised and which is reduced.**

Reduction using carbon

To extract iron...

Iron ore is reduced, which means taking away oxygen.
This is done with a reducing agent.
The reducing agent used is carbon monoxide.
Carbon monoxide is made by heating carbon.

In industry the process has many steps but put simply:

carbon + oxygen → carbon monoxide

iron oxide + carbon monoxide → iron + carbon dioxide

$$2C + O_2 \rightarrow 2CO \quad \text{(i)}$$

$$Fe_2O_3 + 3CO \rightarrow 2Fe + 3CO_2 \quad \text{(ii)}$$

2 **In equation (i) explain which substance has been oxidised.**

3 **In equation (ii) explain which substance has been oxidised and which substance has been reduced.**

4 **The ore cuprite contains copper (II) oxide (CuO) and can be smelted with carbon. The ore is mixed with coal and roasted in a furnace. Copper metal and carbon monoxide, CO, are produced.**

a **Write a balanced equation for the reaction.**

b **State what is happening to the copper(II) oxide and carbon.**

HIGHER TIER ONLY

Ionic equations

In the extraction of zinc, the equations are:

zinc sulfide + oxygen → zinc oxide + sulfur dioxide

then

$$2ZnO + C \rightarrow 2Zn + CO_2$$

The half equation for zinc ions converting to zinc is:

$$2Zn^{2+} + 4e^- \rightarrow 2Zn$$

In the extraction of iron, the half equation for the final reaction (ii) is:

$$2Fe^{3+} + 6e^- \rightarrow 2Fe$$

5 **Write a full balanced symbol equation for the conversion of zinc sulfide, ZnS, to zinc oxide, ZnO, and sulfur dioxide, SO_2.**

6 **The iron half equation is derived from reaction (ii) for the extraction of iron. Write this half equation.**

REMEMBER!

You do not need to know the details of zinc extraction or details of the processes used in the extraction of iron. You need to know about oxidation and reduction.

Figure 6.4 Molten iron extracted from iron ore

KEY INFORMATION

This is done by reacting the iron ore 'haematite', which is iron oxide, with carbon monoxide.

DID YOU KNOW?

Most iron extracted is turned into steel, by adding carbon and other elements. Steel, an alloy, is less easily corroded.

Using electrolysis to extract metals

Learning objectives:

- explain the process of the electrolysis of aluminium oxide
- explain why a mixture is used and the anode needs constant replacement
- write half equations for the reactions at the electrodes.

KEY WORDS

aluminium
bauxite
carbon anode
cryolite

Aluminium is a very useful metal due to its low density. It is used in alloys for the aircraft industry and in the drinks industry for drinks cans. It is very expensive to manufacture but easy to recycle.

Extracting the more reactive metals

Metals such as iron and zinc can be extracted from their ores using reduction by carbon. However, some metals are too reactive to be extracted by reduction with carbon or the metal reacts with carbon. These metals are found towards the top of the reactivity series.

These metals can be extracted from molten compounds using electrolysis.

Large amounts of energy are used in the extraction process to melt the compounds and to produce the electrical current for electrolysis. As a result, these metals can be very expensive.

Aluminium is one of the metals extracted in this way. Although it is very abundant in the Earth's crust it is very expensive to extract. This is why there are so many recycling schemes to encourage us to recycle aluminium cans.

Figure 6.5 Recycling aluminium

1 **Explain why sodium has to be extracted by electrolysis rather than by reduction with carbon.**

Manufacturing aluminium by electrolysis

Aluminium is manufactured by the electrolysis of its ore, **bauxite**.

The ore bauxite is a form of aluminium oxide. As the purified ore melts at a very high temperature it is mixed with **cryolite**, as the mixture melts at a lower temperature. It is melted in a steel electrolysis cell.

The molten mixture of aluminium oxide and cryolite is electrolysed in a cell that is a carbon lined cathode with suspended **carbon anodes**.

Molten aluminium oxide contains positive aluminium ions. The positive ions migrate towards the negative electrode (cathode) and are discharged at the electrode making a molten metal.

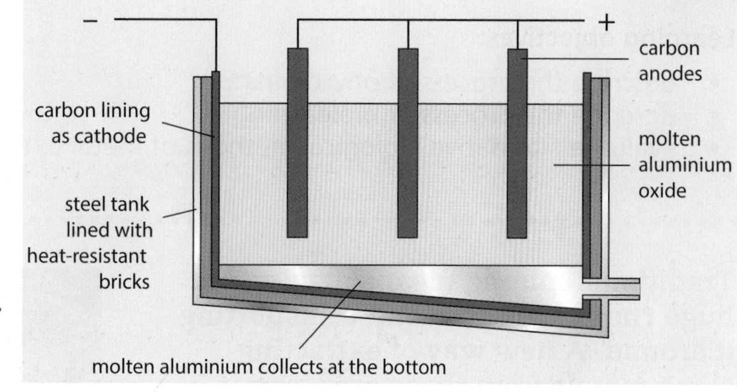

Figure 6.6 Manufacturing aluminium by electrolysis

The oxide ions are negative ions which migrate to the positive electrode (anode). The oxide ions are discharged as oxygen molecules. The positive electrode (anode) is made of carbon. The anodes, which are suspended in the melt, react with the oxygen evolved to produce carbon dioxide. The carbon anodes are continually reacting and so must be regularly replaced.

2. **Describe what happens at the two electrodes during the electrolysis of aluminium oxide to make aluminium.**
3. **The cost of making aluminium is very high. Suggest a reason, other than the cost of materials, why it is so expensive.**

HIGHER TIER ONLY

Electrode half equations

$2Al_2O_3 \rightarrow 4Al + 3O_2$

Aluminium ions are positive, Al^{3+}, so they migrate to the negative cathode. Here they gain electrons.

$4Al^{3+} + 12e^- \rightarrow 4Al$

or per ion is:

$Al^{3+} + 3e^- \rightarrow Al$

Oxide ions are negative O^{2-}, so they migrate to the positive anode. Here they transfer electrons to the anode.

$6O^{2-} - 12e^- \rightarrow 3O_2$

As the anodes are made of carbon operating at high temperature they react with the oxygen being evolved to make carbon dioxide.

$3C + 3O_2 \rightarrow 3CO_2$

4. **Describe the flow of electrons as aluminium oxide is electrolysed.**
5. **Write two half equations for extraction of another reactive metal, calcium, from calcium chloride $CaCl_2$ by electrolysis.**
6. **Write the overall equation and the two half equations for the electrolysis of molten potassium fluoride.**

KEY INFORMATION

Aluminium makes a 3^+ ion and oxide a 2^- ion, so the compound is Al_2O_3

Alternative methods of metal extraction

KEY WORDS

phytomining
bioleaching
hyperaccumulators
toxic metals

Learning objectives:

- describe the process of phytomining
- describe the process of bioleaching
- evaluate alternative biological methods of metal extraction.

Traditional mining means digging out huge chunks of rocks and transporting it around. A new way of extracting some metals from these ores is to use bacteria. Is this feasible? What if we could clean our discarded sites from the toxic metals left behind? What if we could use plants to do this as well?

Figure 6.7 Traditional forms of mining need huge movements of rock

HIGHER TIER ONLY

Phytomining

The Earth's resources of metal ores are limited. Copper ores are becoming scarce and the price of this valuable metal is rising sharply. Copper's electrical conductivity is high, so we use a lot it for wiring and also in alloys. We do recycle it but we can also extract smaller amounts using other methods.

The new ways of *extracting copper* from *low-grade* ores include **phytomining** and **bioleaching**. These methods avoid the traditional mining methods of digging, moving and disposing of large amounts of rock.

Phytomining uses *plants* to absorb metal compounds. The plants are harvested and then burned to produce ash that contains the metal compounds. Plants will naturally absorb compounds from the soil through their root system. We need some of these chemicals for our good health. There are certain metals that plants will absorb that are poisonous to them and they will die. But there are other plants that will collect these metals in their leaves.

Figure 6.8 Some special plants can be used in phytomining

These plants are called **hyperaccumulators**. They concentrate the **toxic metals** in their tissue.

Nickel is very difficult to extract. We use it a lot, in stainless steel, alloys for coins and in batteries. However, it can be absorbed and stored by a particular *hyperaccumulator*.

If a hyperaccumulator could be planted in soil that had nickel in it and grown so that the metal was absorbed, then it could be harvested and the metal extracted from its leaves.

1. **Explain three reasons why it is important that new ways of extracting copper ore are developed.**
2. **Explain why plants are important in the technique of phytomining.**

Bioleaching

Bioleaching uses *bacteria* to produce *leachate* solutions that contain metal compounds. These can then be processed to obtain the metal. For example, copper compounds in leachate solution can be used to obtain pure copper by two methods:

- displacement using scrap iron
- by electrolysis.

Bioleaching is a cleaner process than the traditional leaching that uses cyanide.

Figure 6.9 Copper compounds extracted by bioleaching can be purified by electrolysis

DID YOU KNOW?

Metals such as iron and copper that have been mined since ancient times leached into the river in Spain that became known as Rio Tinto.

Figure 6.10 The Rio Tinto river

3. **Explain how scrap iron and bacteria can be used to extract copper from waste copper ore.**
4. **Bioleaching is a slow process and a byproduct is sulfuric acid. Explain how this might influence whether or not bioleaching is used.**

Evaluating production

Some bacteria can convert metal sulfides to metal sulfates, which is a technique used for recovering mainly copper. However, the costs are still quite high and do not compare with smelting processes yet. So, although there may be some environmental advantages, the process will also need to have economic advantages too before it is used on a large scale in remote areas.

However, bioleaching could be used for cleaning up toxic metals from old industrial sites that have been contaminated so that new building sites can be claimed back from 'brown sites', instead of using green fields. This would help the sustainability cycle.

KEY SKILLS

You will need to know how to read data tables that will not always give a clear indication of the best method of metal extraction to use.

5. **Suggest why mining and smelting may still be a more cost-effective way of producing copper.**
6. **The annual world production of copper is 1.9×10^7 tonnes. About 20% of this comes from bioleaching. Bioleaching is up to 90% efficient. Calculate the mass of copper produced annually by bioleaching.**

The Haber process

Learning objectives:

- apply principles of dynamic equilibrium to the Haber process
- use graphs to explain the trade-off between rate and equilibrium
- explain how commercially used conditions relate to cost.

KEY WORDS

compromise
Haber process
equilibrium
reversible reaction
optimum conditions

Fertilisers that provide nitrogen to the soil are needed for intensive farming to feed the global population of 7 billion.

The manufacture of ammonia

The **Haber process** is used to manufacture *ammonia*, which can be used to produce nitrogen-based fertilisers. The raw materials for the Haber process are *nitrogen* and *hydrogen*. The purified gases are passed over a *catalyst* of iron at a *high temperature* and a *high pressure*. Some of the hydrogen and nitrogen reacts to form ammonia.

$$\text{nitrogen} + \text{hydrogen} \rightleftharpoons \text{ammonia}$$

The ammonia is cooled. It liquefies and is removed. The remaining hydrogen and nitrogen are *recycled* to start again.

DID YOU KNOW?

Ammonia is needed as the basis for making nitrogenous fertilisers. The Haber process makes industrial ammonia from nitrogen and hydrogen. Is there a more sustainable way to make ammonia?

1 **Name the raw materials needed for the Haber process.**

2 **Describe what happens to the raw materials from start to finish.**

Commercially used conditions

Nitrogen is obtained from the air and hydrogen obtained from natural gas or other sources. The nitrogen has to be extracted from the air by cryogenic distillation (distillation by making the mixture colder, not heating it). This requires a *large amount of energy*. There must be no oxygen left in the nitrogen as it damages the catalyst, so the separation has to be done several times – *this means more energ*y and so more cost.

Hydrogen is produced from natural gas and liquefied petroleum gas after the distillation of crude oil. There must be no sulfur in the feedstock as it damages the catalyst too, so it must be removed. The conversion to hydrogen is done by reacting methane with steam. This takes a *great deal of energy too* and so more cost.

Is there a better way of making ammonia that is more sustainable? Could the hydrogen be generated by electrolysis? Is this efficient? Would it only be cost-effective if the electricity was generated by solar or hydroelectric methods? Can hydrogen be obtained from organic waste at less cost? (It may only be 20% of the cost of electrolysis and solve a problem of waste going to landfill).

3 **Describe how the nitrogen for the Haber process is obtained.**

4 **Describe how hydrogen is obtained for the Haber process.**

5 **Suggest, with a reason, a substance that could be electrolysed to produce hydrogen.**

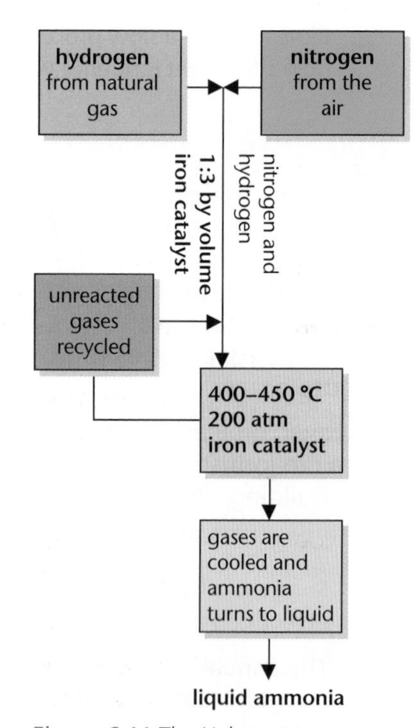

Figure 6.11 The Haber process

Equilibrium versus rate

The reaction for making ammonia is **reversible** so some of the ammonia produced breaks down into nitrogen and hydrogen:

$$N_2 + 3H_2 \rightleftharpoons 2NH_3$$

The reaction requires four 'volumes' of gas to become two 'volumes' of gas so the gases need to be 'pushed together'. To *maximise the yield* a high pressure is needed. Maintaining this pressure requires even more energy and so costs are higher.

The reaction is *exothermic* (energy is transferred out), so a lower temperature favours the forward reaction and a higher temperature favours the backward reaction. To make ammonia a low temperature is preferred.

However, a low temperature would make the reaction too slow, even with a catalyst, so a **compromise** temperature is chosen.

A high pressure (about 200 atmospheres) moves the **equilibrium** position to the right and increases the *rate*.

A high temperature moves the *equilibrium* position to the left but increases the *rate of reaction*. So a higher *compromise* temperature (about 450 °C) is chosen.

The Contact process

The Contact Process is another example of a reversible reaction used in industry.

The Contact Process for making sulfuric acid has three stages:

In the first stage sulfur is burned to make sulfur dioxide.

In the second stage sulfur dioxide reacts with oxygen to make sulfur trioxide:

sulfur dioxide + oxygen $\rightleftharpoons$ sulfur trioxide

$$2SO_2(g) + O_2(g) \rightleftharpoons 2SO_3(g)$$

This reaction is reversible. It uses a catalyst, pressures of 1 to 2 atmospheres and a high temperature (around 450 °C). The lower the temperature the more complete the reaction is, but the longer it takes. These conditions give a conversion rate of over 95% in a short enough time to make the process economic.

In the third stage, sulfur trioxide reacts with water to make sulfuric acid.

6. **Look at Figure 6.13. Explain the trends for the percentage of ammonia made with changes in:**

   a pressure **b temperature.**

7. **Suggest two reasons why a pressure of 200 atmospheres is used in the Haber process and not a much higher pressure.**

DID YOU KNOW?

450 million tonnes of fertiliser is produced by the Haber process each year. Around four times as much land would be needed to feed the population if there were no industrial fertilisers.

Figure 6.12 Separating nitrogen from the rest of the air

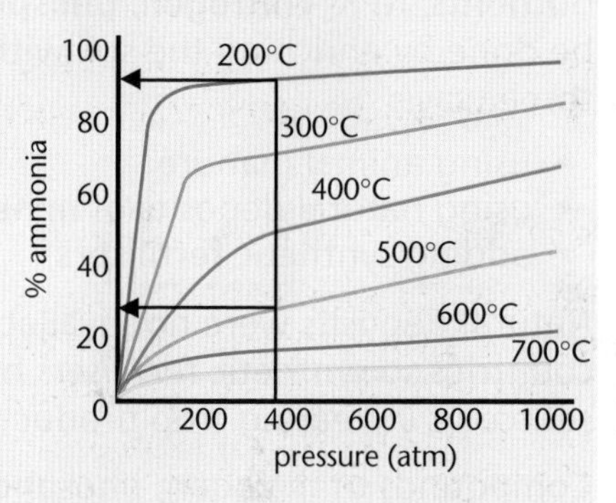

Figure 6.13

REMEMBER!

Conditions used commercially depend on:

- **the availability and cost of raw materials and energy supplies**
- **the control of equilibrium position and rate.**

KEY INFORMATION

This sign: $\rightleftharpoons$ means a reversible reaction.

Production and use of NPK fertilisers

Learning objectives:

- describe how to make a fertiliser in the laboratory
- explain how fertilisers are produced industrially
- compare the industrial production of fertilisers with laboratory preparation.

KEY WORDS

fertiliser
nitrogen
phosphorus
potassium

With the rise in population at the beginning of the 20th century, it became clear that synthetic fertilisers would be needed to prevent the famines that were taking place. How would these fertilisers be made? Would they have any negative effects? Would the population grow even bigger? Was increased agriculture sustainable?

Figure 6.14 Natural potash deposits containing potassium

Laboratory preparation of fertilisers

To feed the world population crops need to be grown successfully. To do this the growing crop needs the right nutrients, NPK – **nitrogen**, **phosphorus** and **potassium**. This can be done by enriching the soil with these nutrients in one of three ways:

- using animal manure
- using natural deposits of minerals
- adding synthetic **fertilisers**

KEY SKILLS

Remember that you will need to be able to carry out this procedure yourself.

Natural deposits were not available to all countries so synthetic *fertilisers* became the source of increasing crop production.

Compounds of nitrogen, phosphorus and potassium can be made in the laboratory. An acid is added to an alkali so that they are neutralised. The solution is evaporated and the crystals are collected.

Ammonia (NH_3) can be added to phosphoric acid (H_3PO_4) to make ammonium phosphate [$(NH_4)_3PO_4$]. This compound contains both the nitrogen and phosphorus that are needed in a fertiliser.

1. Use a measuring cylinder to pour alkali into a conical flask.

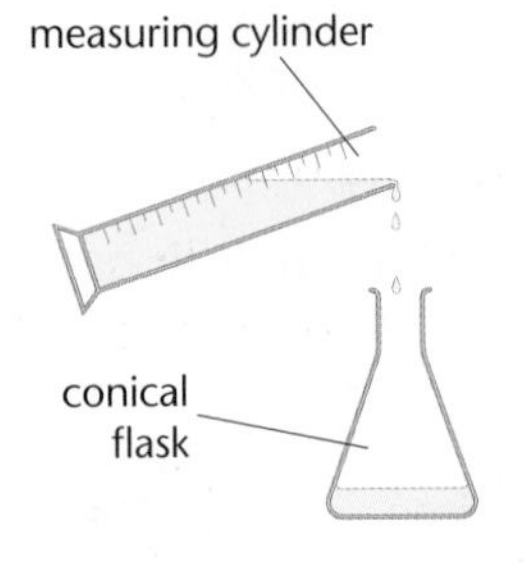

2. Add acid to the alkali until it is neutral.

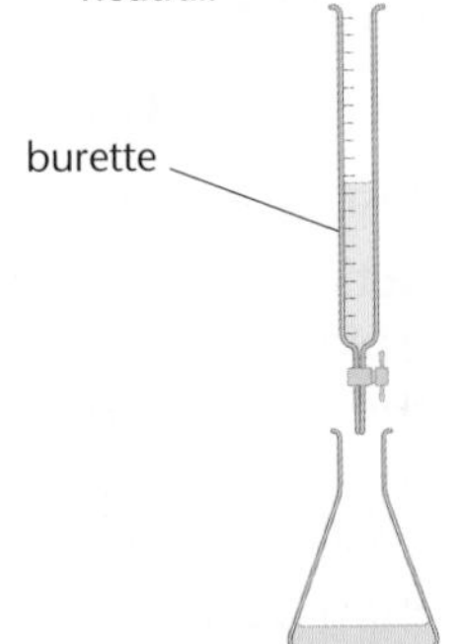

3. Evaporate.

evaporating basin

crystals begin to form

4. Filter off the crystals.

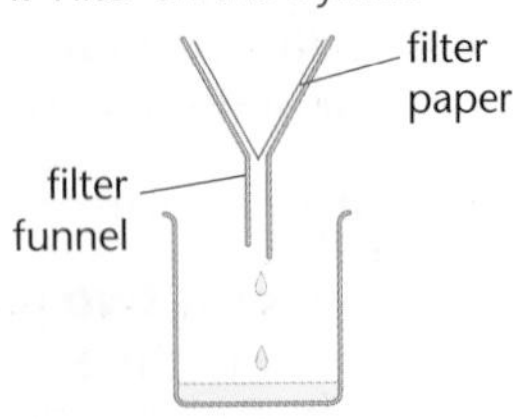

Figure 6.15 Steps in making a fertiliser in the laboratory

1 **Describe the apparatus used in making a neutral solution of fertiliser from ammonia and sulfuric acid.**

2 **Name the fertiliser made from**

a **potassium hydroxide and phosphoric acid.**

b **ammonia and sulfuric acid.**

Industrial production of fertilisers

NPK fertilisers are *formulations* of various salts containing appropriate percentages of the elements. On the packet of a fertiliser it will state the ratio or the percentage of NPK. Different *ratios* are needed for different plants and stages of growth.

Industrial production of NPK fertilisers can be achieved using a variety of raw materials in several *integrated* processes.

a Ammonia (made by the Haber process) can be used to manufacture *ammonium salts* and *nitric acid* (HNO_3).

Potassium chloride, potassium sulfate and *phosphate rock* (calcium phosphate) are obtained by mining. Phosphate rock cannot be used directly as a fertiliser because it is *insoluble*.

b Phosphate rock is treated with *nitric acid* to produce *phosphoric acid* and *calcium nitrate*.

c Phosphoric acid is neutralised with *ammonia* to produce *ammonium phosphate*.

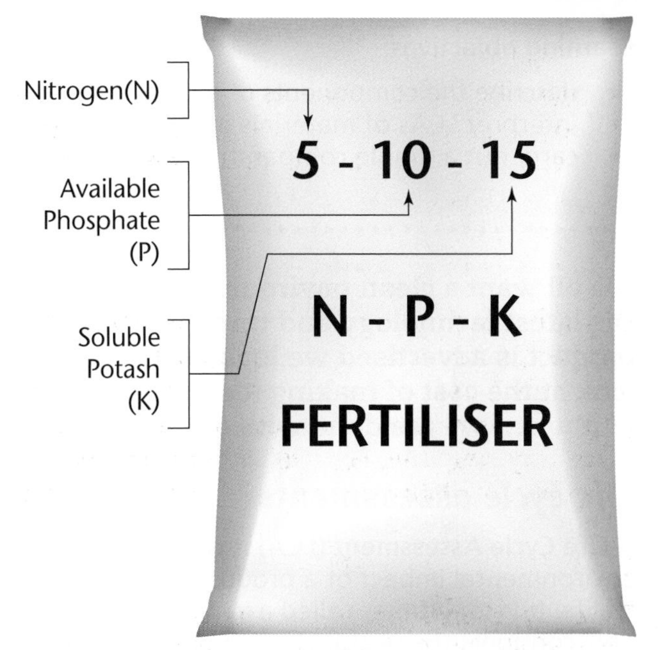

Figure 6.16 Fertilisers have different ratios of N, P and K

Phosphate rock is treated with *sulfuric acid* to produce *single superphosphate* (a mixture of calcium phosphate and calcium sulfate) or with *phosphoric acid* to produce *triple superphosphate* (calcium phosphate).

3 **Write out the word equations for reactions b and c above. Explain how the processes depend on each other.**

4 **Suggest why some scientists say that the reserves of phosphate rock may run out in the next 50–100 years.**

KEY INFORMATION

You may need to be able to compare the industrial production of fertilisers with laboratory preparations of the same compounds from data.

Comparing fertilisers

NPK fertilisers contain compounds of all three elements to improve agricultural productivity. They contain a mixture of products in particular ratios to get the best ratio for the job.

Ammonium phosphate is one of the products that can be made both in the laboratory and in industry.

$$NH_3 + H_3PO_4 \rightarrow (NH_4)_3PO_4$$

The processes used are similar. Acid is added to alkali.

Another product is calcium nitrate. The processes used are different. In the laboratory calcium nitrate is made by adding nitric acid to an alkali or a carbonate. In industry calcium nitrate is made by adding acid to phosphate rock.

5 **Write a balanced equation for the reaction between calcium carbonate and nitric acid.**

6 **Write a balanced equation for the reaction between calcium phosphate $Ca_3(PO_4)_2$ and nitric acid.**

DID YOU KNOW?

There are three stages to making nitric acid:

$4NH_3 + 5O_2 \rightarrow 4NO + 6H_2O$ oxidation

$2NO + O_2 \rightarrow 2NO_2$ further oxidation

$2NO_2 \longleftrightarrow N_2O_4$ reversible reaction

$3N_2O_4 + 2H_2O \rightarrow 4HNO_3 + 2NO$ absorption

Life cycle assessment and recycling

Learning objectives:

- describe the components of a Life Cycle Assessment (LCA)
- interpret LCAs of materials or products from information
- carry out a simple comparative LCA for shopping bags.

KEY WORDS

life cycle assessment
extracting
manufacture
disposal

We all want a clean environment but we also want the latest technology and transport. Each time a new product is advertised we look at the cost of it, but do we look at the cost of making it and the effects of disposing of it when we are done with it?

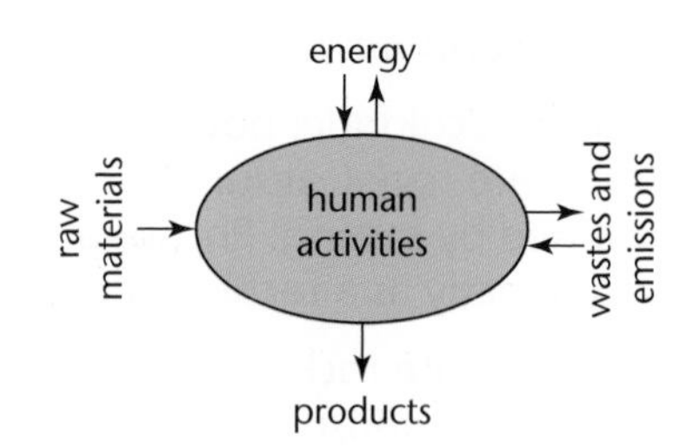

Figure 6.17 A Life Cycle Assessment diagram for a product

Life cycle assessments

A **Life Cycle Assessment (LCA)** is carried out to assess the environmental impact of a product in each of these stages of its life. It is sometimes called a 'cradle to grave' analysis. It not only considers the useful life of a product but also the materials and energy it takes to make the product and what happens to it on its **disposal**.

The stages in a simplified LCA are to consider:

- **extracting** and *processing* the *raw materials*
- **manufacturing** the product and *packaging*
- the *use* and *operation* of the product during its lifetime
- *disposal* at the end of its useful life

Transport and distribution and *waste* are included at each stage. Let's look at shopping bags made of plastic and paper.

Stage	Plastic	Paper
Raw materials	Processing crude oil	Made from wood
Manufacturing	Do both use these the same processes?	
Use and operation	Which lasts longer?	Which is stronger?
Disposal	How do you recycle, reuse or dispose of plastic or paper?	
Transport	Is this the same for both?	
Emissions and waste	Do the manufacturing processes give off emissions? What happens to waste plastic?	Do the manufacturing processes give off emissions? What happens to waste paper?

There are a number of processes in this comparison that are the same. You need to identify the processes and outputs that are different and compare these. There is quite a lot of difference in the raw materials used and the disposal in this example.

Figure 6.18 Comparing the LCA of a plastic bag and a paper bag

1. **Suggest which raw material is the most sustainable in this example.**
2. **Explain what happens at the end of the useful life of each of these bags. Suggest which one is easier to dispose of.**
3. **Suggest what the term 'cradle-to-cradle life cycle assessment' means.**

REMEMBER!

You only need to be able to compare the impact on the environment of the stages in the life of a product by description.

Objectivity and subjectivity

This exercise is done partly as a *description* and partly with numerical data. A description is called *qualitative* information.

Use of water, resources, energy sources used (*consumption*) and *production* of some wastes can be fairly easily quantified (put into *numerical* terms). This is called *quantitative* data.

It is much less easy to allocate numerical values to pollutant effects and requires *value judgements*. These are *subjective* (swayed by the person's viewpoint). So making LCAs is not a purely *objective* process (one where the data can only be interpreted in one way).

Figure 6.19 Considering objective data

4. **Identify whether the following statements are qualitative, quantitative, subjective or objective.**
   - **a** **Plastic bags are more environmentally friendly than paper bags.**
   - **b** **It takes 0.48 MJ of energy to make a plastic bag and 1.6 MJ to make a paper bag.**
5. **Two students have different viewpoints on how much CO_2 goes into the air from the making of paper or plastic. They can find no data on it. Explain why their LCAs may be different.**

DID YOU KNOW?

You only need to be able to use quantified data for energy, water, resources and wastes when it is readily available.

Advertising

It is important to identify the source of the LCA. It could be unbiased from an impartial agency, a reputable company or a protesting group that is fair. It *may* have some bias as a company may want to sell a lot of their product or because a group does not want the product to go ahead. LCAs can be devised that are *selective* or *abbreviated* when evaluating a product and be misused to reach pre-determined conclusions, for example in support of claims for advertising purposes. It is important to be critically aware of bias or non-bias.

Figure 6.20 Have they done an LCA? Are we persuaded?

6. **Draw up a table for the LCA for bag use as on the previous page. Answer the questions in the table.**
7. **It takes an average of 119 000 MJ of energy to make a typical car.**
   - **a** **Suggest how this sort of data can affect the accuracy of an LCA.**
   - **b** **State one factor that is not taken into account in an LCA and how it might affect the manufacturer and consumer.**

Ways of reducing the use of resources

Learning objectives:

- describe ways of recycling and reusing materials
- explain why recycling, reusing and reducing are needed
- evaluate ways of reducing the use of finite resources.

KEY WORDS

recycling
finite resource
reduction of use
reuse of resources

Reduce, reuse, recycle. Is that what we should be doing? The population is growing and we have limited resources on the planet for everyone both now and for future generations. What will it be like for our grandchildren if we take all the finite resources? What can we do about it?

Figure 6.21 Building materials are made from resources that often have to be mined or quarried

Recycling and reusing

Metals, glass, building materials, and clay ceramics are produced from limited raw materials that mostly have to be dug up from the ground.

Obtaining raw materials from the Earth by *quarrying* and *mining* causes *environmental impact*.

Some products, such as glass bottles, can be reused.

Figure 6.22 Reusing or recycling materials can cut down the need for quarrying or mining which have large environmental impacts

Other products cannot be *reused* and so are *recycled* for a different use.

Glass bottles can also be recycled. They can be crushed and melted to make different glass products.

The amount of separation required for **recycling** depends on the material and the properties required of the final product.

Metals can be recycled by melting and recasting or reforming into different products.

For example, some scrap steel can be added to iron from a blast furnace to reduce the amount of iron that needs to be extracted from iron ore.

1. **Describe two environmental impacts that a quarry has on local residents.**
2. **Explain why it is important to recycle metals and glass.**
3. **Some plastic bags are strong and can be repeatedly reused. Suggest why it is better to reuse than recycle.**

DID YOU KNOW?

Milk used to be distributed in glass bottles that could be reused then recycled.

Reducing the use of finite resources

Most plastics are made from crude oil, which is a **limited**, finite **resource**. It is not as easy to recycle as it is to recycle metal or glass.

Plastics do not degrade easily and take up valuable land when dumped in landfill sites that need to last for decades.

Some plastics are used in small quantities and greatly enhance our lives, but do we really need to use plastic bags?

The same limited resource – crude oil – is needed to produce the *energy* used in the *extraction of resources*, such as metal ores, and in the *manufacturing processes* of materials such as metal and glass, and objects such as cars and plastic furniture.

Figure 6.23 Metals can be re-melted with metal from new ore to reduce the amount of new ore needed

4. **Explain why it is important to use crude oil wisely.**
5. **Suggest objects made in plastic that could be made by materials that will degrade.**
6. **It is common to see the slogan 'Reduce, reuse, recycle'. Discuss how this slogan can be applied to a typical car.**

End use reduction

We have seen that the reuse and recycling of materials *reduces*:

- the use of limited resources
- use of energy sources
- waste
- the environmental impact.

We could also do one more thing. That is not to use so much in the first place.

The **reduction of use**, by *end users*, would have a massive impact on the amount of resources that we need to take out of the Earth at the start of the process of manufacturing.

Figure 6.24 Discarded plastic takes up valuable land and does not degrade easily

REMEMBER!

In complex arguments like this where there are a multitude of factors to take into account, you may not always be right. It is the thinking that is important.

7. **It is possible to assemble cars a long distance from or near to the point where they will be sold (the market). Explain which way may save more limited resources.**
8. **An industry produces metal parts for precision hospital equipment. They are discussing how to set up an assembly line so that waste is reduced. Explain two things they could do.**
9. **The annual use of plastic carrier bags in the UK was 7.6×10^9 (7.6 billion). On the introduction of a charge, use fell by 80%.**

   a Describe the environmental impact of the reduction.

   b Calculate the number of plastic bags used after the reduction.

Alloys as useful materials

Learning objectives:

- describe the composition of common alloys
- interpret the composition of other alloys from data
- evaluate the uses of other alloys.

KEY WORDS

alloy
brass
bronze
steel

Why make alloys? Most metals are not quite suitable for the uses that we need. They are often too soft, too prone to corrosion and their melting point is too high. We can make them harder and more resistant. What other uses can metal alloys have?

Figure 6.25 A bronze statue

Common alloys

Most metals in everyday use are **alloys**. Pure copper, gold, iron and aluminium are too soft for many uses and so are mixed with other metals to make them harder.

Bronze is an alloy of copper and tin. It is used to make statues and decorative objects.

Brass is an alloy of copper and zinc. It is used to make water taps, door fittings and musical instruments. It has higher *malleability* than bronze.

Both these copper alloys can be separated for *recycling* as they are not attracted to the powerful magnets that are used to separate iron-based (ferrous) metals. This allows the valuable copper resources to be reused.

Gold used as jewellery is usually an alloy. The proportion of gold is measured in carats. 24 carat gold is 100% (pure gold). An alloy of 18 carat is 75% gold and 25% other metals. The alloys of gold use silver, copper and zinc.

Steels are alloys of iron that contain specific amounts of carbon and other metals. High-carbon steel is strong but brittle. Low-carbon steel is softer and more easily shaped.

Stainless steels contain chromium and nickel. They are hard and resistant to corrosion.

Aluminium alloys have a low density and are used in the aerospace industry.

Figure 6.26 Objects made from stainless steel are resistant to rusting

1 **Three types of cast iron, X, Y and Z, contain carbon at X 3.4%, Y 2.5% and Z 3.0%.**

Determine which is:

a **most brittle** **b** **most easily hammered**

2 **The three iron samples in Q1 have tensile strengths of 35, 50 and 27 (units). Match the strengths to the samples X, Y and Z.**

Other alloys

Alloys are made so that physical properties are altered. Some properties may not be very different to the original metals in the alloy, such as density, but others may be very different, such as tensile strength. *Solder* is one alloy of tin and lead used to join other metals together. The mixture has a lower melting point than pure tin or pure lead.

% of tin	60	10	50
% of lead	40	90	50
Use	Electrical soldering	Car radiators	Plumbing and gas meters

Lead is toxic and so some traditional uses of the alloy are no longer in practice.

Phosphor bronze is made from 94.8% copper, 5% tin and 0.2% phosphorus. It has a *range of uses* because of different *physical properties.* It is used for springs and bolts due to its resistance to fatigue and corrosion.

It is also used for cryogenics (working at very low temperatures) as it has a low thermal conductivity.

A third use is for cymbals, trumpets and saxophones because of its ability to sustain sound and produce a spectrum of tones.

3 **The mass of a baritone saxaphone made out of phosphor bronze is 5.6 kg. Work out the mass of each component in grams.**

4 **Explain why alloys with the same constituents are made but with different percentages.**

Evaluating uses

Pewter is a very ancient alloy that is very malleable.

% Tin	99	96	85	94	97.5
% Copper	1	0	0	1	1
% Lead	0	4	15	0	0
% Antimony	0	0	0	5	1.5

Different percentage compositions allow different uses of the metal. The 99% tin and 1% copper alloy was used for the finest tableware. Higher proportions of tin make the alloy softer and more malleable.

5 **Explain which pewters are no longer used for making drinking cups.**

6 **Suggest why higher proportions of tin make pewter softer.**

7 **Nitinol typically is composed of 44% titanium, which is a very reactive metal. The other component is nickel. The alloy is made under vacuum.**

a **Suggest why nitinol is made under vacuum.**

b **Calculate the mass of nickel in 500 g of nitinol.**

KEY SKILLS

You do not need to remember the percentages of metals in an alloy but you do need to be able to read data from a table and evaluate uses.

Figure 6.27 Alloys of aluminium are used for aircraft due to their low density

Figure 6.28 Pewter is a malleable alloy

DID YOU KNOW?

Some metal alloys have 'shape-memory' properties and can be squeezed to a shape but will return to its original shape on heating. Nickel–titanium alloy, called nitinol, is in common use now, but it was first discovered by accident in 1932 in Sweden when experimenting with other alloys.

Corrosion and its prevention

Learning objectives:

- show that air and water are needed for rusting
- describe experiments and interpret results on rusting
- explain methods for preventing corrosion.

KEY WORDS

corrosion
sacrificial protection
galvanising
electroplating

The Titanic sank almost 100 years ago but it is rusting only slowly. Why is that?

It sank in water that is 2.5 miles deep so there is very little oxygen dissolved, which is why it is rusting so slowly. What is rusting, why is oxygen needed and how can we prevent it?

What is needed for rusting?

Sam and Alex set up an experiment to find out which conditions are needed for rusting. They set up three boiling tubes and put two shiny iron nails in each.

They intend to have one tube open to the damp air, one tube to have air but no water and the last tube to have water but no air.

They put calcium chloride in a glass drying tube attached to the test tube, open to the air so that no water can enter.

They boil water to remove the air and put the water in the tube with a layer of oil on it to prevent air getting to the nail.

They leave them for one week. Which do you predict will rust?

These are their results:

Conditions	air and water	air but no water	water but no air
Appearance	rust develops	no rust develops	no rust develops

Their conclusion is that both air and water are needed for rusting to occur.

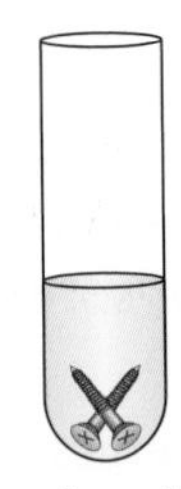

air but no water

Figure 6.29 Experiment to see if air, water or both are needed for rusting

Figure 6.30 Painting can protect objects from rusting

1 **A car left in the Sahara desert does not rust as quickly as a car in the UK. Suggest why.**

2 **Explain why ships that sink in shallow water rust quite quickly.**

The prevention of corrosion

Rusting is an example of **corrosion**.

Corrosion is the destruction of materials by chemical reactions with substances in the environment.

If we are trying to use less resources and less energy for a more sustainable way of life, then it makes sense to reduce the amount of corrosion.

Figure 6.31 Electroplating with another metal to prevent corrosion

Corrosion can be prevented by having a barrier in place. This can be done by applying a synthetic coating or encouraging a natural barrier.

Iron can be protected with barrier methods such as greasing, painting or **electroplating**.

Aluminium does not corrode as it forms a natural oxide coating that protects the metal from further corrosion.

3. **Explain why painting or greasing protects iron from rusting.**
4. **Explain why iron that has been painted to prevent corrosion has to be repainted every so often.**

KEY INFORMATION

Remember, **reduction** and **oxidation** happening at the same time is called a *redox* reaction.

Electrons in corrosion

When metals corrode they are taking part in a redox reaction. Rusting iron is just one example of this.

Rusting is a redox reaction where iron is oxidised to hydrated iron(III) oxide.

iron + oxygen + water → hydrated iron(III) oxide

$$Fe - 3e^- \rightarrow Fe^{3+}$$

Some coatings are not barriers, but are reactive. They may contain *corrosion inhibitors* or a more reactive metal.

Figure 6.32 Magnesium is used as a sacrificial metal

Zinc is a more reactive metal and is used to *galvanise* iron. **Galvanising** protects iron from rusting by covering it with a layer of zinc. The layer of zinc stops water and oxygen from reaching the surface of the iron. It also acts as a sacrificial metal. When the surface is scratched it is the zinc that reacts and provides protection to the iron as it is more reactive than iron. The zinc surface is sacrificed to protect the less reactive iron.

A ship's surface is not covered in zinc but instead blocks of magnesium (which are more reactive than steel) can be attached to steel ships to provide **sacrificial protection**.

Sacrificial protection works better where the metal to be sacrificed is much more reactive than the metal to be saved. It is the *relative* reactivity of the two metals that is important.

Underground pipelines and the metal legs of seaside piers also often have lumps of magnesium attached at intervals.

DID YOU KNOW?

Another way to prevent corrosion is by alloying. Stainless steel does not rust like iron as it has chromium, carbon and other metals within the alloy.

5. **Explain why magnesium is more effective than zinc at protecting iron.**
6. **Describe briefly two methods for protecting a bridge from corrosion. Explain how each method works.**
7. **The magnesium in the blocks attached to ships is oxidised to Mg^{2+} ions. This causes the Fe^{3+} ions to be reduced back to Fe.**
   - **a** **Write an equation for the oxidation of Mg to Mg^{2+}.**
   - **b** **Explain in terms of electrons how this prevents Fe from oxidising.**
   - **c** **Suggest what happens to the magnesium block over a period of time.**

DID YOU KNOW?

Salt speeds up the process of rusting iron. Salt doesn't make iron rust in the first place.

Ceramics, polymers and composites

Learning objectives:

- compare quantitatively the properties of materials
- compare glass, ceramics, polymers, composites and metals
- explain the difference between thermosoftening and thermosetting polymers by their structures
- select materials by relating their properties to uses.

KEY WORDS

thermosoftening
thermosetting
composite
reinforcement

The oldest piece of ceramic in the world is a small female figure around 29 000 years old. People have been using fire to make ceramics since then and later for glass and metalworking. New composites and polymers have recently revolutionised the objects we can make. Where will materials technology take us next?

Figure 6.33 Blowing soda-lime glass into shape

Glass, ceramics and composites

Most of the glass we use is *soda-lime glass*. This is made by heating a mixture of sand, sodium carbonate and limestone.

Borosilicate glass is made from sand and boron trioxide and melts at higher temperatures than soda-lime glass.

Ceramics, including pottery and bricks, are made by shaping wet clay and then heating in a furnace.

Most **composites** are made of two materials. A matrix or binder surrounds and binds together the fibres or fragments of another material. The other material is called the **reinforcement**. Concrete is made of cement with stones (aggregate) in it. Other examples of composites include wood and fibreglass.

Some advanced composites are made using carbon fibres or carbon nanotubes instead of glass fibres.

Figure 6.34 Borosilicate glass withstands heating much better than soda-lime glass

Figure 6.35 Bricks and pottery are ceramics made from clay

1. **Describe the difference between soda-lime glass and borosilicate glass.**
2. **Explain why fibreglass is a composite.**

Polymers

The properties of *polymers* depend on which monomers they are made from and the conditions under which they are made.

Low-density (LD) and high-density (HD) poly(ethene) are both produced from ethene.

To make their properties different they are made using different catalysts and different reaction conditions.

Both types of poly(ethene) materials are **thermosoftening** polymers.

DID YOU KNOW?

The older glassware under the name Pyrex® was made from borosilicate glass but is now made from tempered soda-lime glass.

Thermosoftening polymers consist of individual, tangled polymer chains and melt when they are heated. **Thermosetting** polymers consist of polymer chains with cross-links between them and so they do not melt when they are heated.

Thermosoftening	Thermosetting
Flexible or can be made into fibres.	Hard and brittle
Polymer chains move more freely over each other when heated	Polymer chains are linked across chains by cross-links so do not re-soften on heating
↑ small force ↑ large force	

3 **Melamine resin used on worktops is a thermoset plastic. Suggest what the structure of the melamine polymer looks like.**

4 **Suggest why you would use a thermosoftening polymer to make a drinks bottle.**

Figure 6.36 Sports equipment made of carbon fibre

Evaluating materials

Material	Tensile strength (MPa)	Flexibility (MPa)	Electrical conductivity (S/m)	Strength to density ratio	Relative transparency (0-1 scale)
glass	33	50	1×10^{-11}	n/a	1
ceramic	30	37	1×10^{-18}	0.2	0
wood	40	n/a	1×10^{-16}	n/a	0
glass fibre	2600	17	1×10^{-15}	1.6	0.2
polymer	80	4	1×10^{-21}	0.3	0.2
steel	2700	200	7×10^{6}	0.7	0
graphene	130000	1050	1×10^{8}	n/a	n/a

Look at the table above. It contains some of the many properties of materials that can be measured for comparison.

If a particular property is needed for a new object or part of a complex object, the data can be compared and the best material selected for trial. The materials are evaluated for their usefulness for particular functions. Compositions of alloys, ceramics, polymer or composites can be changed to match the exact requirements.

Figure 6.37 Low density poly(ethene) is used to make milk cartons and plastic bags. High density poly(ethene) is used to make water pipes.

5 **Explain why neither the polymer nor the ceramic could be used for the body work of a car.**

6 **A part for a machine needs to have high tensile strength, be flexible and have a relatively low density. Justify which material you choose.**

KEY SKILLS

You need to be able to consider more than one property for a use when evaluating materials.

MATHS SKILLS

Translate information between graphical and numerical form

Learning objectives:

- to represent information from pie charts numerically
- to represent information from graphs numerically
- to represent information from numeric form graphically.

KEY WORDS

graph
numerical
scale
pie chart

Is it easier to read lists of numbers or to see those numbers visually represented? It often depends on the context. What is important is the ability to change between one form and another and be able to read data tables, histograms and graphs. Can you choose the best scales to construct graphs?

Representing information with pie charts

The Haber process is an example of a synthesis of a chemical (ammonia) needed to synthesise other chemicals (fertilisers). How big is that sector of the chemical industry compared to other chemicals that are made?

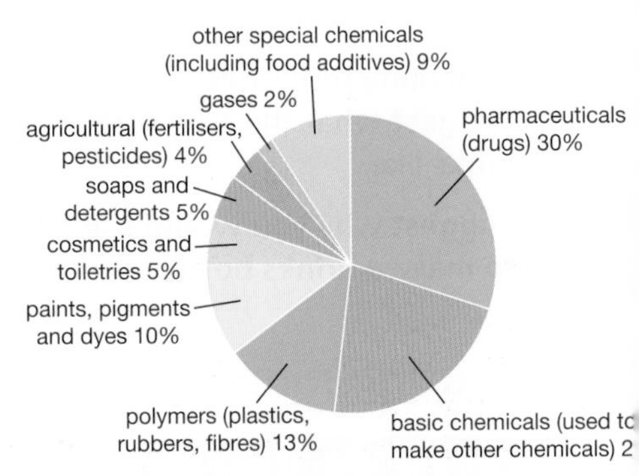

Figure 6.38 The chemical industry

Look at the **pie chart.** What percentage of the industrial chemicals made are used in the agriculture sector? The answer is 4%.

Is that more or less than the chemicals used for soaps and detergents? The answer is less, as 5% are used for soaps and detergents.

What is the biggest sector of use of chemicals for synthesis? 30% are used in the synthesis of drugs and medicines. These are called pharmaceuticals.

How do you know how big to draw each sector? If the numbers are given as percentages then the pie chart is divided equally into 100 segments. As there are 360° in a circle each segment takes up 3.6°. What size, in degrees, is the paints, pigments and dyes sector? The answer is 36° as the sector represents 10%, so the size is $10 \times 3.6 = 36°$.

1. **Determine the percentage of chemicals used to synthesise polymers.**
2. **How many degrees does the cosmetics and toiletries sector measure?**

KEY INFORMATION

Can you see that, in this case, the pie chart starts at '12 o'clock' and the sectors are drawn in order of size going clockwise. The first sector is 30%, whereas sector number two represents only 22%. The sectors go all the way round to gases which represents only 2%. Why is the last sector larger? This is because it is an 'other' sector that holds all the smaller contributors together;
it is put at the end.

Reading from graphs

The Haber process is the reversible reaction between nitrogen and hydrogen producing ammonia. Only a small percentage of ammonia is made and the reaction is affected by both temperature and pressure.

To show the effects the pressure is increased by steps and the percentage of ammonia is measured.

The pressure is the *independent variable*, so it is marked along the *x*-axis.

The percentage of ammonia depends on the pressure, so it is the *dependent variable*. It is marked along the *y*-axis.

Look at the graph line (Figure 6.39) of the reaction at 200°C.

What is the percentage of ammonia formed at 400 atmospheres? The answer is 92%.

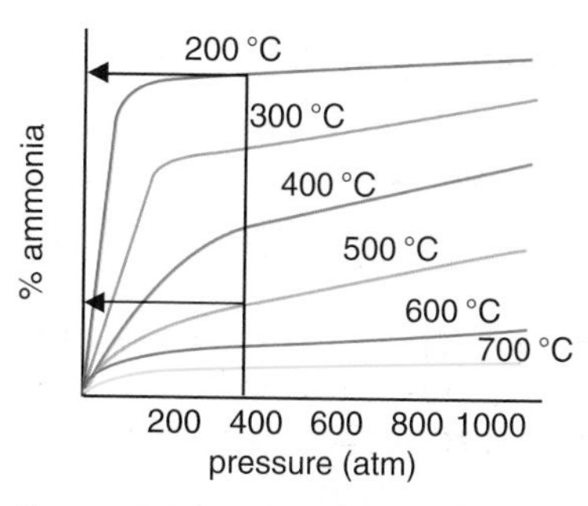

Figure 6.39 Haber process the effect of increasing pressure at different temperatures

3 **Look at the graph line of the reaction at 400°C. Estimate the percentage of ammonia formed at 600 atmospheres.**

4 **Estimate the pressure needed to make 40% ammonia at 400°C.**

Drawing graphs from data

First the data must be recorded so that the independent variable is measured at regular intervals. The dependent variable is recorded against this.

Once the data has been collected you need to look at the maximum values of the independent variable and choose a **scale** to write along the *x*-axis, with even steps and values written against each major gridline. For example, if the first value is 0 and the last one is 70 then the scale will be every 10 units to each of the 7 darker gridlines.

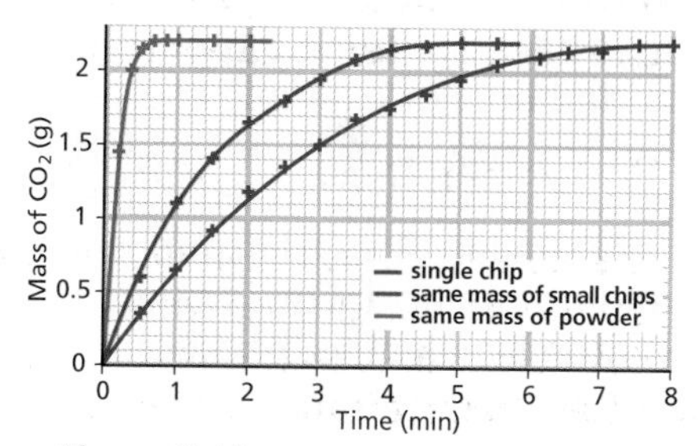

Figure 6.40

In the graph in Figure 6.40, graph the first value is 0 and the last one is 8.

The reaction gave off between 0 and 2.5g of CO_2 so the scale chosen is 0.5 units to every major gridline. The *x*-axis is labelled 'mass of CO_2 in g'.

The graph in Figure 6.41 shows the volume of a gas plotted against its temperature. It has been extrapolated through the negative values of temperature so is drawn in two quadrants.

DID YOU KNOW?

If a graph line is extended beyond the data collected this is called extrapolation. If data points are read between plotted points this is called interpolation.

5 Draw a pie chart to represent the percentage use of chlorine in other products using this data:

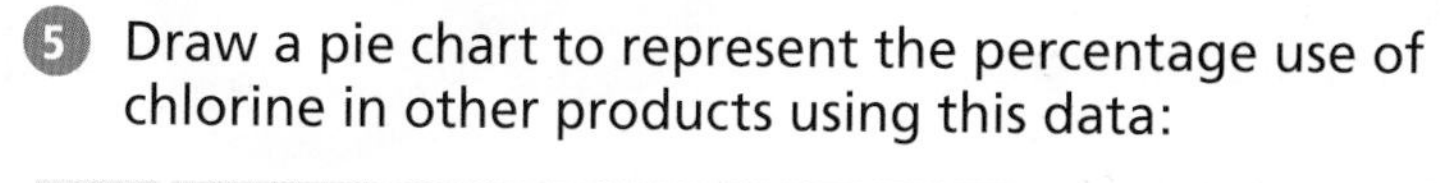

PVC	Other polymers	Disinfectants and cleaners	Solvents	Water treatment	Paper treatment	Other compounds
36%	20%	14%	6%	5%	5%	14%

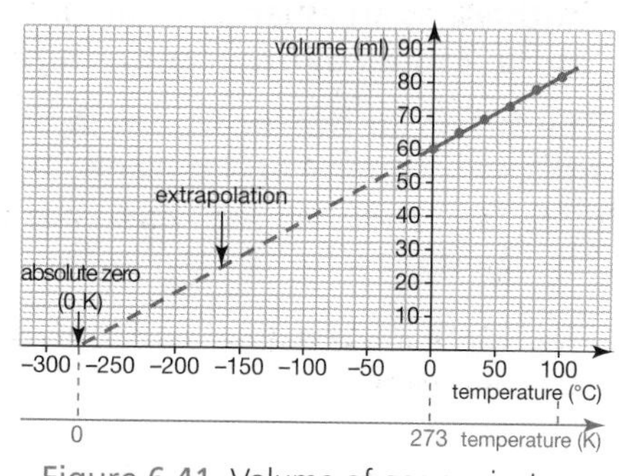

Figure 6.41 Volume of gas against temperature

6 Draw a line graph of this data:

Time in minutes	0	2	4	6	8	10
Mass of CO_2 in g	0	0.2	0.38	0.42	0.42	0.42

Functional groups and homologous series

Learning objectives:

- identify the first four hydrocarbons in the alkane series
- name the first four compounds in homologous series
- identify the functional group of a series.

KEY WORDS

Hydrocarbon
Homologous
Functional group
Alkane

Organic compounds are made of carbon atoms and hydrogen atoms joined with atoms of a few other elements such as oxygen and nitrogen. Carbon atom form four bonds and can make a vast array of compounds.

Backbones and functional groups

Carbon compounds are often formed in series of compounds that have the same basic structure but more carbon atoms added on. For example the simplest alcohol is methanol CH_3OH, but the next alcohol is C_2H_5OH. Another C atom has been added and two more H atoms but otherwise the structure is very similar. The OH group is the group that makes it an alcohol and not an acid. This is called the **functional group**. The C and H atoms make the 'backbone' or 'skeleton' of the molecules. There can be any number of them and they make a series of the compounds called a **homologous series**.

(a)

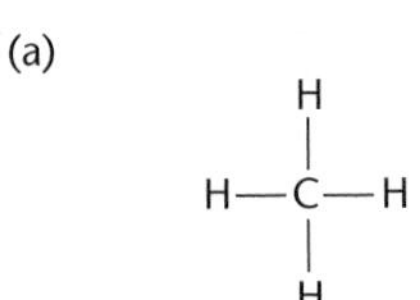

(b)

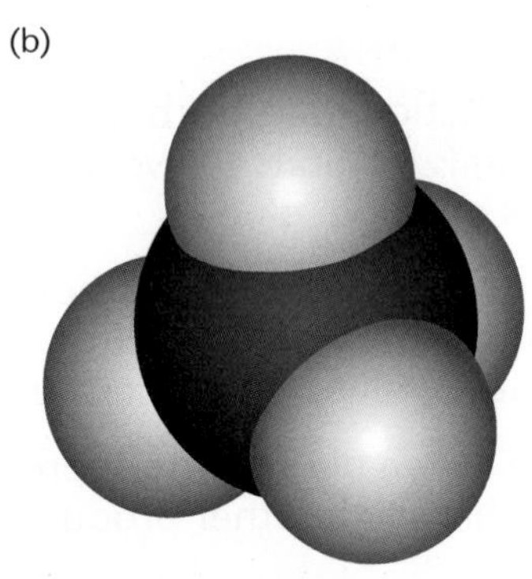

Figure 6.42 Two ways to represent the molecule of methane CH_4

1. **Look at Figure 6.42a. What is the formula of the simplest compound of carbon and hydrogen?**
2. **How many bonds does carbon always make?**

Hydrocarbons

Hydrocarbons are molecules made from carbon and hydrogen *only*. The smallest hydrocarbon is called methane. One atom of carbon and four atoms of hydrogen chemically combine to make it.

Methane is an **alkane**. Alkanes are one series of hydrocarbons. You will meet another series later.

The first four hydrocarbons in the alkane series are methane, ethane, propane and butane.

Name	Number of carbon atoms	Formula	Displayed formula
Methane	1	CH_4	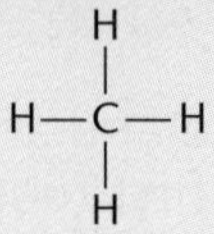
Ethane	2	C_2H_6	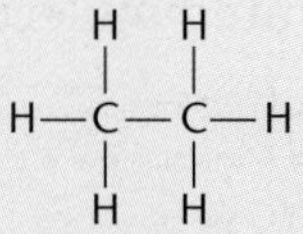
Propane	3	C_3H_8	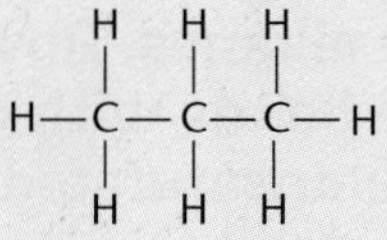
Butane	4	C_4H_{10}	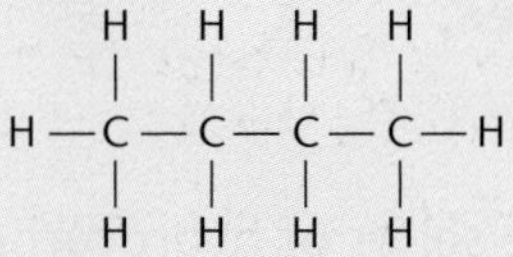

3 **Explain why the formula for the next one in the series is C_5H_{12}.**

4 **Predict the formula for the alkane that has eight carbon atoms.**

Homologous series

All the alkanes have the same general formula, so they form one series of compounds. The general formula is C_nH_{2n+2}.

One carbon atom forms four bonds. Hydrogen has one bond, so four hydrogen atoms bond to one carbon atom to make methane.

All the bonds in alkanes are single bonds. If one carbon bonds to another carbon atom to make ethane, the carbon–carbon bond is a single bond. Each carbon atom has three bonds left to make bonds with hydrogen atoms. The formula can be written as CH_3CH_3.

Look back at the displayed formula of butane in the table. How many hydrogen atoms are bonded:

- to each of the carbon atoms at either end?
- to each of the carbon atoms at in the middle?

You can also write the formula for butane as $CH_3CH_2CH_2CH_3$.

5 **Explain how to use the general formula to work out the formula of a hydrocarbon with 18 carbon atoms.**

6 **Write the formula of C_7H_{16} in two other ways, remembering to only use single bonds.**

7 **Alkanes can also form branched chains. Draw the displayed formula for the straight chain and two branched chain alkanes for pentane, C_5H_{12}.**

DID YOU KNOW?

It is easy to predict the name of an alkane if the number of carbon atoms is five or more. C_5H_{12} is *pen*tane. C_8H_{18} is *oct*ane. What do you think is the name for $C_{10}H_{22}$?

KEY INFORMATION

Remember that you can work out any formula of an alkane if you know the number of carbon atoms (n) using the formula C_nH_{2n+2}.

Structure and formulae of alkenes

Learning objectives:

- describe the difference between an alk*a*ne and an alk*e*ne
- draw displayed structural formulae of the first four members of the alkenes
- explain why alkenes are called unsaturated molecules.

KEY WORDS

alkenes
double bond
homologous series
unsaturated

Carbon atoms that bond using four single bonds form a series of hydrocarbons called the alkanes. However, carbon atoms can also form double bonds with other carbon atoms. These molecules are unsaturated molecules. Series of unsaturated hydrocarbons exist; one of these is the alkene series.

The alkene series

We have seen that alkanes are one series of hydrocarbons. Ethane is an alkane with the formula C_2H_6. All of the bonds *within* the molecule are single bonds.

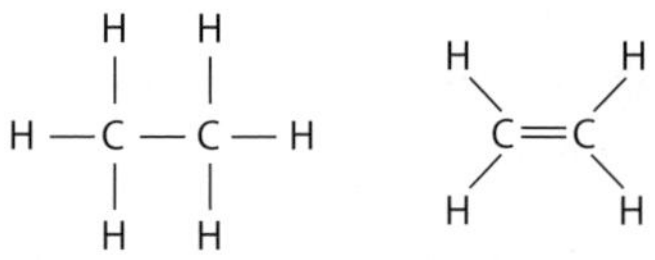

Alkenes are another series of hydrocarbons. Ethene is an alkene with the formula C_2H_4. Look at Figure 6.43. The alkenes are molecules with a double carbon–carbon bond.

ethene

Figure 6.43 (a) Ethane has a single carbon-carbon bond. (b) Ethene has a double carbon-carbon bond.

The first four members of the series of alkenes are ethene, propene, butene and pentene.

Number of carbon atoms	Name	Formula	Displayed formula
2	Ethene	C_2H_4	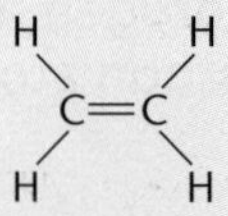 $H_2C{=}CH_2$
3	Propene	C_3H_6	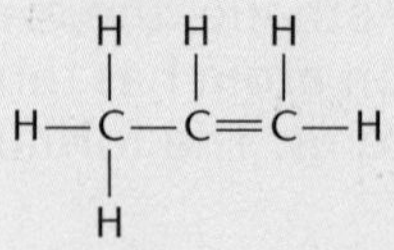 $CH_3{-}CH{=}CH_2$
4	Butene	C_4H_8	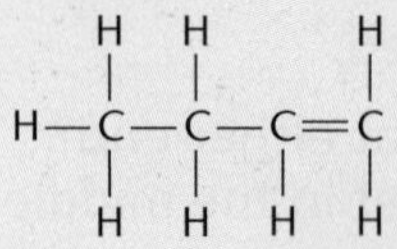 $CH_3{-}CH_2{-}CH{=}CH_2$
5	Pentene	C_5H_{10}	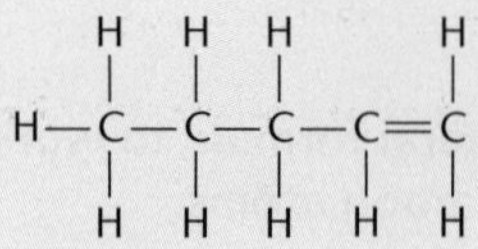 $CH_3{-}CH_2{-}CH_2{-}CH{=}CH_2$

1. **Predict the name, formula and structure of the next alkene in the series.**
2. **Compare the bonding in an alkene and an alkane.**

Unsaturated molecules

Ethene has the formula C_2H_4. Ethane has two more hydrogen atoms than ethene. This is because ethane has a single carbon–carbon bond and ethene has a carbon–carbon double bond. This means that each carbon atom is already using two of its four bonds in ethene. Look at Figure 6.43 a and b to see this. Only two hydrogen atoms can bond to each carbon atom, not three.

Ethene is an **unsaturated** molecule. It has a **double bond**.

Alkene molecules are unsaturated because they contain two fewer hydrogen atoms than the alkane with the same number of carbon atoms. They are not 'full up' with hydrogens.

Alkenes have a general formula of C_nH_{2n}. (Remember we saw that alkanes are C_nH_{2n+2}.)

We have already seen that bromine water is used to test for unsaturated molecules. Bromine water is an orange solution. When an alkene is added the orange solution turns colourless.

3 Explain why the formula for the molecule $C_{16}H_{32}$ shows that it is an alkene.

4 Predict which of these molecules will turn bromine water colourless:

a $C_{16}H_{32}$ **b** $C_{16}H_{34}$ **c** C_9H_{20} **d** C_7H_{14}

Homologous series

Like alkanes, alkenes also have a general formula, as they form one series of compounds with similar structures.

The general formula for the **homologous series** of alkenes is C_nH_{2n}.

The alkene series is made up of carbon atoms joining together with *double* bonds and hydrogen atoms bonding to the carbon atoms so that each carbon atom has four completed bonds.

Look at the displayed formula of pentene in the table. How many hydrogen atoms are bonded:

- to each of the carbon atoms at either end
- to each of the carbon atoms bonded by single bonds
- to each of the carbon atoms bonded by double bonds?

You can also write the formula for pentene as $CH_2CHCH_2CH_2CH_3$

5 Write the formula of C_8H_{16} in two other ways.

6 Write the formula of the first four alkenes in the series and their reactions with hydrogen.

7 Ring alkenes are called cycloalkenes. The formulae for the first four members of the series having one carbon–carbon double bond are: C_3H_4, C_4H_6, C_5H_8, C_6H_{10}.

a Determine a general formula for cycloalkenes.
b Give the formula of the cycloalkene with 10 carbons.

DID YOU KNOW?

The term 'saturated' comes from the same idea as being saturated when you are caught out in very heavy rain without a coat and you can't get any wetter.

DID YOU KNOW?

Hydrocarbons that have at least four carbon atoms and that have a double bond can have the double bonds in different places. Look at the displayed formula of butene in the table. You can also have the double bond between carbons 1 and 2 instead of between carbons 2 and 3. These molecules are called *isomers*. $CH_2CHCH_2CH_3$ instead of $CH_3CHCHCH_3$. Draw out the two structures and see.

KEY INFORMATION

Always remember that carbon forms FOUR bonds. They can be single, double or triple bonds. If a triple bond is formed then the carbon atom can only form *one* more bond to make FOUR bonds altogether.

Reactions of alkenes

Learning objectives:

- describe the addition reactions of alkenes
- draw the full displayed structural formulae of the products that alkenes make
- explain how alkenes react with hydrogen, water and the halogens.

KEY WORDS

addition
carbon–carbon double bond
finely divided nickel catalyst
incomplete combustion

Alkenes are more reactive than alkanes as they have a reactive double bond within the carbon hydrogen skeleton. They react by the addition of molecules across this double bond, so that the carbon to carbon bond becomes a single bond. This means the unsaturated molecule becomes saturated.

Incomplete combustion

We have seen with hydrocarbon fuels that if they burn without enough oxygen then **incomplete combustion** happens.

Alkenes react with oxygen in the same way as other hydrocarbons, but they tend to burn in air with smoky flames because of incomplete combustion.

Ethene is an alkene with the formula C_2H_4.

Complete combustion is shown by the balanced equation:

$$C_2H_4 + 3O_2 \rightarrow 2CO_2 + 2H_2O$$

In the reactants there are *six* oxygen atoms.	In the products of complete combustion there are *four* in the carbon dioxide plus *another two* oxygen atoms in the water.

KEY INFORMATION

The products are water, carbon monoxide (CO) and carbon (C). Carbon is noticed as soot and the flame burns *yellow*. Carbon monoxide is a highly toxic gas that has no smell. It causes illness and death by preventing oxygen being carried in the red blood cells.

With less oxygen there is incomplete combustion, as this equation shows:

$$C_2H_4 + 2O_2 \rightarrow 2CO + 2H_2O$$

In the reactants there are now only *four* oxygen atoms.	In the products of incomplete combustion(there are only *two* in the carbon monoxide plus another *two* oxygen atoms in the water

1. **Write another balanced symbol equation for the incomplete combustion of ethene, but this time only one oxygen molecule in the reactants.**
2. **Write a balanced symbol equation for the incomplete combustion of butene (C_4H_8).**
3. **Suggest why alkenes tend not to be used as fuels, particularly in boilers in the home.**
4. **Predict, with a reason, which of the following are likely to incompletely combust: C_5H_{10}, C_5H_{12}, C_6H_{14}.**

Functional groups

Functional groups are usually attached to a carbon skeleton.

However, alkenes are hydrocarbons with the functional group C=C in the skeleton. This is a **carbon to carbon double bond**.

Alkenes react across the carbon–carbon double bond so that the double bond becomes a single carbon–carbon bond.

The reaction is an **addition** of atoms to the carbon atoms on either side of the double bond.

For example a bromine molecule (Br_2) will add across the double bond of ethene (C_2H_4).

$$H_2C{=}CH_2 + Br_2 \longrightarrow CH_2Br{-}CH_2Br$$

In this *addition* reaction ethene reacts with bromine to produce *dibromoethane.*

Bromine solution will undergo an *addition* reaction with any alkene molecule. As it adds across it will form a dibromo-compound with the alkene. It will therefore lose its orange/brown colour as it is no longer bromine but part of a new compound.

In general the addition of any halogen (Cl_2, Br_2 or I_2) to an unsaturated alkene molecule produces a *saturated* compound with two halogen atoms in the molecule. They will be on the carbon atoms at the two ends of the original double bond.

KEY INFORMATION

Make sure you draw all the bonds in a fully displayed structural formula. Check that each carbon atom has four bonds and when drawing the product of an addition reaction that it is fully saturated.

5. **Draw the displayed structural formula of the addition of bromine to propene.**
6. **Predict the product of the reaction between chlorine and pentene, $CH_3CHCHCH_2CH_3$.**

Other addition reactions

Ethene reacts with hydrogen through an addition reaction across the carbon–carbon double bond. This takes place over a heated **catalyst of finely divided nickel** at a temperature of 300 °C.

$$C_2H_4 + H_2 \rightarrow C_2H_6$$

Ethene also reacts with water through an addition reaction. Ethene and steam are passed over a catalyst of hot phosphoric acid to produce ethanol.

$$C_2H_4 + H_2O \rightarrow C_2H_5OH$$

This is an addition reaction that takes place on an industrial scale to produce ethanol. The ethanol produced is very pure (unlike the fermentation process) but the yield for each cycle is very low.

DID YOU KNOW?

The addition of hydrogen reaction is used to harden the oils used to make margarine. Hydrogen and the unsaturated vegetable oils are passed over a catalyst of finely divided nickel so that the addition reaction takes place. Hydrogen adds across the carbon–carbon double bonds of the oils to make more saturated compounds.

7. **Predict the product made when hydrogen is added to butene $CH_3CHCHCH_3$.**
8. **Draw the displayed formula for the product of the reaction between but-1-ene, $CH_3CH_2CH{=}CH_2$ and chlorine.**
9. **Draw the products made with the first four alkenes and their reactions with I_2.**
10. **Propene will react with steam to form an alcohol, which contains an –OH connected to a carbon. Draw the displayed formula for the two possible products of this reaction.**

Alcohols

KEY WORDS

alcohol
fermentation
oxidation
enzymes

Learning objectives:

- recognise alcohols from their names or from given formulae
- describe the conditions used for fermentation of sugar using yeast
- write balanced chemical equations for the combustion of alcohols.

Alcohols are flammable. You may have seen this in pictures of Christmas pudding alight with brandy or flaming crepes served in French restaurants. Alcohol for drinking has been made since ancient times by fermenting fruit, grain and rice with yeast.

Figure 6.44 Alcohol is sometimes used in giving a 'flame flourish' to puddings

The alcohol series

Alcohols contain the functional group –OH.

We can recognise an alcohol by the –OH group and the fact that its name ends in –ol, for example pentan*ol*.

Ethanol can be represented in the following forms:

CH_3CH_2OH or

```
    H   H
    |   |
H — C — C — O — H
    |   |
    H   H
```

Ethanol is the main alcohol in alcoholic drinks. Aqueous solutions of ethanol are produced when sugar solutions are **fermented** using yeast.

Methanol, ethanol, propanol and butanol are the first four members of a series of alcohols.

DID YOU KNOW?

The equation for the fermentation reaction is
$C_6H_{12}O_6 \rightarrow 2C_2H_5OH + 2CO_2$

Number of carbon atoms	Name	Formula	Displayed formula
1	Methanol	CH_3OH	H | H—C—O—H | H
2	Ethanol	C_2H_5OH	H H | | H—C—C—O—H | | H H
3	Propanol	C_3H_7OH	H H H | | | H—C—C—C—O—H | | | H H H
4	Butanol	C_4H_9OH	H H H H | | | | H—C—C—C—C—O—H | | | | H H H H

Methanol, ethanol, propanol and butanol:

- are used as fuels and solvents
- dissolve in water to form a neutral solution
- react with sodium to produce hydrogen
- burn in air to produce carbon dioxide and water
- can be **oxidised** to produce carboxylic acids.

1. **Predict the structure of hexanol.**
2. **Predict the products of the combustion of butanol.**

Making alcohol

Most ethanol is made from plants, so it is a renewable resource. Ethanol for drinking is made by fermentation. This process uses yeast to turn sugars into alcohol.

glucose → ethanol + carbon dioxide

Fermentation needs sugars from plants and the following conditions:

- water
- **enzymes** from yeast
- a temperature between 25 °C and 40 °C
- an absence of oxygen.

The *enzymes* in yeast are the biological catalyst for the process of fermentation. Alcohol will only be made if there is no air present otherwise vinegar (ethanoic acid) will be produced. There needs to be an *optimum temperature* as:

- if the temperature is too cold the enzymes in yeast are inactive
- if the temperature is too hot the *enzymes* in yeast are *denatured*.

The ethanol produced is dilute. There needs to be *distillation* of the dilute liquid to produce almost pure ethanol.

3. **Explain why fermentation in an open flask would produce very little ethanol.**
4. **Describe what you might see as glucose ferments.**

DID YOU KNOW?

Fermentation is done in enclosed tanks. These tanks are temperature-controlled.

DID YOU KNOW?

Pure alcohol is very poisonous. The strongest whisky is about 50% water. Any stronger and it would do people serious harm.

Figure 6.45 Sodium reacts with ethanol to produce hydrogen

Reactions of alcohols

Sodium reacts with alcohols, producing hydrogen.

ethanol + sodium → sodium ethoxide + hydrogen

Ethanol burns in air to produce carbon dioxide and water.

$$C_2H_5OH + 3O_2 \rightarrow 2CO_2 + 3H_2O$$

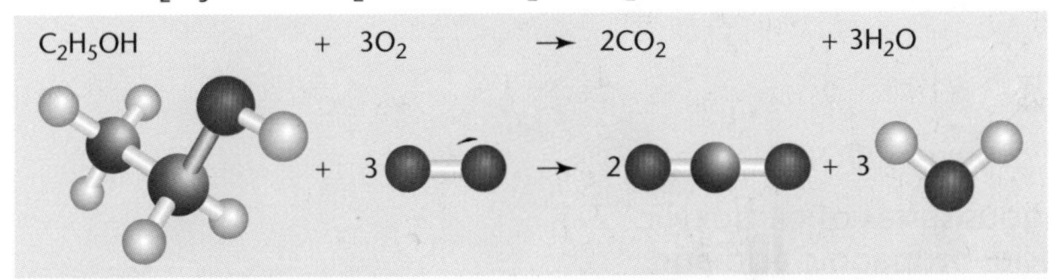

Alcohols react to produce carboxylic acids. To do this they need an oxidising agent.

$$RCH_2\text{-}OH \xrightarrow{O} RCOOH + H_2O$$

Alcohols are *oxidised* to form carboxylic acids.

5. **Write the balanced equation for the complete combustion of butanol.**

KEY INFORMATION

One oxidising agent used is potassium dichromate(VI). As it is the oxidising agent, it is itself reduced. Orange dichromate(VI) ions are reduced to a green solution containing chromium(III) ions.

Carboxylic acids

Learning objectives:

- describe the reactions of carboxylic acids
- recognise carboxylic acids from their formulae
- explain the reaction of ethanoic acid with an alcohol
- understand why carboxylic acids are weak acids.

KEY WORDS

ionises
incomplete ionisation
ester
acid catalyst

Vinegar is made when wine goes sour. The important component of vinegar is ethanoic acid. Ethanoic acid is a carboxylic acid that is used as a flavouring. There are many other carboxylic acids. When some react with alcohols they make molecules that form the basis of the perfume industry.

Behaving as an acid

Carboxylic acids are compounds that contain carbon, hydrogen and oxygen atoms but can still behave as acids.

As they behave as acids, carboxylic acids:

- dissolve in water to produce acidic solutions
- react with carbonates to produce carbon dioxide.

So ethan*oic* acid will react with sodium carbonate to produce carbon dioxide, water and a salt called sodium ethan*oate*.

They turn universal indicator yellow/orange.

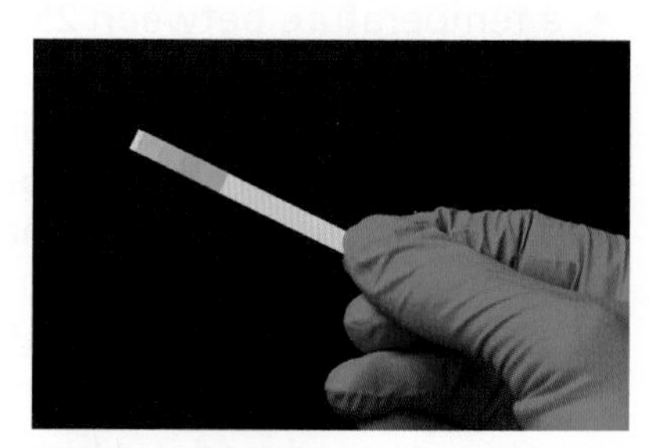

Figure 6.46 Universal indicator (UI) turns yellow/orange with ethanoic acid.

1. **Describe what you would see when ethanoic acid reacts with sodium carbonate.**

2. **Predict the salt that is made when ethanoic acid reacts with potassium carbonate.**

Carboxylic acid series

Ethanoic acid is a carboxylic acid with the formula CH_3COOH which displays as

```
      H
      |     O
      |    //
  H — C — C
      |    \
      H     O — H
```

Carboxylic acids have the functional group – COOH.

The first four members of a homologous series of carboxylic acids are methanoic acid, ethanoic acid, propanoic acid and butanoic acid.

Name	Methanoic acid	Ethanoic acid	Propanoic acid	Butanoic acid
Number of carbon atoms	1	2	3	4
Formula	HCOOH	CH_3COOH	C_2H_5COOH	C_3H_7COOH
Displayed formula	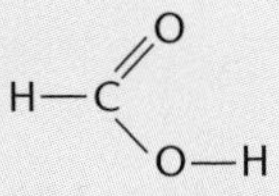 H—C(=O)—O—H	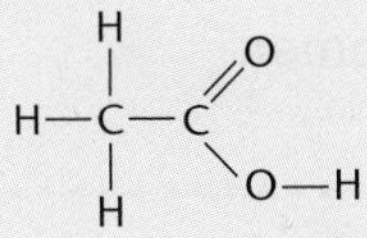 H—C(H)(H)—C(=O)—O—H	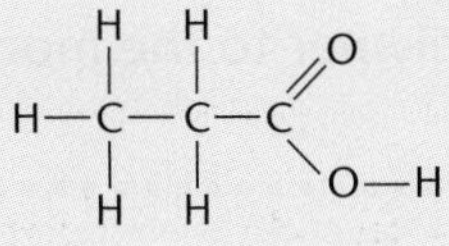 H—C(H)(H)—C(H)(H)—C(=O)—O—H	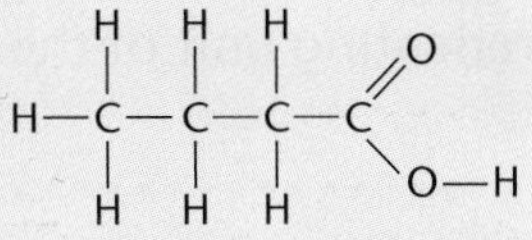 H—C(H)(H)—C(H)(H)—C(H)(H)—C(=O)—O—H

An alcohol reacts with a carboxylic acid to make an ester. (An **acid catalyst** is needed.)

Alcohol	Acid	Ester	Displayed formula
ethanol	ethanoic acid	ethyl ethanoate	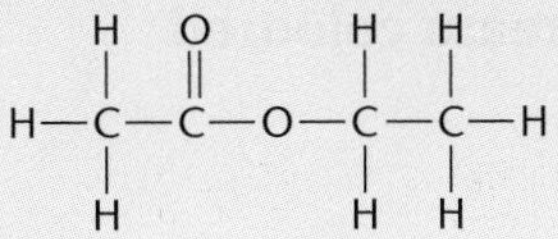 H—C(H)(H)—C(=O)—O—C(H)(H)—C(H)(H)—H

So ethanoic acid will react with ethanol to form an ester. The ester is ethyl ethanoate.

3 **Describe what kind of chemical butanoic acid will make when it reacts with an alcohol.**

alcohol + acid → ester + water

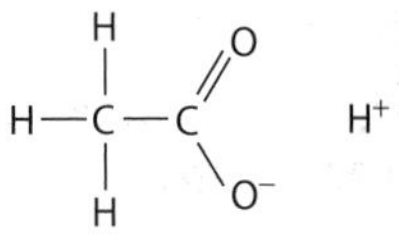

H—C(H)(H)—C(=O)—O^- H^+

Figure 6.47 H^+ ions are 'given up' from the ethanoic acid molecule

HIGHER TIER ONLY

Carboxylic acids as weak acids

Carboxylic acids are weak acids as they do not **ionise** completely when dissolved in water.

Carboxylic acids make two ions in solution. For example, a molecule of ethanoic acid dissolves to make a hydrogen ion and an ethanoate ion.

The H^+ ion is the ion that makes the solution acidic.

However, not all the molecules of ethanoic acid lose their hydrogen ions at once. So ethanoic acid is a *weak* acid. Carboxylic acids in solution make *some ions* but *some molecules* remain. *Ionisation* is not complete.

4 **Explain why propanoic acid is a weak acid but hydrochloric acid is a strong acid.**

5 **Predict the ions present in a solution of hexanoic acid.**

DID YOU KNOW?

A strong acid loses or gives up all its H^+ ions at once.

REMEMBER!

You are not expected to write balanced chemical equations for the reactions of carboxylic acids.

Addition polymerisation

Learning objectives:

- recognise addition polymers and monomers from diagrams
- draw diagrams of the formation of a polymer from an alkene monomer
- relate the repeating unit of the polymer to the monomer.

KEY WORDS

monomer
polymer
polymerisation
repeating unit

Many of the materials that we depend on for our modern lifestyle are polymers. Although they are very common now, polymers have only been made in the last century when crude oil was extracted on a large scale. The simplest polymer poly(ethene) was made by accident in 1933 in Cheshire where they made a cream coloured walking stick as the very first item.

Polymers

A **polymer** is a:

- very big molecule
- very long chain molecule
- molecule made from many small molecules called **monomers**.

Alkenes can be the monomers used to make polymers such as poly(ethene) and poly(propene).

When many monomers are joined to make a polymer the reaction is called **polymerisation**.

In polymerisation reactions, many small molecules, called monomers, join together to form very large molecules, polymers.

A small letter '*n*' is used in science to mean 'lots of'. So a polymer made from one monomer [X] is written -[X]-_n and can be represented like this:

$n(\square) \rightarrow [\square\text{-}\square\text{-}\square\text{-}\square\text{-}\square\text{-}\square\text{-}\square\text{-}\square\text{-}\square]_n$

The dashed line means 'and longer'.

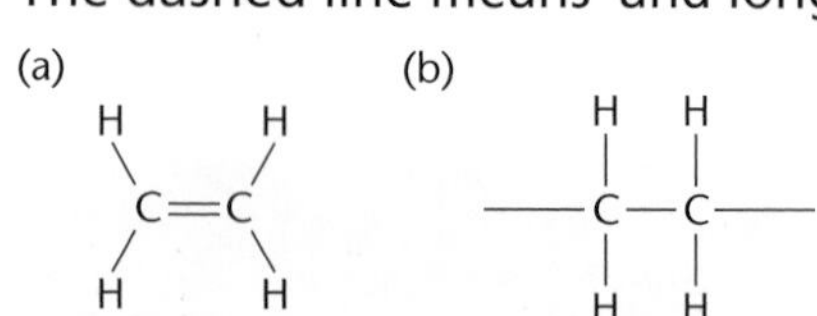

These are the displayed formulae for ethene and for ethene making a polymer.

$\left[\text{—C(H)(H)—C(H)(H)—} \right]_n$

This is the displayed formula for the polymer of ethene, which is poly(ethene).

Figure 6.48 Raincoat made from poly(ethene) and rope made from poly(propene)

The displayed formulae of polymers are written as **repeating units** with a square bracket at each end. A small '*n*' follows the last bracket to show that there could be more repeating units and the chain could go on.

1 **If O represents a monomer, draw the formula for a polymer showing 3 repeat units.**

KEY SKILLS

You should be able to draw diagrams to represent the formation of a polymer from a given alkene monomer.

Addition polymers

There are two ways polymers can be made:

- by addition
- by condensation.

In this topic we shall look at addition polymers.

Alkenes are able to make polymers by *addition* reactions.

Alkenes are able to be monomers because of the C=C double bond. Addition can take place across this double bond.

2 **Explain why butene is able to function as a monomer for an addition polymer.**

3 **Draw the three repeating units of poly(propene) made from the monomer propene shown on the right.**

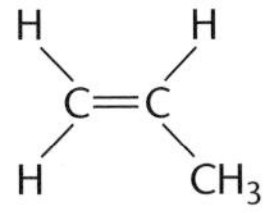

4 **Predict the monomer that reacts to produce the polymer shown on the right.**

H H
C=C
H CH_3

H H H H H H
—C—C—C—C—C—C—
H Cl H Cl H Cl *n*

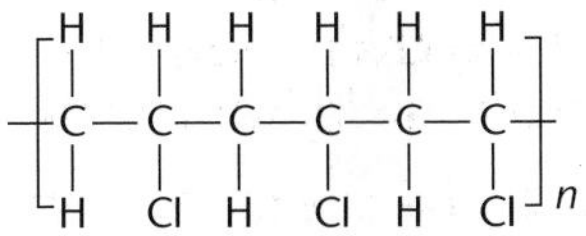

Industrial addition polymerisation

Poly(ethene) and poly(propene) are made by addition polymerisation.

The conditions needed for these reactions on an industrial scale are a high pressure and a catalyst.

Addition polymerisation is the reaction of many monomers that have double bonds to form a polymer that has single bonds.

In addition polymers, the repeating unit has the same atoms as the monomer as no other molecule forms in the reaction.

If the displayed formula of an addition polymer is known, the displayed formula of its monomer can be worked out by looking at its repeated units.

DID YOU KNOW?

Many of the advances in healthcare and communication technology would not have been possible without the development of polymers.

5 **Draw three repeating units of the polymer made from the monomer shown on the right.**

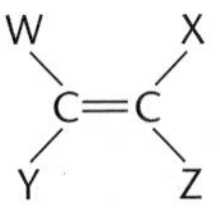

6 **Deduce the monomer that reacts to produce the polymer shown on the right.**

W X
C=C
Y Z

A B A B A B
—C—C—C—C—C—C—
D E D E D E *n*

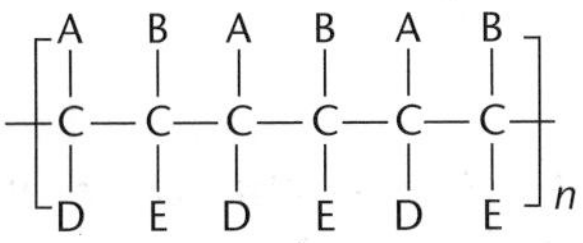

7 **Draw the repeat units of the polymers made from the following monomers:**

a

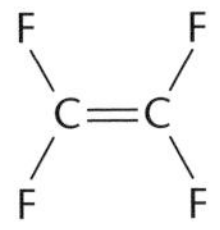

F F
C=C
F F

b

H H
C=C
H C≡N

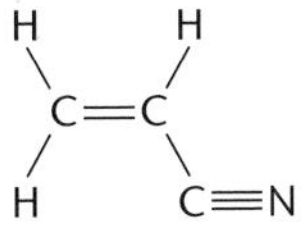

Condensation polymerisation

Learning objectives:

- explain the basic principles of condensation polymerisation
- explain the role of functional groups in producing a condensation polymer
- explain the structure of the repeating units in condensation polymers.

KEY WORDS

condensation
polyester
ethanediol
hexanedioic acid

Fabrics for clothes, paint for cars and cases for computers are all made from different polymers. Polymers such as polyester and nylon are made from two component molecules on an industrial scale. They are synthetic polymers made by condensation reactions.

HIGHER TIER ONLY

Condensation reactions

We have already seen that polymers can be made in one of two ways by addition reactions and by **condensation** reactions.

We saw that addition polymers are made with monomers that have at least one double bond. Usually there is only one type of monomer so the addition polymer has the same empirical formula as the monomer.

In this topic we shall look at the other way that polymerisation occurs, which is by condensation reactions.

Some polymers such as **polyesters** are not made with just one monomer. They need at least two monomers to join together.

The polymer structure is then represented as: $-\!\!\!+[X][Y]+\!\!\!-_n$

The [X] is one type of monomer and [Y] is another type of monomer.

When two of these types of monomers react they join together, usually losing small molecules such as water. This is why these reactions are called condensation reactions.

Examples of condensation polymers are nylon, Kevlar and polyesters.

1 Explain the difference between the structure of a condensation polymer and an addition polymer.

KEY INFORMATION

Just as before in addition polymerisation;

- a small letter '*n*' is used in science to mean 'lots of'
- the dashed line means 'and longer'.

Condensation polymerisation

Condensation polymerisation involves monomers that have different *functional* groups. For example the functional group on the monomers could be an alcohol (–OH), a carboxylic acid (–COOH) or an amine (–NH_2).

In condensation reactions, two functional groups react together to produce a new functional group plus a small molecule such as water.

For example, an –OH group from alcohol and a –COOH group from carboxylic acid react together to form an ester plus water.

alcohol + carboxylic acid → ester + water

R–OH + R′–COOH → R′COOR + H_2O

This type of single condensation reaction can be used multiple times to form long chains of condensation polymers with repeat units containing the joined functional groups.

The simplest polymers are produced from two different monomers with different functional groups. Each monomer has *two* of the same functional groups, one on each end of the monomer. (di- means two)

For example: ethane**diol** (two OH groups) and hexane**dioic acid** (two COOH groups) polymerise to produce a polyester.

2 Explain why ethanediol and hexanedioic acid can join together to make a polymer.

Predicting the products of polymerisation

All condensation polymers are made from two functional groups joining together with the loss of a small molecule.

3 Draw two repeat units of the polymer made from ethanediol, HO–CH_2–CH_2–OH, and ethanedioic acid, HOOC–COOH.

4 The monomers used to make Terylene, a polyester, are shown below. Draw the repeat unit.

HO–CH_2–CH_2–OH HOOC–C_6H_4–COOH

5 3-hydroxypropanoic acid has the structure HO–CH_2CH_2–COOH. Explain whether it can be made into a polymer.

6 The repeat unit of nylon-6,6, which is a polyamide, is shown below. Amides contain the functional group –CONH–.

$$\left[-N(H)-(CH_2)_6-N(H)-C(=O)-(CH_2)_4-C(=O)- \right]$$

a Explain whether nylon-6,6 is a condensation or addition polymer.

b Amides are made when a carboxylic acid reacts with an amine. Amines contain the functional group –NH_2. Draw the monomers from which nylon-6,6 is made.

DID YOU KNOW?

Polyester was first invented in 1941 in the UK.

DID YOU KNOW?

Nylon was the first condensation polymer to be made. It was developed in the USA in 1935. Its name is made up from New York and London. Nylon is made by the condensation reaction of a diamine and a dioic acid.

Amino acids

Learning objectives:

- describe the functional group of an amine
- identify the two functional groups of an amino acid
- explain how amino acids build proteins.

KEY WORDS

amino acid
glycine
peptide
protein

Amino acids are the building blocks of peptides which are the building blocks of proteins. Proteins are essential for all animals and plants to enable healthy growth and repair of the body. You cannot live without amino acids. There are some that the body cannot make so it is important that we eat foods that contain them.

Amines

Amino acids have *two* functional groups. One of these groups is the amine group. An amine is a carbon compound containing an NH_2 group. For example, methylamine CH_3NH_2.

Again, these compounds form a homologous series.

Number of carbon atoms	Name	Formula	Displayed formula
1	Methylamine	CH_3NH_2	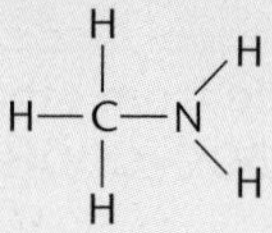 H–C(H)(H)–N(H)(H)
2	Ethylamine	$C_2H_5NH_2$	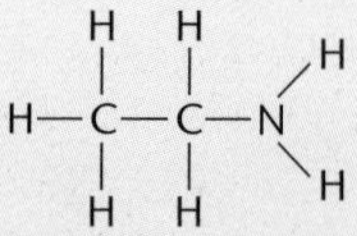 H–C(H)(H)–C(H)(H)–N(H)(H)
3	Propylamine	$C_3H_7NH_2$	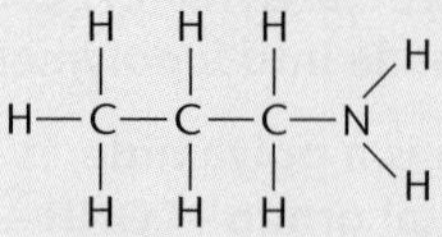 H–C(H)(H)–C(H)(H)–C(H)(H)–N(H)(H)
4	Butylamine	$C_4H_9NH_2$	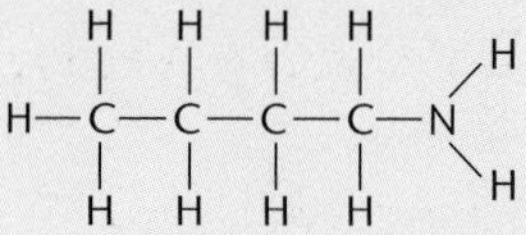 H–C(H)(H)–C(H)(H)–C(H)(H)–C(H)(H)–N(H)(H)

1 **Predict the name and formula of the eighth member of the series.**

2 **Predict the number of carbons, *x*, in an amine with the formula $C_xH_{15}NH_2$.**

DID YOU KNOW?

The amine group can be attached to more than one carbon atom in the molecule of propylamine. For example the three carbon atom skeleton can form two types of amine. These are called isomers.

$CH_3CH_2CH_2NH_2$

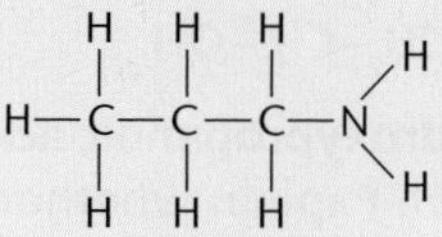

$CH_3CHNH_2CH_3$

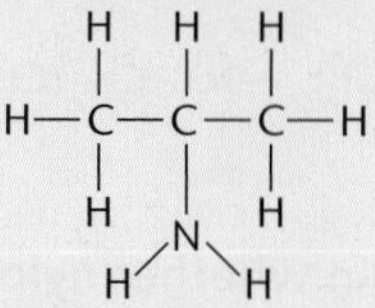

Figure 6.49 Two isomers of propylamine (primary and secondary)

Amino acids

Amines have one functional group, R–NH_2. Carboxylic acids have one functional group, R–COOH.

Amino acids, however, have two functional groups, H_2N–R–COOH.

The simplest amino acid is H_2NCH_2COOH This is known as **glycine**.

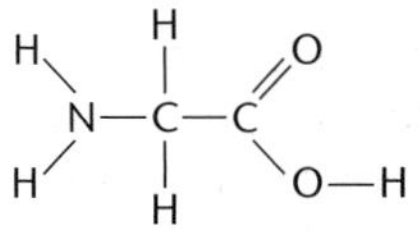

H H O
N—C—C
H H O—H

Figure 6.50 Displayed formula of glycine

Number of carbon atoms	Name	Formula	Displayed formula
1 + 1	Glycine	H_2NCH_2COOH	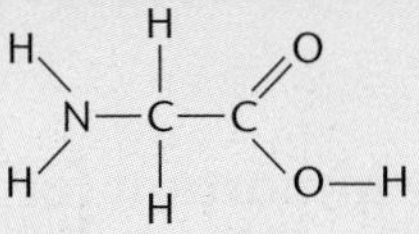 H H O N—C—C H H O—H
2 + 1	Alanine	$H_2NC_2H_4COOH$	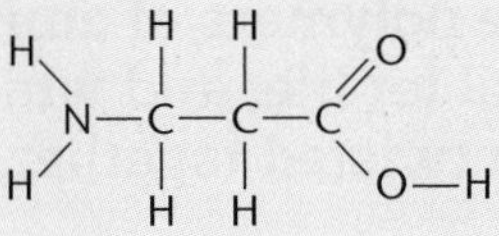 H H H O N—C—C—C H H H O—H
3 + 1	Valine	$H_2NC_3H_6COOH$	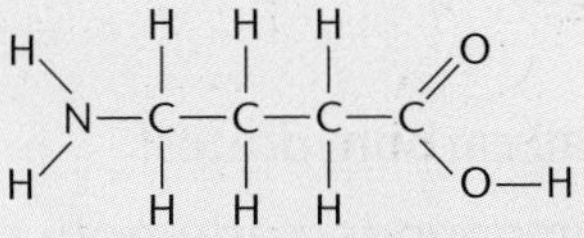 H H H H O N—C—C—C—C H H H H O—H
4 + 1	Leucine	$H_2NC_4H_8COOH$	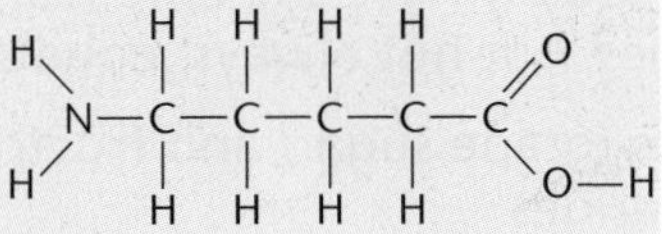 H H H H H O N—C—C—C—C—C H H H H H O—H

KEY INFORMATION

An amine and a carboxylic acid joined together without a carbon skeleton in between is H_2NCOOH carbamic acid. It is a molecule needed in the production of urea.

Four simple amino acids are shown in the table. The number of carbon atoms increases for these simple examples. Valine has a three carbon atom skeleton and one carbon atom as the carboxylic acid functional group. However, there are many different types of amino acids and they do not form one homologous series. They can have different elements in them, such as S or P. Importantly, 20 of them are used to build proteins in the body.

3 **Describe the structure of an amino acid.**

Polypeptides and proteins

Amino acids react by *condensation polymerisation* to produce polypeptides.

For example: glycine polymerises to produce

$$n\ HN_2CH_2COOH \rightarrow (\text{-}HNCH_2COO\text{-})_n + n\ H_2O$$

Different amino acids combine to make different **polypeptides**.

Different polypeptides join to make **proteins**.

So different amino acids combine to produce proteins.

4 **Use the table to predict the repeat unit of the polypeptide made by condensation polymerisation of alanine.**

DID YOU KNOW?

There are 9 essential amino acids and 11 non-essential amino acids.

- **Essential means that these amino acids *have to be taken in from food* as the body cannot make them.**
- **Non-essential means that they can be made in the body from other molecules.**

REMEMBER!

You only need to know glycine as an example of an amino acid.

DNA and other naturally occurring polymers

Learning objectives:

- describe the components of natural polymers
- explain the structures of proteins and carbohydrates
- explain how a molecule of DNA is constructed.

KEY WORDS

cellulose
deoxyribonucleic acid
DNA bases
protein

Many biological molecules that are important for life are naturally occurring polymers. Starch and cellulose are polymers of sugars. Proteins are polymers of amino acids. Another large molecule essential for life, and which is made up of two polymer chains twisted together, is DNA (deoxyribonucleic acid).

Figure 6.51 Sugar, starch and cellulose are all carbohydrates

Starch and cellulose

Sugar, starch and **cellulose** are all carbohydrates.

Carbohydrates are biological compounds made from carbon, hydrogen and oxygen. They usually have the empirical formulae $C_m(H_2O)_n$, (where *m* sometimes, but not always, equals *n*).

Simple sugars such as glucose (grape sugar) and fructose (fruit sugar) are called *mono*saccharides.

If monosaccharides join together they can form very long chains, which are natural *polymers*. These are called *poly*saccharides and include starch and cellulose.

Starch is produced from glucose as an energy store by most green plants. Starch is the most common carbohydrate consumed by humans, mostly from wheat, potatoes, rice, maize and cassava.

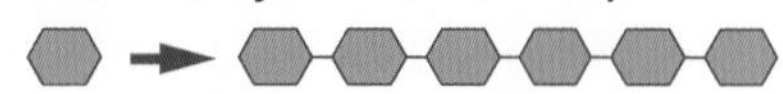

glucose
a monosaccharide

starch
a polysaccharide

Figure 6.52 Glucose forms the polymer starch

Cellulose is another polymer of glucose that is a component of the cell wall of green plants.

DID YOU KNOW?

If glucose and fructose react together they form sucrose. Sucrose is a *di*saccharide and is the sugar from sugar cane and sugar beet. Lactose (milk sugar) and maltose (malt sugar) are also disaccharides.

1. **Draw the other strand of DNA that would pair with the four nucleotides in Figure 6.54 on the next page.**

DNA

DNA encodes genetic instructions for both the development of living organisms and viruses and their functioning.

Most DNA molecules are two polymer chains, made from monomers called *nucleotides*. These are joined together in two long chains in the shape of a *double helix*.

Each *nucleotide* is made of three parts: a sugar molecule, a phosphate group and a **base**.

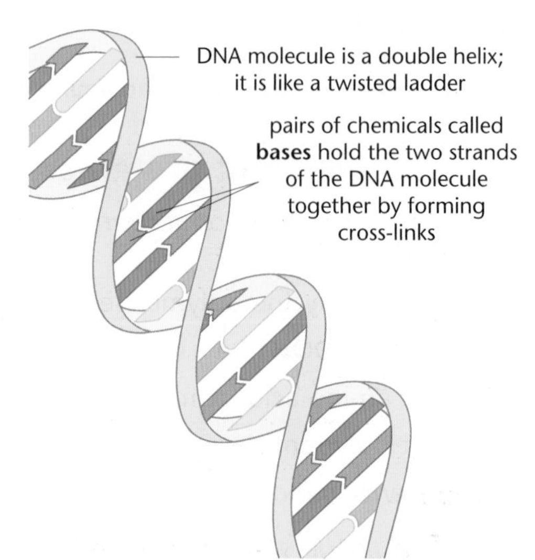

Figure 6.53 DNA is a molecule in the form of a double helix

The sugar molecule and the phosphate group are always the same. They form the backbone of the polymer chains.

There are four different bases.

They are: A (adenine) T (thymine) G (guanine) and C (cytosine) so there are four different nucleotides.

The bases always pair with each other in the same way. A with T and C with G. It is the order of the bases within the polymer chains that encodes the genetic information.

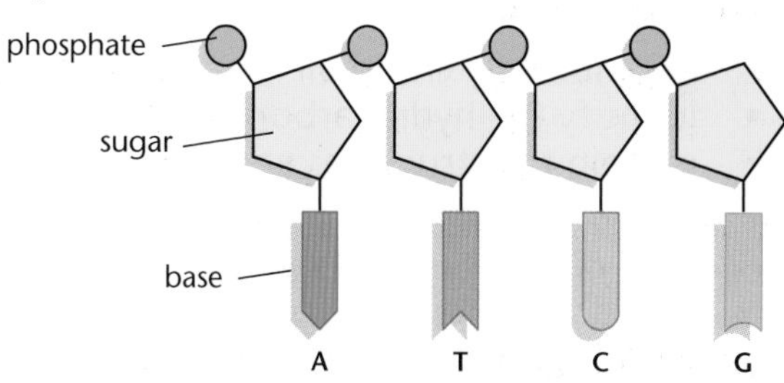

Figure 6.54 Four nucleotides made from alternating sugar and phosphate groups with a different base attached each time.

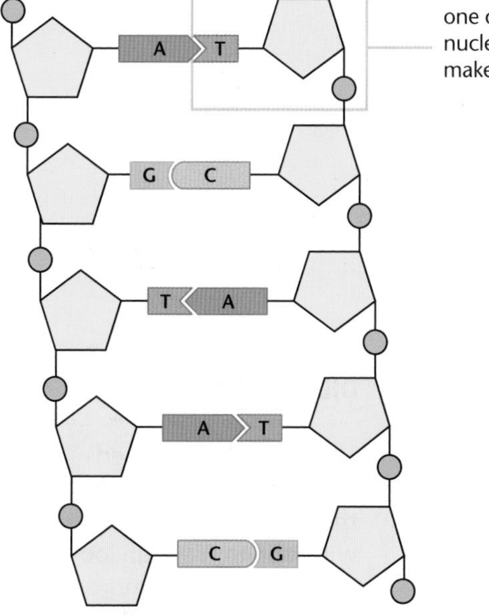

Figure 6.55 Five nucleotides, in random order, as part of a DNA polymer

DID YOU KNOW?

Polysaccharides also include *glycogen* (the long term energy store molecule for humans) and *chitin*, which is used to make the exoskeletons of crabs, lobsters and shrimps.

2 **Draw a diagram of the whole nucleotide on the opposite helix to the nucleotide with a guanine base.**

HIGHER TIER ONLY

Proteins

Amino acids also form natural polymers that are called **proteins**. Proteins differ from one another according to the sequence of the amino acids that formed them. There are 20 different amino acids that can be arranged in almost any sequence. Many proteins act as enzymes, can be part of structures such as hair, nails and collagen and are essential to many biochemical functions, such as transport of oxygen by haemoglobin. Proteins have to be part of an animal's diet as animals cannot make amino acids.

All of these large biological molecules are called macromolecules.

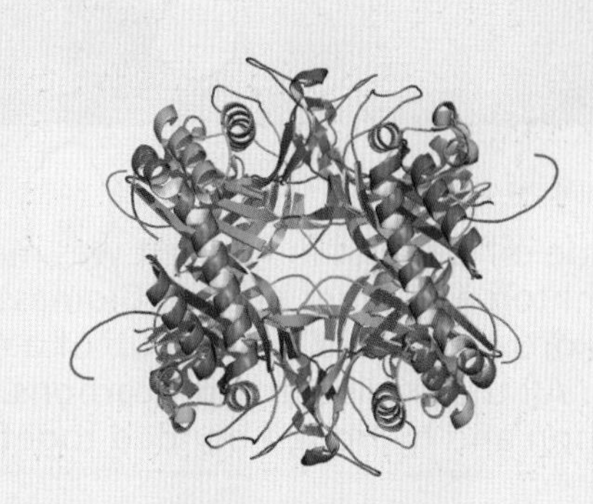

Figure 6.56 Amino acids make polymers, which are proteins

3 **Explain the function of an enzyme.**

4 **Explain why enzymes are polymers.**

Crude oil, hydrocarbons and alkanes

Learning objectives:

- describe why crude oil is a finite resource
- identify the hydrocarbons in the series of alkanes
- explain the structures and formulae of alkanes.

KEY WORDS

alkane
crude oil
finite resource
hydrocarbon

Crude oil is a smelly, yellow-to-black liquid that we can change to provide us both with fuels we use everyday and the chemicals needed to make synthetic fabrics, medicines and dyes. How do we do this? The mixture has to be separated by boiling and collecting the different parts.

Crude oil

Crude oil is a fossil fuel. It was made millions of years ago. Fossil fuels are **finite resources** because they are no longer being replaced. The conditions on Earth are not the same as they were millions of years ago. They are called non-renewable sources, as they cannot be made again.

DID YOU KNOW?

There is a theory called 'peak oil' that talks about the oil running out and what will happen. You can look this up to find out more.

Figure 6.57 Drilling for crude oil

Crude oil is found in rocks. It was made from the remains of plankton and other ancient biomass compressed in mud over millions of years. It is made up of a mixture of many types of oils. All these oils are **hydrocarbons**. Hydrocarbons are made of carbon and hydrogen bonded together.

1. **Name one other fuel that is a fossil fuel.**
2. **Suggest why alternative energy sources such as wind power are being developed.**

Products from crude oil

The mixture of oils in crude oil is separated into 'fractions' by **fractional distillation**. Then each fraction can be processed to produce fuels and feedstocks for the petrochemical industry.

Many of the useful materials on which we depend for our modern lifestyle are produced from crude oil.

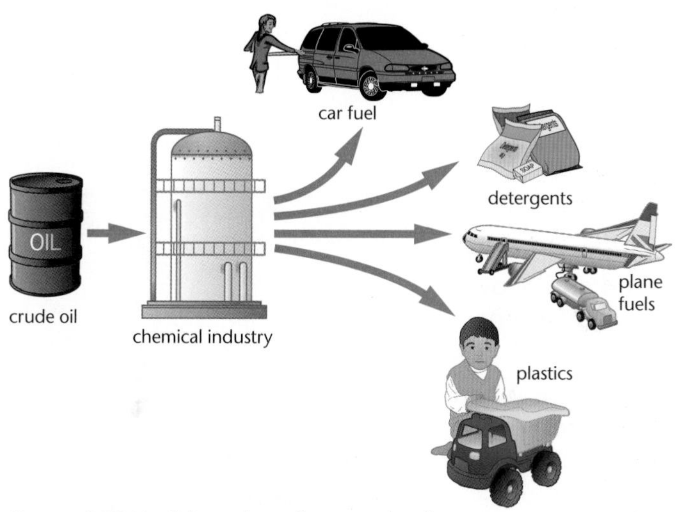

Figure 6.58 Useful products from crude oil

- Fuels, such as **liquefied petroleum gases (LPG)**, petrol, kerosene, diesel oil, heavy fuel oil
- petrochemicals, such as solvents, lubricants, polymers, detergents.

The huge number of natural and synthetic carbon compounds occurs because of the ability of carbon atoms to form 'families' of similar compounds using its four bonds.

3 **Explain the difference between a fuel and a petrochemical.**

Molecular size

The boiling point, viscosity and flammability of hydrocarbons depend on the size of their molecules. Look at the table in topic 6.23. Petrol has 6–10 carbon atoms in its molecule. Diesel has 9–16 carbon atoms so its molecules are longer.

The boiling point of petrol is lower than the boiling point of diesel. So from these two data points we could say that the larger the molecule is the higher the boiling point.

Two data points is not enough for a trend. Look at all the data from the two columns. Is the predicted trend correct? Yes – from six data points we can say the larger the molecule is the higher the boiling point.

4 **Use data from the table to describe the trends in:**
a viscosity
b flammability
as the number of carbon atoms increases.

5 **Predict the boiling point of the fraction of hydrocarbons with 28–32 carbon atoms.**

(a)

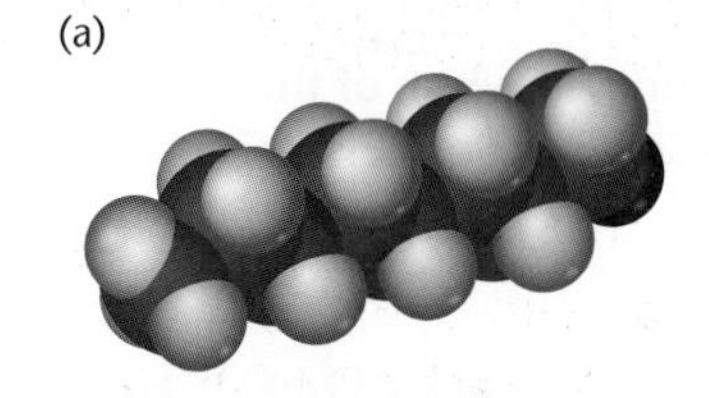

(b)

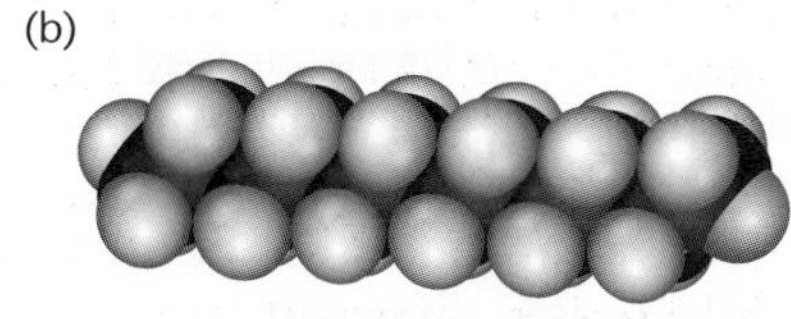

Figure 6.59 Models of one of the molecules found in (a) petrol (b) diesel. Molecule (a) is shorter than molecule (b)

Fractional distillation and petrochemicals

Learning objectives:

- describe how crude oil is used to provide modern materials
- explain how crude oil is separated by fractional distillation
- explain why boiling points of the fractions are different.

KEY WORDS

boiling point
fractional distillation
hydrocarbon
liquid petroleum gas

Separating fractions

Fractional distillation is the process of separating the mixture of **hydrocarbons** in crude oil. The oil is evaporated and allowed to cool and condense. The different fractions have different **boiling points** and so will condense at different temperatures.

Crude oil is heated at the bottom of a tower, so it is hot when it enters the bottom of the fractionating column. The top of the tower is colder. The column has a temperature gradient, hotter at the bottom, cooler at the top.

- Most fractions boil and their gases rise up the tower towards the colder top.
- Fractions such as liquid petroleum gas (LPG) have low boiling points. They stay as a gas and exit at the top of the tower.
- Fractions with slightly higher boiling points, such as petrol and kerosene, boil and stay as a gas until they reach a point in the tower where it is cold enough that they condense.
- Fractions with high boiling points, such as heavy fuel oils, boil but then condense first and exit near the bottom of the tower.
- Bitumen has such a high boiling point that it does not boil at all and sinks as a thick liquid to the bottom of the tower. This fraction is used to make tar for road surfaces.

Each fraction from the distillation contains a mixture of hydrocarbons with similar boiling points. Each fraction contains molecules with a similar number of carbon atoms.

1. **Explain why crude oil itself does not have any uses.**
2. **Suggest which of the following hydrocarbons are likely to be in the same fraction: C_2H_6, C_6H_{14}, C_7H_{16}, $C_{11}H_{24}$, $C_{20}H_{42}$. Justify your answer.**

KEY INFORMATION

Liquids boil at their boiling point and the hot gases condense back to liquids at *that same temperature.* When liquid petrol has turned to hot petrol vapour it travels up the tower at a temperature higher than its boiling point. When it reaches a colder part of the tower that is the same temperature as the boiling point of petrol, the vapour will condense back to liquid.

KEY INFORMATION

Remember that the hot chemicals enter the tower at the bottom. You may get a clearer understanding of this process if you read this part of the text again, starting at the bottom.

Why can fractions be separated?

Crude oil can be separated by fractional distillation because the molecules in different fractions have different numbers of carbon atoms and so have different length chains. This means that the total quantity of the weak forces *between* the molecules is different.

lots of weak intermolecular forces between large molecules

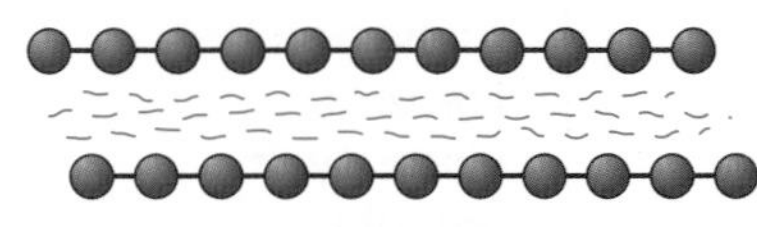

weak intermolecular forces between small molecules

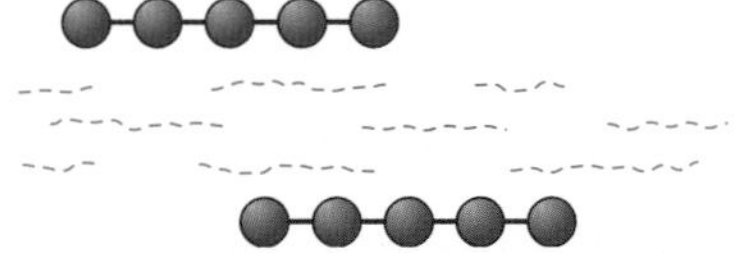

Figure 6.60 Forces *between* molecules. Large molecules such as bitumen have more weak forces of attraction between them (shown by continuous wavy lines). Smaller molecules have fewer weak forces between them (shown by short wavy lines).

These weak forces *between* the molecules are overcome during boiling. When a liquid boils, the molecules separate from each other as molecules of gas.

Large molecules, such as those that make up bitumen and heavy oil, have very long chains so there are lots of weak forces of attraction between the molecules along the chains. This means that they are difficult to separate. A lot of energy is needed to pull each molecule away from another. They have high boiling points.

3 Hydrocarbon X condenses at 68°C and hydrocarbon Y at 98°C. Explain why.

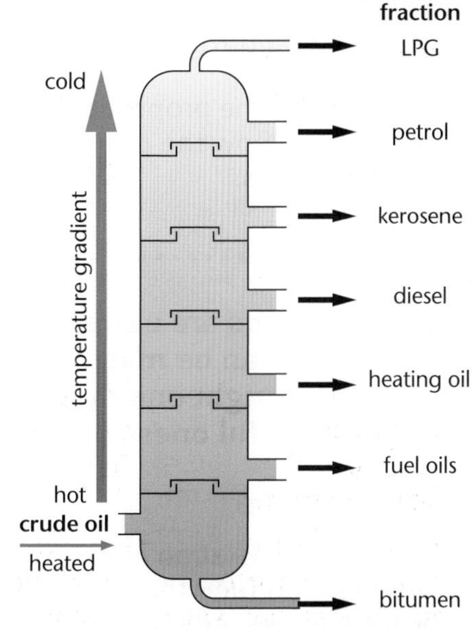

Figure 6.61 A fractional distillation column

> **KEY INFORMATION**
>
> **Forces *between* molecules are called *inter*molecular forces. You could remember this by thinking of *inter*national sporting matches which are matches *between* nations.**

Why trends in properties occur

You can see from the table in topic 6.23 that the boiling point, viscosity and flammability of hydrocarbons depend on the size of their molecules.

Boiling point and viscosity *increase* with increasing molecular size. Flammability *decreases* with increasing molecular size.

Many of the hydrocarbons in crude oil fractions are long chains. The larger the hydrocarbon chains the more of the weak forces *between* the molecules, so more energy is needed to separate them. This means that more energy is needed for the molecules to be free enough to form a vapour.

So the *larger* the molecule is the *higher* the boiling point.

4 Explain why the larger the hydrocarbon molecule is the more viscous the liquid. Use ideas about forces between molecules.

5 Figure 6.62 shows the melting points of alkanes.

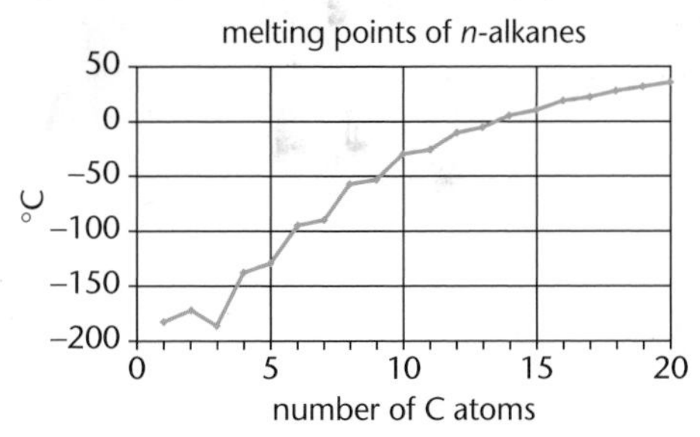

Figure 6.62

a Explain the general trend in melting points.

b The graph is not smooth. Identify a pattern between the number of carbons and melting point that takes this into account.

lots of weak intermolecular forces between large molecules

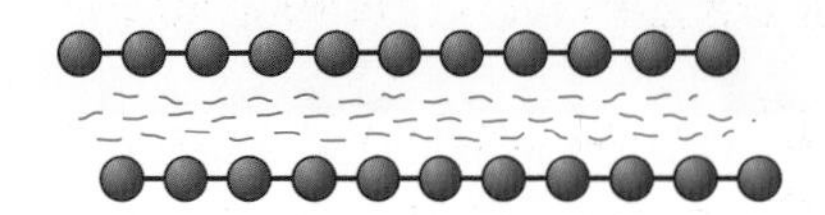

weak intermolecular forces between small molecules

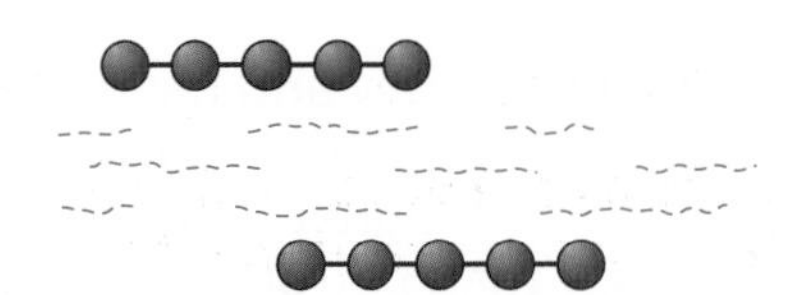

Figure 6.63 Larger molecules have stronger forces of attraction *between* them than the forces between smaller molecules

Properties of hydrocarbons

Learning objectives:

- describe how different hydrocarbon fuels have different properties
- identify the properties that influence the use of fuels
- explain how the properties are related to the size of molecules.

KEY WORDS

flammability
ignite
viscosity
volatile

Hydrocarbons are extremely useful because their properties can be matched to applications. How do we choose the right one though, and what can we do with the less useful ones?

Using fuels

Fractional distillation of crude oil results in many types of hydrocarbon fuels that have different properties. Their properties influence *how* they are used as fuels.

Figure 6.64 Different uses need different fuels

Petrol and diesel are used as fuels in car engines. Petrol and diesel are both flammable so they both burn to transfer energy out energy. They are both liquids so are transported more easily than a gas.

Petrol flows easily and has a fairly low boiling point so it vaporises easily. This means it ignites easily. Diesel is more viscous, has a higher boiling point so is less **volatile**. It **ignites** less easily than petrol.

The properties that are related to the use of a fuel are their boiling point, **viscosity** and **flammability**.

DID YOU KNOW?

Diesel has a higher freezing point than petrol. This is why diesel cars and lorries can have trouble starting in very cold winters. As the temperature falls, diesel can stop being a liquid at a higher (i.e. less cold) temperature than petrol. You could have problems with a diesel engine at around –15 °C but a petrol engine would still work well at –40 °C.

Fuels	Number of carbon atoms	Boiling point °C	Viscosity Relative scale 1 low–10 high	Flammability Relative scale 1 poor–10 good	Use
LPG	3–4	Below 30	Gas	10	Camping stoves
Petrol	6–10	80–120	2	8	Cars
Diesel oil	9–16	110–190	3	6	Cars and lorries
Kerosene	12–19	130–230	6	5	Aircraft
Heavy fuel oil	20–27	240–340	8	3	Ships

1. **An engine requires a fuel with a boiling point of 180°C but with low viscosity. Suggest which should be used.**

Supply and demand

It is difficult for industry to match supply with the demand for petrol. For example more petrol is needed, but less kerosene is needed than crude oil can provide. The solution to the problem of **supply and demand** is **cracking**. Industrial cracking plants are built near to refineries so that large hydrocarbons can be cracked to make smaller, more useful molecules. The supply of petrol can then be matched more closely with demand as more petrol is produced.

The table here contains an example of the data that a cracking plant could collect.

2. **Look at the table. At this cracking plant what do they need to do to match supply with demand?**

Product	Supply in tonnes	Demand in tonnes
Petrol	100	300
Diesel	200	100
Kerosene	250	50

Cracking and alkenes

Learning objectives:

- describe the usefulness of cracking
- balance chemical equations as examples of cracking
- explain how modern life depends on the uses of hydrocarbons.

KEY WORDS

alkene
bromine
catalyst
cracking
supply and demand

The fractional distillation of crude oil provides petrol and many other useful fractions containing hydrocarbons for our modern lifestyle. But too much kerosene is produced and not enough petrol. This can be solved by turning the kerosene and other large hydrocarbons into smaller molecules. This process is called cracking.

Cracking

Kerosene can be broken down or 'cracked' into petrol. This can be done both on an industrial scale and also in the laboratory.

Figure 6.65 Cracking alkanes on an industrial scale

To crack a liquid alkane you need a high temperature and a catalyst.

This process involves heating long-chain hydrocarbons to vaporise them. The vapours are either passed over a hot **catalyst** or mixed with steam and heated to a very high temperature so that thermal decomposition reactions then occur. These processes are called catalytic cracking or steam cracking.

Large molecules are not so useful. Large hydrocarbon molecules can be broken down or 'cracked' to produce smaller, more useful molecules. So cracking makes more petrol and some of the shorter alkane products of cracking are also useful as fuels.

KEY INFORMATION

Remember you have learned that thermal decomposition means breaking down by using thermal energy.

However, the **alkenes** that are also produced can be used to produce polymers and as starting materials for the production of many other chemicals.

1 **State two conditions that are needed for cracking to take place.**

Testing for products of cracking

The products of cracking include alkanes and another type of hydrocarbon called *alkenes*.

This can be represented by an equation such as:

$$\underset{\text{long-chain alkane}}{C_{16}H_{34}} \rightarrow \underset{\text{shorter alkane}}{C_8H_{18}} + \underset{\text{alkene}}{2C_3H_6} + \underset{\text{alkene}}{C_2H_4}$$

Bromine water is used to test for alkenes. It is an orange solution. When an alkene is added the orange solution turns colourless.

When an alkane is added to the bromine water it remains orange because an alkane does not react with bromine.

Figure 6.66 Two test tubes of bromine solution one with an alkane added to it and one with an alkene added to it. Which tube had the alkene added to it?

Alk*en*es are more reactive than alk*an*es and react with bromine water, turning it from orange to colourless. The test tube on the right is the alkene as it has reacted with the bromine water.

2 **Bromine solution is used to test two hydrocarbons, pent*ane* and pent*ene*. Predict which one will turn the solution from orange to colourless.**

3 **Balance the cracking equation by working out the numbers D, E, F and G.**

$$\mathbf{D}\ C_{14}H_{30} \rightarrow \mathbf{E}\ C_7H_{16} + \mathbf{F}\ C_3H_6 + \mathbf{G}\ C_2H_4$$

4 **Write a balanced equation for the cracking of $C_{10}H_{22}$ that produces two products.**

DID YOU KNOW?

Zeolites are now used as modern catalysts for cracking. They have an open 'cage-like' structure.

KEY INFORMATION

You will find out later that an alkene has a *double* bond. Bromine solution is used to test for a double bond in alkenes. The orange bromine solution turns colourless because the bromine has reacted with the alkene and formed a new compound. The new compound is a dibromo compound, which is colourless.

Cells and batteries

Learning objectives:

- explain how a voltage can be produced by metals in an electrolyte
- evaluate the use of cells
- interpret data for the relative reactivity of different metals.

KEY WORDS

cell
electrolyte
electrode
rechargeable

Cells and batteries are useful to provide small voltages for mobile devices. The original cells were based on two dissimilar metals dipped in an electrolyte. These were obviously difficult to carry around. Many advances have been made in the structure of cells. Rechargeable cells and tiny cells are now in everyday use.

Simple cells

You already know that if a more reactive metal is put into a solution of ions of a less reactive metal a displacement reaction occurs.

For example, if magnesium is put into a solution of zinc ions, zinc metal will be deposited and magnesium metal will become magnesium ions in solution.

$$Mg + Zn^{2+} \rightarrow Mg^{2+} + Zn$$

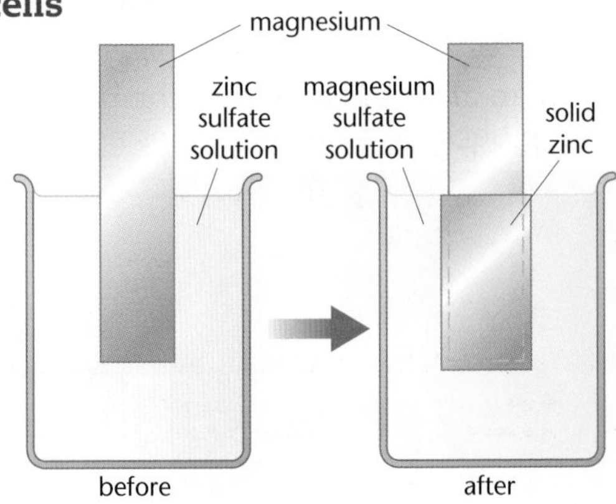

Figure 6.67 Magnesium and zinc

This happens because magnesium metal forces the zinc ions to accept two electrons.

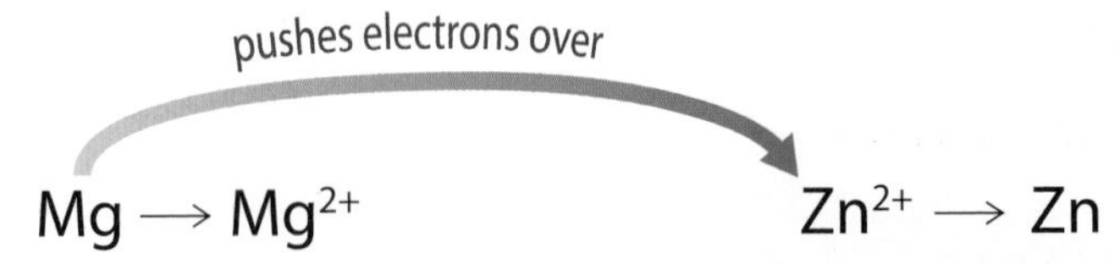

Figure 6.68 Magnesium pushes electrons over to zinc.

$Mg - 2e^- \rightarrow Mg^{2+}$ (oxidation) $Zn^{2+} + 2e^- \rightarrow Zn$ (reduction)

Magnesium is the more reactive metal.

If these two metals are put into an **electrolyte**, joined by a wire through a lamp, the lamp will light because of the *difference* in 'reactivity'.

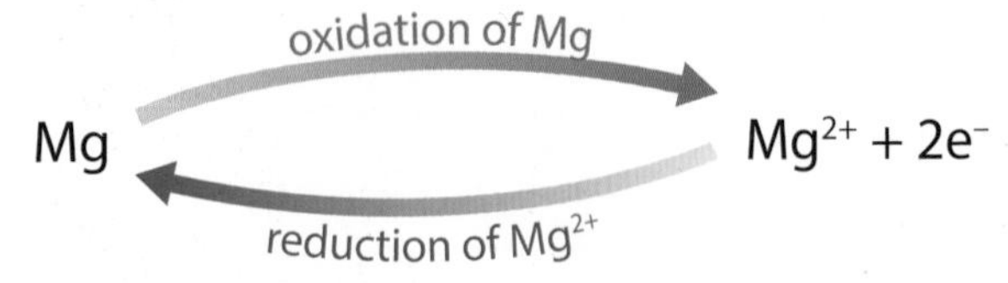

Figure 6.69 Oxidation of magnesium

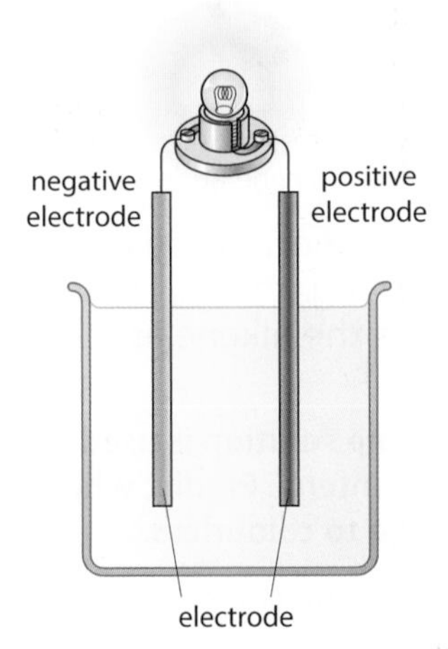

Figure 6.70 Two metals in an electrolyte

If instead of a lamp, they are joined through a voltmeter, a voltage will be noted, about 1.6 V.

This is the basis of a simple **cell**.

If two other metals are put into an electrolyte a different voltage will be noted.

The voltage produced by a cell depends on a number of factors including the type of **electrode** and electrolyte.

1 Explain how a cell between copper and zinc works.

DID YOU KNOW?

Two or more cells make a battery. Some of the batteries that you buy in a shop are actually cells.

Cells and batteries

Batteries consist of two or more cells connected together in series to provide a higher voltage.

In non-rechargeable cells and batteries the chemical reactions stop when one of the reactants has been used up. These have to be recycled as they contain heavy metals, which can cause contamination of soil and water in landfill sites.

Alkaline batteries are non-rechargeable.

Rechargeable batteries are more expensive to begin with but can be recharged easily and used many times.

Rechargeable cells and batteries can be recharged because the chemical reactions are reversed when an external electrical current is supplied.

2 Describe the advantages of using rechargeable batteries.

3 Three cells were tested for use in a small toy car.

Cell	A	B	C
Voltage volts	4	12	24
Mass g	40	150	200
Type	Rechargeable	Non-rechargeable	Lead-acid

Evaluate the data and explain which you would choose for the toy.

Comparing voltages

These voltages are based on the 'relative' reactivity of metals and measured but not under 'standard' conditions.

Metals	Magnesium	Nickel	Silver
Magnesium	0.0 V		
Nickel	–2.1 V	0.0 V	
Silver	–3.2 V	–1.2 V	0.0 V

REMEMBER!

It is possible to measure 'standard' voltages but for now you only need to use voltages based on the 'relative' reactivity of metals.

4 From these voltages, identify which metal is least reactive.

5 A cell was set up with Ni electrodes in nickel(II) sulfate and silver electrodes in silver sulfate. Write half equations for the reactions that occur at the electrodes. Ni forms Ni^{2+} ions and silver forms Ag^{+}.

Fuel cells

Learning objectives:

- describe how a fuel cell works
- explain the processes in the energy conversions
- evaluate the use of hydrogen fuel cells in comparison with rechargeable cells and batteries.

KEY WORDS

fuel cell
less pollution
oxidation
oxygen

Hydrogen fuel cells offer a potential alternative to rechargeable cells and batteries. Fuel cells are a very efficient way of producing electric current because they have no moving parts. Until recently their main use was in spacecraft. However, they are increasingly used in small vehicles. Fuel cells are mobile energy sources.

Figure 6.71 Some electric cars use fuel cells instead of batteries

What is a fuel cell?

Fuel cells are a special type of electric cell. They do not need replacing or recharging like ordinary cells and batteries. Instead, fuel cells have a fuel tank that needs refilling every now and then.

The fuel in the fuel cell is hydrogen. It reacts with **oxygen** from the air. This is an exothermic reaction so you would expect heat to be released. However, the energy released in the reaction is converted into electrical energy. This is a very efficient process.

Figure 6.72 Hydrogen fuel cells

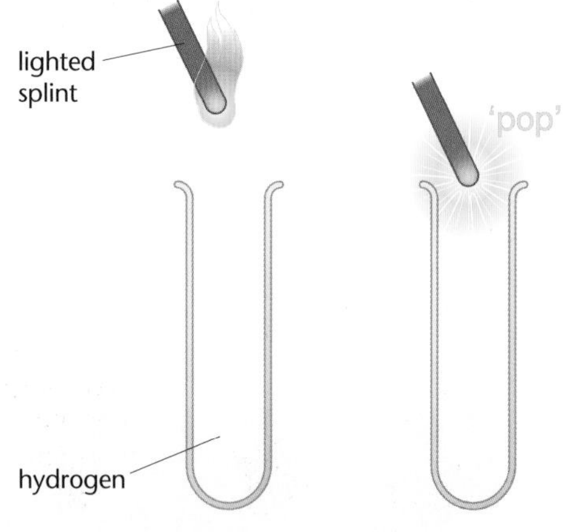

Figure 6.73 Remember the squeaky pop?

Hydrogen is a pollution-free fuel because when hydrogen reacts with oxygen, it only makes water.

hydrogen + oxygen → water

1. **Write a balanced symbol equation for the reaction of hydrogen with oxygen.**
2. **Explain why hydrogen and oxygen do not explode in a fuel cell.**

DID YOU KNOW?

Fuel cells are used in spacecraft because:

- they are efficient – they waste very little energy
- the water produced is not wasted – the astronauts drink it
- they are lightweight – so the spacecraft can carry a bigger payload
- they are compact and there are no moving parts
- they can be used continuously – they do not need to be recharged
- they do not need a special fuel – the spacecraft already has hydrogen and oxygen for its engines.

Advantages of fuel cells

The main product of a hydrogen-powered fuel cell is water, which is not a pollutant.

The advantages of a fuel cell are:

- Direct energy transfer. Chemical energy into electrical energy. Energy does not have to be converted into heat first.
- Fewer transfer stages.
- **Less polluting**, as water is the product.
- Fuel cells last longer than conventional rechargeable batteries.

The overall reaction in a hydrogen fuel cell involves the **oxidation** of hydrogen to produce water.

$2H_2 + O_2 \rightarrow 2H_2O$

When hydrogen reacts with oxygen by burning, the chemical energy is given out as heat. It is an exothermic reaction.

This is shown in an *energy level* diagram.

A fuel cell however, converts the chemical energy directly into electrical energy – there is no heat transferred out.

3 Suggest some disadvantages of using fuel cells to power cars.

KEY INFORMATION

You may remember the reaction between hydrogen and oxygen that you have tested before. Look at Figure 6.71. Do you remember the squeaky pop?

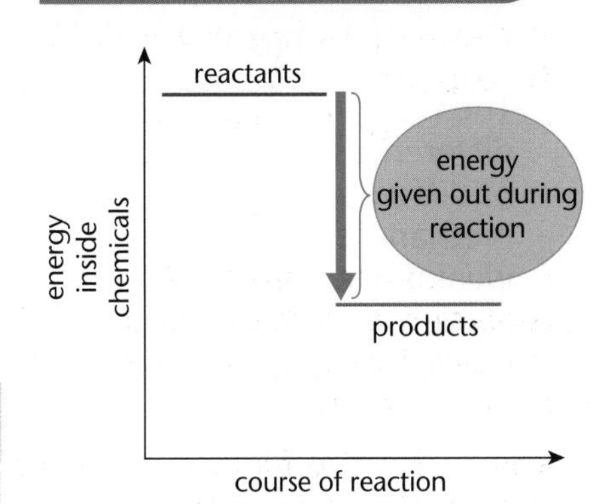

Figure 6.74 Energy level diagram for the fuel cell

HIGHER TIER ONLY

Redox reaction

Fuel cells are supplied by an external source of fuel (for example, hydrogen) and oxygen or air. The fuel is *oxidised electrochemically* within the fuel cell to produce a *potential difference*.

The reaction in the fuel cell is a redox reaction as shown in the half equations at the electrodes:

At the *negative* electrode		At the *positive* electrode
The hydrogen forms an H^+ ion at the catalyst releasing an electron	→ The electrons move through the wire, while the ions travel through the electrolyte.	The ions are reduced while reacting with oxygen in the fuel cell, which takes in the electrons
$2H_2 \rightarrow 4H^+ + 4e^-$		$4H^+ + O_2 + 4e^- \rightarrow 2H_2O$
Oxidation		*Reduction*
	A reaction where electrons are gained and lost is a *redox* reaction.	
	The overall reaction is $2H_2 + O_2 \rightarrow 2H_2O$	

4 Explain why the fuel cell reaction is a redox reaction.

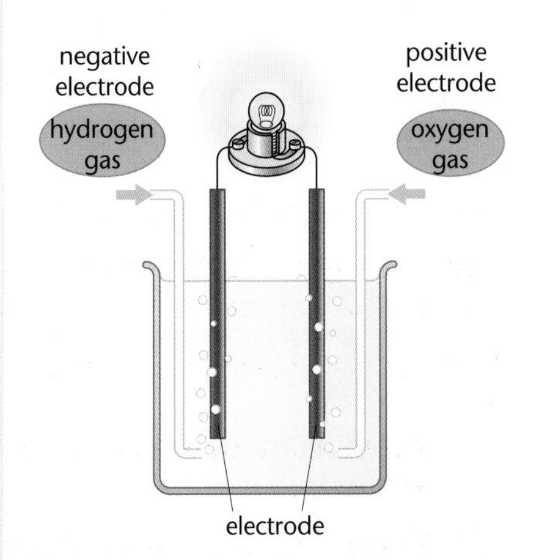

Figure 6.75 The negative and positive electrodes of a fuel cell

KEY CONCEPT

Intermolecular forces

Learning objectives:

- recognise the strong covalent bonds within molecules
- recognise the weak intermolecular forces between molecules
- describe the effects of weak intermolecular forces on properties of substances.

KEY WORDS

intermolecular
covalent
force of attraction
intramolecular

Forces between molecules

The weak forces between molecules are the cause of many differences in the way molecules and materials behave and of differences in their physical properties. The reason why methane is a gas and petrol is a liquid is, in part, due to these weak forces. The reason why poly(ethene) is a solid that can be made into stretchy film is also linked to weak forces between poly(ethene) molecules. These weak forces are *intermolecular* forces.

Bonds within molecules

Carbon can make four strong bonds. Carbon atoms can bond with other carbon atoms, with some other non-elements and also with a few metals.

Carbon atoms can bond by using single bonds, double bonds and triple bonds. For example:

C–C C=C C≡C C–H C–O
C=O C–N C–Cl

These (covalent) bonds are *very strong* and do not break easily. These bonds help to make compounds such as those shown in these formulae.

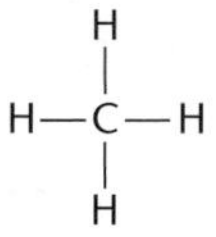

methane 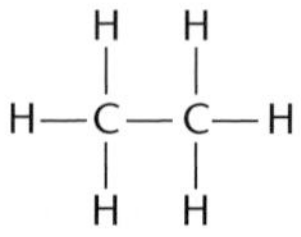 ethane 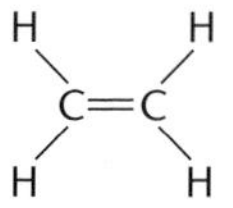 ethene 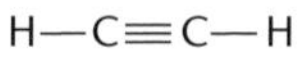 ethyne

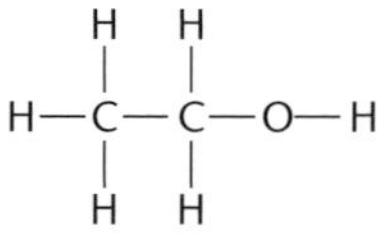

ethanol 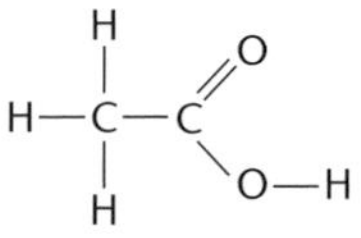 ethanoic acid 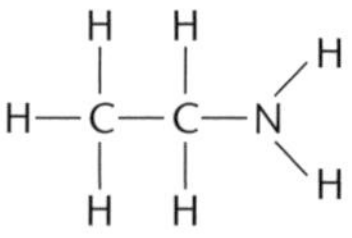 ethylamine

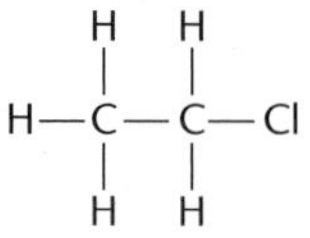

chloroethane 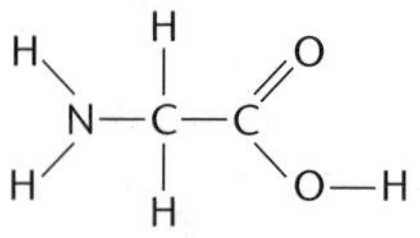 glycine 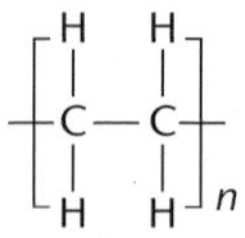 poly (ethene)

1 **Identify two molecules from those shown on the right that have double bonds.**

Forces between molecules and changes of state

A molecule of methane does not break down easily. The four bonds *within* it are very strong. If two molecules of methane are side by side at room temperature there is a very weak force of attraction *between* the two molecules.

If there are lots of methane molecules there are more of the weak forces between them. The molecules move rapidly and randomly as a gas. When the temperature reduces to –161 °C the molecules have reduced their energy of movement and are closer together. The forces of attraction are greater so the gas condenses to a liquid.

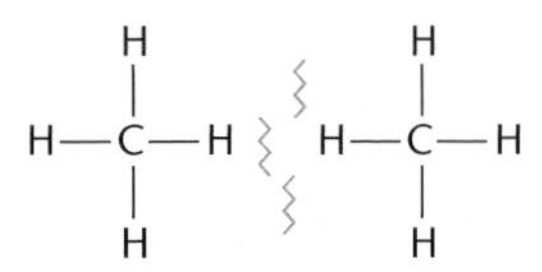

weak force of attraction between methane molecules

If the liquid is heated up again the molecules have more energy and the forces between them reduce once more so that the molecules can move freely as a gas again. They can continue to be heated but the molecules themselves will not break down. The bonds holding the atoms in each molecule together are so much stronger and do not break. These are strong bonds *within* molecules. These bonds are called *intramolecular* bonds.

The intermolecular forces between molecules are very weak. Those between methane molecules, CH_4, are weaker than those between octane molecules, C_8H_{18}. Octane is found in petrol.

Octane is a liquid at room temperature because there are more of the weak forces along the longer chains. Because there are more forces the longer chains need more energy to separate. They do not move from being part of a liquid to part of a gas as easily, so the temperature at which they can separate is higher, so the boiling point is higher.

2 **Explain why decane has a higher boiling point than octane**

(a)

methane

(b)

octane

(c)

poly (ethene)

Diagrams of (a) methane molecules, CH_4, (b) octane molecules, C_8H_{18}, and (c) poly(ethene) $–[CH_2CH_2]–_n$ (where *n* can be greater than 10 000). Methane is a gas at room temperature, octane is a liquid and poly(ethene) is a solid.

Intermolecular forces, stretching and rigidity

The size of the methane molecules CH_4 (found as a gas) are smaller than those of the octane molecules (found in liquid petrol) C_8H_{18}, but these are both much shorter than polymer chains such as poly(ethene) $–[CH_2CH_2]–_n$ which is a solid. The reason they are found at different states at room temperature is because the longer the molecule chains the greater number of weak intermolecular forces between the molecules holding them together. So the long chains of polymer are held together as a solid.

The chains in poly(ethene) are only loosely held together so they can 'slide over' one another. This allows the material that is made up from the chains to stretch easily.

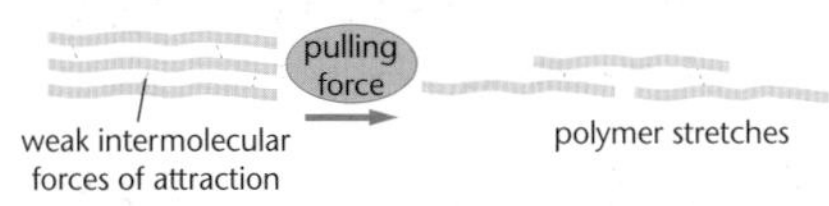

Weak intermolecular forces; the polymer can stretch.

3 **Propene has a melting point of –185°C and poly(propene) a melting point of 68°C. Explain why.**

4 **Melamine resin is used on worktop surfaces. Explain how this polymer will respond to a stretching force.**

5 **Ethene, CH_2=CH_2, has a boiling point of –104°C. Ethane also has a two-carbon chain but has a boiling point of –89°C. Explain why.**

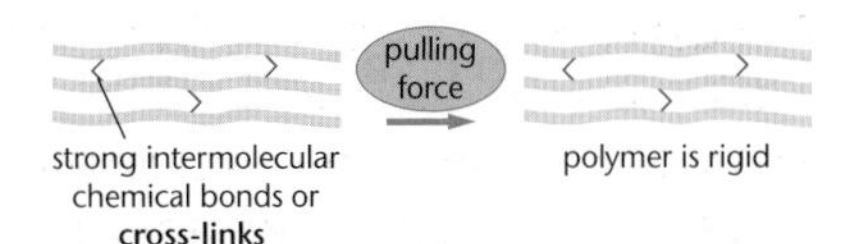

Strong intermolecular forces; the polymer cannot stretch

MATHS SKILLS

Visualise and represent 3D models

Learning objectives:

- use 3D models to represent
  - › hydrocarbons
  - › polymers
  - › large biological models.

KEY WORDS

3D model
representations
tetrahedron
helix

We use carbon compounds every day as fuels, foods and materials. Carbon atoms are able to form a huge number of combinations with hydrogen, oxygen and other atoms to make up millions of different carbon compounds. We need to have a system of representing these compounds. Can you use 3D models to represent molecules of carbon compounds and then draw them?

DID YOU KNOW?

The series of similar carbon compounds are called homologous series. Each carbon atom makes 4 bonds, but the bonds can be single, double or triple bonds.

Models of hydrocarbons

Carbon atoms are able to make four bonds, so can make whole series of similar compounds by adding another carbon atom to a chain. We use **3D models** to help us **represent** the compounds. They are especially useful because carbon atoms form 4 bonds in the shape of a **tetrahedron** (also called a triangular-based pyramid), so it is useful to see the direction of the bonds.

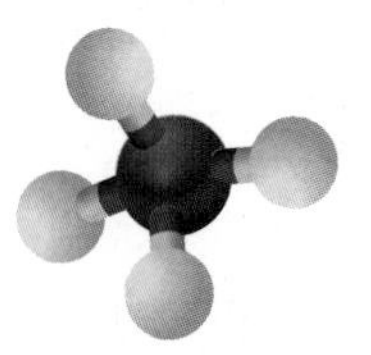

Figure 6.76 3D model of methane CH_4

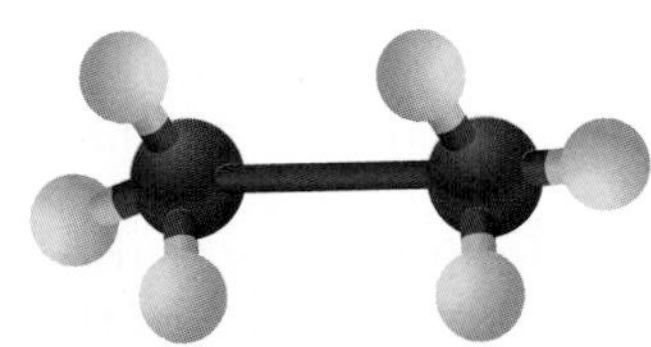

Figure 6.77 3D model of ethane C_2H_6

These can be represented by 2D 'atom and bond' diagrams that show the type of bond but not the direction of it.

```
    H
    |
H — C — H
    |
    H
```

CH_4

Figure 6.78 2D diagram of CH_4

```
    H   H
    |   |
H — C — C — H
    |   |
    H   H
```

C_2H_6

Figure 6.79 2D diagram of C_2H_6

If 5 carbon atoms join together the compound, in the same series as methane and ethane, is known as pentane, (the compound with 8 carbon atoms is known as octane).

A compound in a different series but with 5 carbon atoms is known also as 'pent', such as pentanol which is an alcohol in the same series as methanol and ethanol.

1 **Draw a 2D diagram of a molecule of propane C_3H_8.**

2 **What do these models represent:**

a

```
    H   H   H
    |   |   |
H — C — C — C — O — H
    |   |   |
    H   H   H
```

b heptane, decane or dodecane?

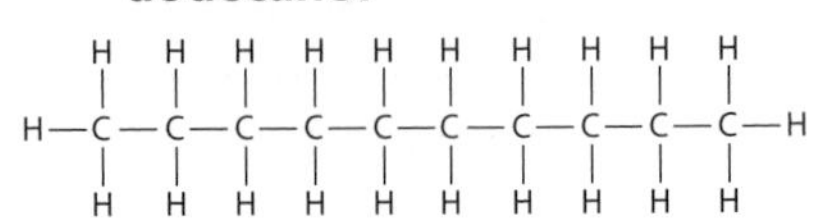

REMEMBER!

Carbon always makes 4 bonds.

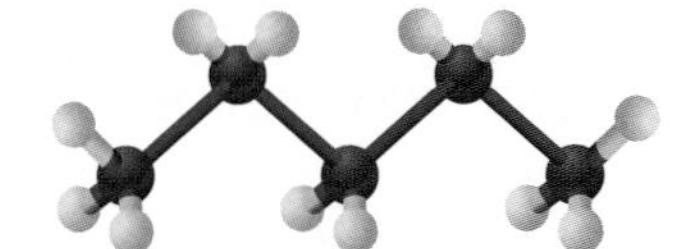

Figure 6.80 3D model of pentane C_5H_{12}

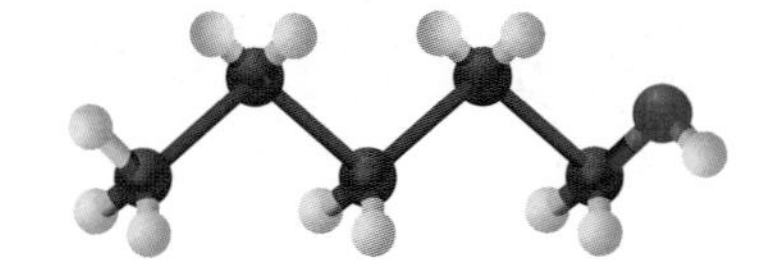

Figure 6.81 3D model of pentanol $C_5H_{11}OH$

```
     H   H   H   H   H
     |   |   |   |   |
 H — C — C — C — C — C — H
     |   |   |   |   |
     H   H   H   H   H
```

Figure 6.82 2D diagram of pentane $C_{45}H_{12}$

```
     H   H   H   H   H
     |   |   |   |   |
 H — C — C — C — C — C — O — H
     |   |   |   |   |
     H   H   H   H   H
```

Figure 6.83 2D diagram of pentanol $C_5H_{11}OH$

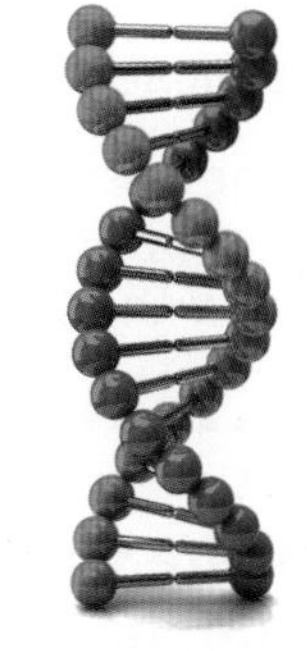

Figure 6.84 3D model of DNA

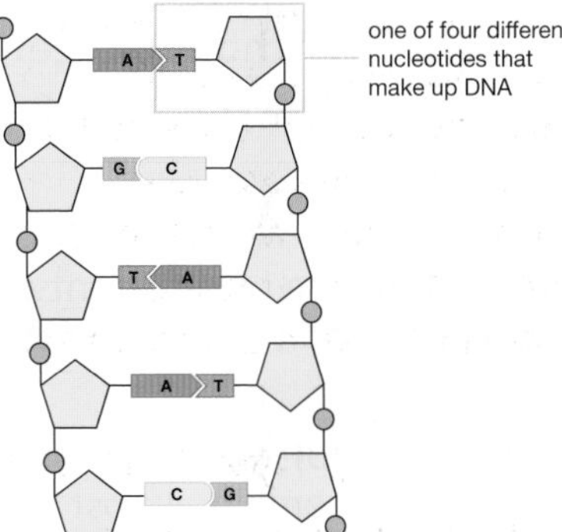

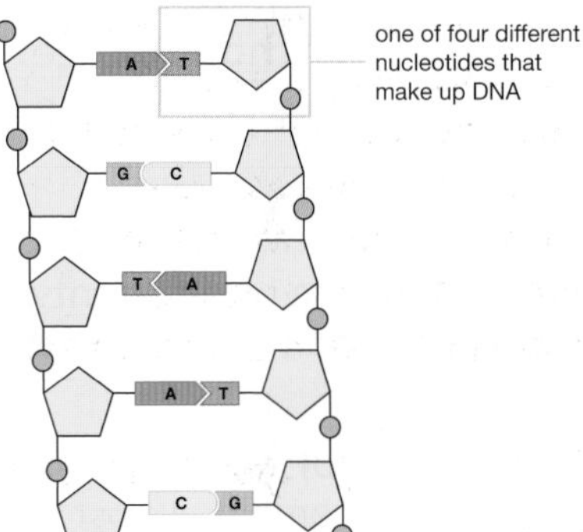

Figure 6.85 2D model of DNA

Models of larger structures

Carbon atoms can form very long chains, which are called polymers. Although the polymer is made of the same small molecule (called a monomer) added together, the number of monomers in each chain may vary. So we write the model slightly differently. We say that there are '*n*' monomers in the polymer, where '*n*' can be any number (e.g. 100 or 20,000)

```
 ( H   H )
 ( |   | )
—( C — C )—
 ( |   | )
 ( H   H )n
```

Three of these monomers joined will be:

```
  ( H   H   H   H   H   H )
  ( |   |   |   |   |   | )
 —( C — C — C — C — C — C )—
  ( |   |   |   |   |   | )
  ( H   H   H   H   H   H )n
```

This is poly(ethene).

```
 H       Cl
  \     /
   C = C
  /     \
 H       H
```

The monomer is ethene, which has a carbon to carbon double bond.

```
 H       H
  \     /
   C = C
  /     \
 H       H
```

3. **Draw a diagram of 3 monomers joined to make a polymer from CF_2=CF_2.**

4. **Draw the monomer that makes this polymer.**

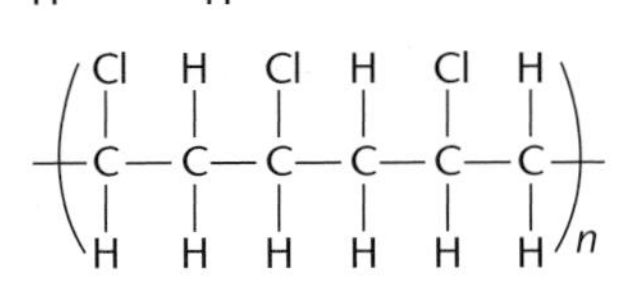

```
  ( Cl  H   Cl  H   Cl  H )
  ( |   |   |   |   |   | )
 —( C — C — C — C — C — C )—
  ( |   |   |   |   |   | )
  ( H   H   H   H   H   H )n
```

Representing large biological molecules

Molecules such as DNA, RNA and haemoglobin are best represented by 3D models as their function not only relies on their chemistry of their *functional groups* but also on the *shape* of the molecule as they 'curl up'.

DNA is a molecule consisting of two strands that makes a double **helix**.

To describe this molecule by a 2D diagram you can only focus on one part of this molecule such as the pairing of bases or the way that it twists as a double helix.

5. **Explain how many bases there are in a molecule of DNA and which bases 'pair up'.**

6. **Explain why it is necessary to use a 3D model to fully explain the structure of DNA.**

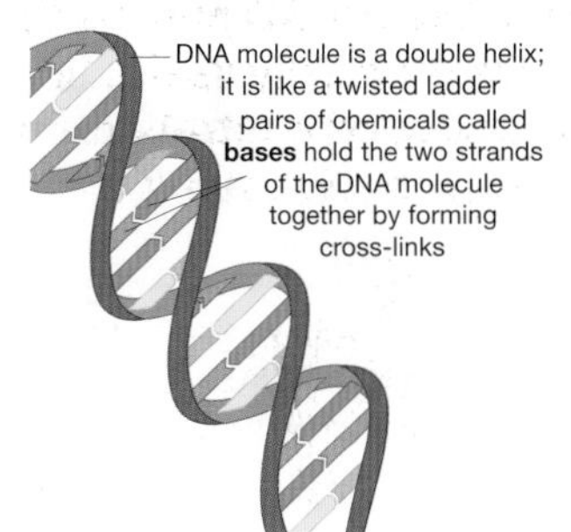

Figure 6.86 Which model gives the clearest information?

DID YOU KNOW?

X-ray diffraction is used to find out the structure of crystals. This was very important in the determination of the structure of DNA. Rosalind Franklin and Maurice Wilkins made diffraction patterns that were key evidence in the search for the structure that was found to be a double helix.

Proportions of gases in the atmosphere

Learning objectives:

- identify the gases of the atmosphere
- recall the proportions of gases
- explain how the balance of the gases is maintained.

KEY WORDS

atmosphere
carbon dioxide
nitrogen
oxygen

The atmosphere was not always made up of the gases in the mixture that we know today. However, for the last 200 million years, the proportions of different gases in the atmosphere have been much the same as they are today. Without oxygen and carbon dioxide life as we know it would not be the same.

KEY INFORMATION

The amount of water in the **atmosphere** in different parts of the world gives rise to dry desert conditions or to warm, wet and humid climates.

Air as a mixture

Air is a mixture of different gases. Clean air contains **nitrogen**, **oxygen**, **carbon dioxide**, water vapour and noble gases, such as argon.

The amount of water vapour changes but the amounts of the other gases remain almost constant.

The amounts of the various gases in clean air remain constant because of two opposing processes:

Photosynthesis	Respiration
plants use carbon dioxide and release oxygen	animals and plants use oxygen and release carbon dioxide

These two processes help to keep the balance of the proportions of gases in the air.

1 **Describe how oxygen is put into the atmosphere.**

2 **Suggest the proportion of water vapour in the atmosphere for the following conditions**

a **dry desert,**
b **hot rain forest, and**
c **cool maritime.**

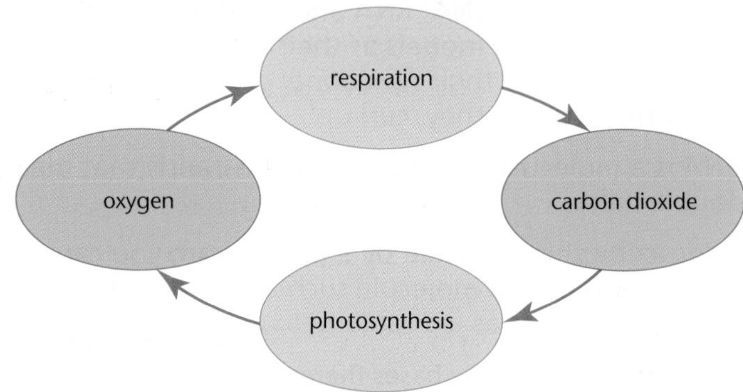

Figure 6.87 Plants photosynthesise, releasing oxygen. Animals and plants respire, releasing carbon dioxide.

Proportions

In the mixture of gases in the air there is:

- approximately 80% nitrogen (about four-fifths)
- approximately 20% oxygen (about one-fifth)
- small proportions of carbon dioxide (approximately 0.040%, often given as 400 parts per million)
- variable amounts of water vapour
- small amounts of noble gases.

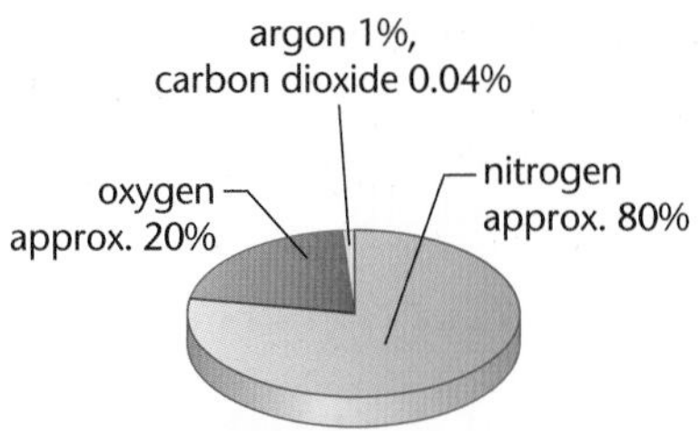

Figure 6.88 Proportion of gases in dry air

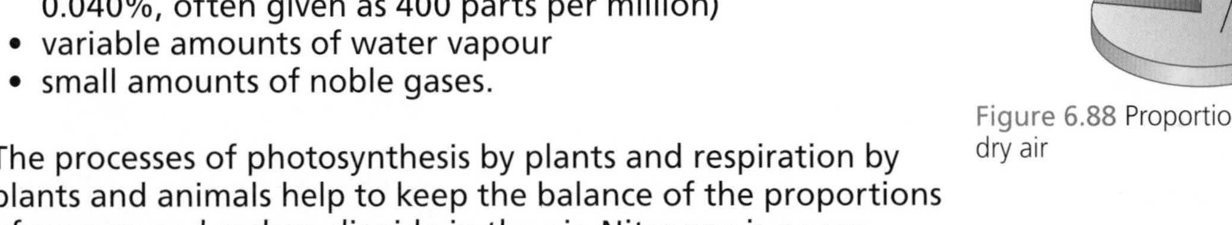

The processes of photosynthesis by plants and respiration by plants and animals help to keep the balance of the proportions of oxygen and carbon dioxide in the air. Nitrogen is a very unreactive gas so its proportion in the air is kept constant.

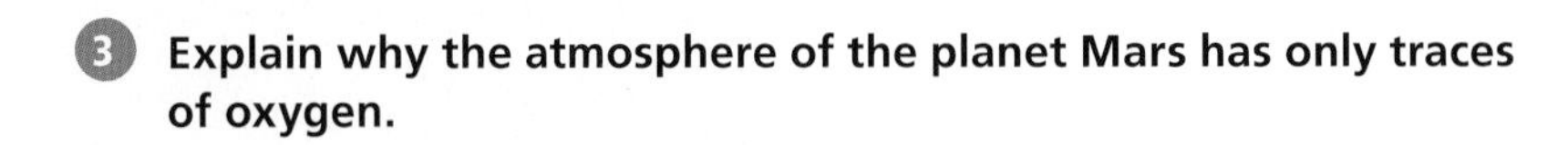

3 Explain why the atmosphere of the planet Mars has only traces of oxygen.

> **DID YOU KNOW?**
>
> **The level of carbon dioxide is constantly monitored at Mauna Loa, a volcano in Hawaii. Measurements are usually given in ppm (parts per million). It is monitored because the rising levels of carbon dioxide are causing great concern. Scientists have been measuring concentrations of CO_2 at Mauna Loa since 1958.**

Balancing the proportions

The percentages of gases in the air do not change very much. This is because there is a balance between processes that use up carbon dioxide and make oxygen and those that use up oxygen and make carbon dioxide.

Plants use carbon dioxide to make glucose through photosynthesis:

carbon dioxide + water → glucose + oxygen

$$6CO_2 + 6H_2O \rightarrow C_6H_{12}O_6 + 6O_2$$

Animals and plants use glucose and oxygen through respiration to release energy:

glucose + oxygen → carbon dioxide + water

$$C_6H_{12}O_6 + 6O_2 \rightarrow 6CO_2 + 6H_2O$$

These two processes are part of the carbon cycle. One other process that is part of this cycle is also part of the balance of the gases. This process is combustion. Combustion uses oxygen and produces carbon dioxide.

fuel + oxygen → carbon dioxide + water

4 Breathing and respiration are different processes. Explain the difference considering the gases involved.

5 Explain why 6 molecules of CO_2 are produced from 1 molecule of glucose.

6 Explain the role of glucose in maintaining the Earth's atmosphere.

The Earth's early atmosphere

Learning objectives:

- describe ideas about the Earth's early atmosphere
- interpret evidence about the Earth's early atmosphere
- evaluate different theories about the Earth's early atmosphere.

KEY WORDS

atmosphere
oceans
sediments
volcanic

The age of the Earth is currently estimated to be between 4.5 and 4.6 billion years. That is a little over four and half thousand million years (written as 4.5×10^9 years). During that time the atmosphere has changed from possibly hydrogen and helium at first, probably followed by gases from volcanic activity to finally an atmosphere rich in oxygen, which is what we have now.

REMEMBER!

You only have to remember this one theory, but there are more theories you can look up and evaluate against evidence.

The Earth's early atmosphere

We *definitely know* we now have an **atmosphere** rich in oxygen but we also *think* we know, from evidence, that oxygen was not in the Earth's early atmosphere. From the evidence we have, scientists have developed theories about the change in atmosphere over the last 4.6 billion years.

One theory suggests that during the first billion years of the Earth's existence there was intense **volcanic** activity that released gases. This process is called degassing and the idea is based on the composition of gases vented out in present-day volcanic activity.

At the start of this period the Earth's atmosphere may have been like the atmospheres of Mars and Venus today, consisting of mainly carbon dioxide with little or no oxygen gas.

Volcanoes also produced nitrogen, which gradually built up in the atmosphere and there may have been small proportions of methane and ammonia.

These gases formed the early atmosphere, along with water vapour, which condensed to form the oceans.

When the **oceans** formed, carbon dioxide dissolved in the water and carbonates were precipitated producing **sediments**, reducing the amount of carbon dioxide in the atmosphere.

Figure 6.89 Volcanic activity releases gases

Figure 6.90 The surface of Mars is being explored by robots but the atmosphere cannot support human life

1. **Explain the scientists' theory of how the first gases became part of the atmosphere for the first billion years.**
2. **Suggest why there was no oxygen in the early atmosphere.**
3. **There are several different theories about the Earth's early atmosphere. Suggest why.**

Evidence for the theories

The theories about what was in the Earth's early atmosphere and how the atmosphere was formed have changed over time.

One set of evidence is gained by measuring carbon and boron isotope ratios in sediments under the sea.

Other models use the composition of gases given out by volcanoes today as evidence. An assumption is made that the same proportion of gases are given out today as were given out by volcanoes billions of years ago.

The question we need to ask is how well do these assumptions fit with evidence gained through other means. We need to *evaluate evidence* to develop new models.

Figure 6.91 Water vapour was produced which condensed to form oceans

4 Scientists use the composition of gases given off by volcanoes to develop models of the Earth's ancient atmosphere. Suggest one reason why.

Evaluating the theories

The evidence for the early atmosphere is limited because of the time scale of 4.6 billion years, which means no direct measurements can be made. Models need to be developed using proxy evidence or by assuming that what happened in the past is still happening today and using that evidence from today (for example the composition of volcanic gases).

Proxy evidence is evidence from one source gathered from ancient times that can be used to make an assumption about another related ancient effect. An example would be counting the number of stomata on fossils of ancient leaves to make assumptions about the levels of carbon dioxide in the atmosphere.

Evidence from direct measurements would seem more valuable but there is not always agreement that the measurements were taken correctly. For example, the measurements taken from isotopes by one team are seen as more reliable if their data compares well with the data from a different site by a different team. The more measurements taken by different teams that provide data that coincide, the more reliable the data. This is why scientists publish data in scientific journals.

5 Scientists have only indirect evidence for the atmosphere of 2 billion years ago. Suggest, with reasons, whether the current composition of volcanic gases gives better or worse evidence than the evidence from counting stomata.

6 Stomata are openings that allow gas exchange in plants. Predict a relationship between number of stomata and atmospheric carbon dioxide levels.

DID YOU KNOW?

Some scientists are now rejecting the early models of the Earth's atmosphere and are studying zircon gemstones for new evidence.

How oxygen increased

Learning objectives:

- identify the processes allowing oxygen levels to increase
- explain the role of algae in the composition of the atmosphere
- recall the equation for photosynthesis.

KEY WORDS

algae
evolve
oxygen
photosynthesis

The ages of the Earth are divided into eons and eras spanning billions of years. The 'Hadean' is the first eon (4.5 to 4 billion years ago). The second eon is the Archaen. During this eon we find the first evidence of organisms that photosynthesise. This material is kerogen, which is the compressed remains of planktons found in shales. When heated, kerogen releases bitumen and crude oil.

The early production of oxygen

We have already seen that the Earth is 4.5 billion years old. In the first billion years the Earth's atmosphere may have been like the atmospheres of Mars and Venus today, consisting of mainly carbon dioxide with little or no **oxygen** gas.

The first life forms appeared over 3.5 billion years ago, but these could live anaerobically, without oxygen. **Algae** first produced oxygen about 2.7 billion years ago.

Soon after the appearance of life forms that used **photosynthesis**, oxygen appeared in the atmosphere. Over the next billion years, plants **evolved** and the percentage of oxygen gradually increased to a level that enabled animals to evolve.

Figure 6.92 Algae started to produce oxygen about 2.7 billion years ago

1. **When did algae first produce oxygen?**
2. **Explain the difference between living anaerobically and living aerobically.**

Algae and photosynthetic production of oxygen

Algae and plants produced the oxygen that is now in the atmosphere. This was produced by photosynthesis.

This process can be represented by the equation:

$$6CO_2 + 6H_2O \rightarrow C_6H_{12}O_6 + 6O_2$$

carbon dioxide + water → glucose + oxygen

You can see that the process uses up carbon dioxide. So, as more plants evolved the levels of carbon dioxide went down and the levels of oxygen went up.

Over billions of years, the percentage of oxygen in the atmosphere increased and the percentage of carbon dioxide decreased, until today's levels were reached.

Figure 6.93 These fossils are of early plants that produced oxygen by photosynthesis

KEY INFORMATION

You can show by experiment that aquatic plants produce oxygen.

The percentage of nitrogen also slowly increased, but since nitrogen is very unreactive, very little nitrogen was removed from the atmosphere.

3. **Oxygen is a product of photosynthesis. What is the other product?**
4. **Suggest why the levels of oxygen increased.**
5. **Explain the role of algae in the increased levels of oxygen.**

REMEMBER!

You do not need to remember the role of nitrogen in the atmosphere but you do need to remember the equation for photosynthesis.

The Great Oxygenation event

There is a theory that the first organisms carrying out photosynthesis, cyanobacteria, produced oxygen as a waste product, but that this oxygen was removed by the oxidation of iron to form iron(III) oxide. The evidence for this is banded iron oxide sediments, which are red and nearly always found in older rocks formed by precipitation in oceans. Examples of these formations can be now found in Minnesota and Western Australia.

The process of iron oxidation captured all the spare oxygen until there was an excess that could no longer be captured. This led to a significant rise in the levels of oxygen in the atmosphere and is known as the Great Oxygenation event. This is a theory based on evidence, which can be interpreted in different ways and combined with other evidence to support the theory. The evidence cannot be gathered by direct measurement.

Figure 6.94 A banded iron formation in Western Australia which originally used up oxygen from the air before the levels rose

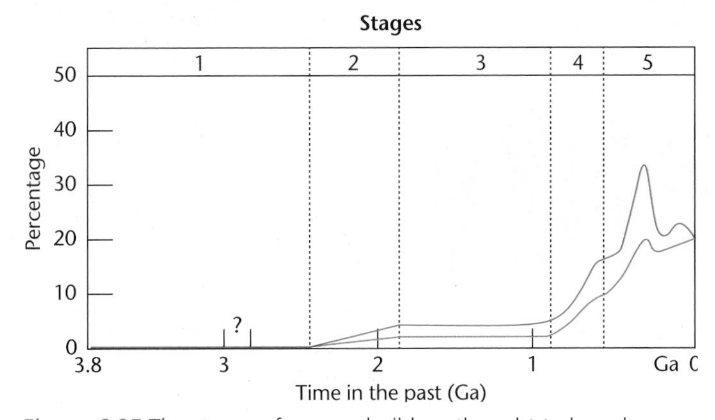

Figure 6.95 The stages of oxygen build-up thought to have happened over billions of years. The unit on the *x*-axis (Ga) means 'billions of years'. The two lines show the range of estimates.

DID YOU KNOW?

This theory has many different names and is an attempt to explain why there was a delay between the appearance of oxygen producing bacteria and increases in oxygen levels.

One piece of evidence scientists used to develop theories about the rise of oxygen levels still persists. This is the presence of microbial mats. They are present in the fossil record of 3500 million years ago but are still present and can produce a thin layer of oxygenated water under thick ice. These microbes do not die in an oxygenated environment or during ice ages.

6. **Suggest why there was a time lag between the appearance of the first organisms producing oxygen and the rise in the level of free oxygen in the atmosphere.**

KEY CONCEPT

Greenhouse gases

Learning objectives:

- describe the greenhouse gases
- explain the greenhouse effect
- explain these processes as the interaction of short and long wavelength radiation with matter.

KEY WORDS

wavelength
radiation
greenhouse
absorb

The average temperature of the Earth is 14°C. This is because we have a blanket of gases as an atmosphere that protects us by keeping the temperatures relatively stable. The greenhouse effect is a natural phenomenon that is beneficial for us. If we are found to be altering it, then that is another story.

Why are greenhouse gases important?

Water vapour, carbon dioxide and methane are the greenhouse gases in our atmosphere. These gases allow **radiation** from the Sun to pass through and warm the Earth. We all feel this on a bright sunny day.

What is less obvious, but very important, is that the warm Earth radiates energy back out into the atmosphere and out into space, otherwise the Earth would get warmer and warmer.

The **greenhouse** gases 'trap' some of this radiation and help to keep us relatively warm.

So greenhouse gases in the atmosphere maintain the temperatures on Earth high enough to support life.

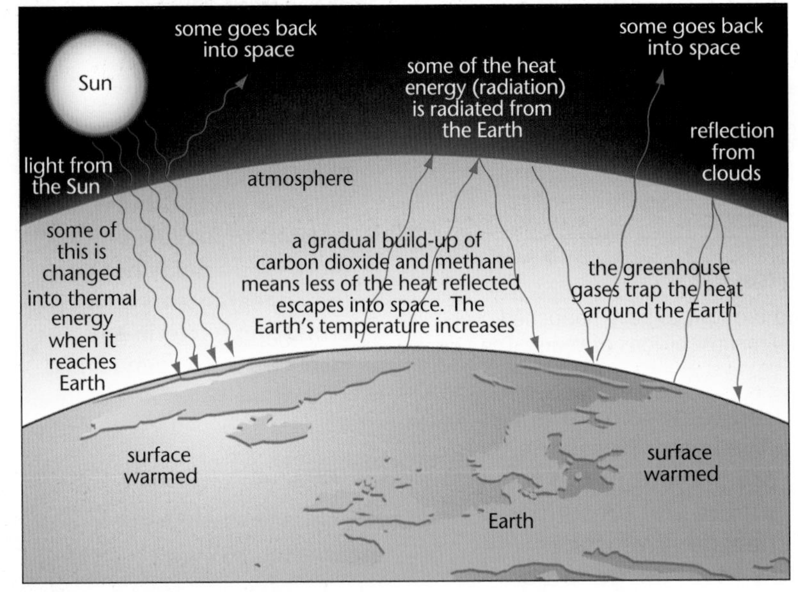

Figure 6.96 Solar radiation reaches Earth. Energy radiated back from the Earth is trapped by the greenhouse gases.

DID YOU KNOW?

The idea of a greenhouse effect was first put forward in 1824 by Joseph Fourier. It is estimated that the greenhouse effect makes the Earth 33°C warmer than it would be otherwise. What is the temperature where you are today? Now reduce that by 33°C ...

1. **Name two greenhouse gases.**
2. **Explain why greenhouse gases are important to life on Earth.**

How does the greenhouse effect work?

Radiation from the Sun is of short **wavelength**. It can pass through the atmosphere and not be **absorbed** by the atmospheric gases.

The gases allow *short wavelength* radiation to pass through the atmosphere to the Earth's surface. The gases then absorb the outgoing *long wavelength* radiation from the Earth, causing an increase in temperature. This temperature increase is beneficial to us as the temperature of the Earth is kept relatively stable.

Even though the temperature at the polar caps gets very low and the temperature at the equator can get very high, between these limits life can be sustained. Both our Moon and other planets experience temperatures way outside these limits. Life, as we know it, would be difficult to sustain elsewhere.

3 Describe the wavelength of incoming solar radiation compared with outgoing radiation from the Earth. Use a diagram to help.

4 Venus has an atmosphere of carbon dioxide. Suggest whether the average temperature will be higher or lower than expected when considering its distance from the Sun. Explain your reasoning.

DID YOU KNOW?

The surface temperature on the Moon varies between –173°C and +127°C. The range on Earth is –89°C and +58°C.

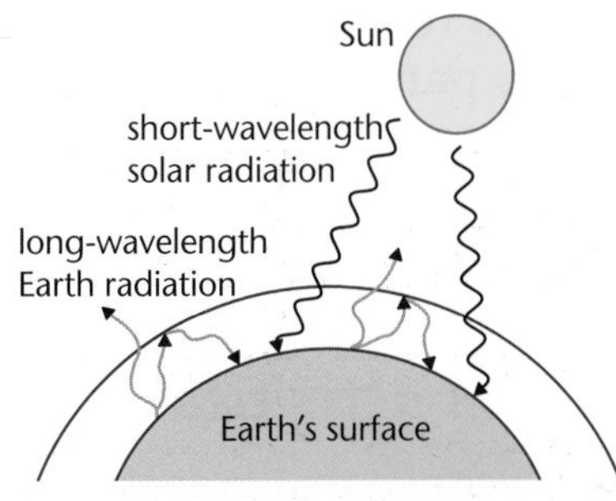

Figure 6.97 Short wavelength radiation from the Sun enters the atmosphere. Long wavelength radiation radiated back from the Earth is absorbed by the greenhouse gases.

What types of radiation are involved?

Radiation from the Sun is of short wavelength such as ultraviolet (UV), light and near infrared (near IR). It can pass through the atmosphere and not be absorbed by the gases. The radiation that comes back from the Earth is of a longer wavelength, which is far IR. This long wavelength IR radiation is absorbed by the molecules of water vapour, carbon dioxide and methane, which are the greenhouse gases.

The electromagnetic spectrum identifies radiation according to its wavelength and/or frequency.

This is an example of the spectrum to show the relative wavelengths of UV, visible light, near IR and far IR.

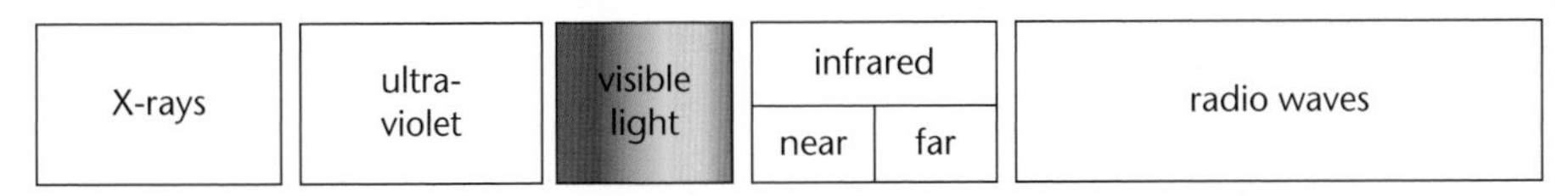

Figure 6.98 The relative wavelengths of UV, near IR and far IR showing how incoming radiation is different to outgoing radiation.

5 About 71 percent of the total incoming solar energy is absorbed by the atmosphere or reaches Earth. Suggest what happens to rest of the solar energy.

6 Suggest what the Earth would be like if we did not have greenhouse gases.

REMEMBER!

You do not need to remember the names of the types of radiation.

Human activities

KEY WORDS

correlation
deforestation
peer review
speculation

Learning objectives:

- describe two activities that increase the amounts of carbon dioxide and methane in the atmosphere
- evaluate the quality of evidence in a report about global climate change
- recognise the importance of peer review of results and of communicating results to a wide range of audiences.

The increasing levels of carbon dioxide from the use of fossil fuels is now concerning world governments. This is particularly because the global population has risen so much in the last few decades, which means even more fuel is being used. However, other human activities are also producing higher levels of greenhouse gases. Do we know what their effect will be?

Human activities

We have seen that greenhouse gases are essential for maintaining stable temperatures on Earth so that life, as we know it, can be sustained. This is a natural phenomenon.

However, as far back as 1896 scientists were concerned that some human activities *increase* the amounts of greenhouse gases in the atmosphere and that we are altering the balance of the greenhouse effect. A Swedish chemist called Svante Arrhenius suggested that doubling the CO_2 level would cause the average temperature on Earth to increase by 5 °C.

The original concern was over the burning of coal, producing carbon dioxide. As more fossil fuels are burned, more carbon dioxide is produced. You can see this in the graph.

The increase in the percentage of carbon dioxide in the atmosphere over the last 100 years **correlates** with the increased use of fossil fuels.

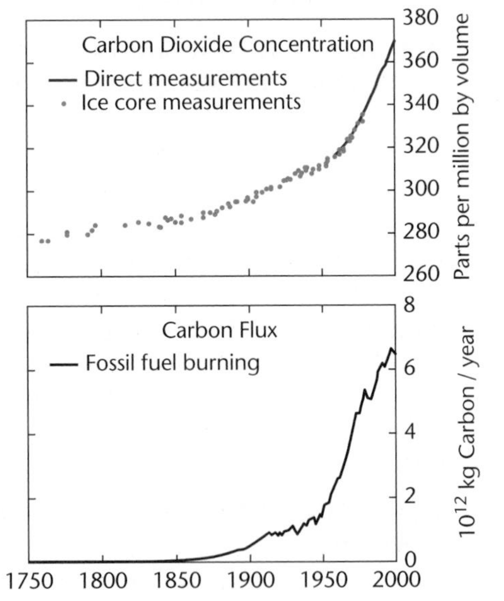

Figure 6.99 Graph showing that fossil fuel use has risen and carbon dioxide levels have risen at the same time. They correlate.

There are now concerns over many other activities, that not only increase the amount of carbon dioxide but also increase the amount of methane.

The human activities that affect these greenhouse gases include more:

- carbon dioxide because of
  1. combustion of fossil fuels
  2. **deforestation**
- methane because of
  1. more animal farming (through digestion and decomposition of waste)
  2. decomposition of rubbish in landfill sites.

1. **Describe two reasons why carbon dioxide levels have increased in the last 200 years.**
2. **Explain why deforestation may lead to increased levels of carbon dioxide.**

Evidence of human activity impact

We can see from ice core measurements of carbon dioxide that scientists believe that the amount of carbon dioxide in the atmosphere has increased significantly since the industrial revolution and the increased use of fossil fuels for transport, electricity production and machinery for manufacturing.

The evidence is **peer-reviewed** but questions asked include: 'Are these measurements accurate? Could the data be interpreted in a different way'? Most scientists agree that CO_2 levels have increased.

However, a few scientists would say that it is not humans that have caused this, but believe it is part of the Earth's longer, natural cycles.

It is a difficult debate, but based on peer-reviewed evidence, many scientists believe that human activities will cause the temperature of the Earth's atmosphere to increase at the surface and that this will result in global climate change.

3. **Explain why it is important for scientific data and interpretation to be peer-reviewed.**

4. **Suggest what will happen in the next 50 years if the average temperature of the Earth goes up by 1.5 to 2 °C.**

Figure 6.100 Human activities that produce more carbon dioxide or methane

Modelling climate change

It is very difficult to model such complex systems as global climate. This is why the Intergovernmental Panel on Climate (IPCC) was set up through the United Nations in 1988. This panel draws on evidence from hundreds of scientists who contribute to papers and peer-review each other's evidence. They try to model what might happen and give probabilities to future events. For example, they *predict* that oceans will warm and that it is *very likely* that Arctic ice cover will decrease.

They produce assessment reports that can be read and evaluated. These reports are written by scientists not only for other scientists but also for policy makers. The scientists need to carefully explain their conclusions to people who may not know the technical language. In these reports they are writing for a different audience.

When the models are presented to the public they are simplified models. This means that **speculation** and opinions may be presented in the media that may be based on only parts of the evidence and part of the complex interpretations. At any stage of any explanation there may also be a biased viewpoint.

5. **There has been an increase in methane from agriculture over the last 50 years. Explain how you would search for evidence of its effect on the temperature of the Earth.**

6. **Explain why a simple model of predicted climate change seen on a 2-minute TV news item may not give quality information.**

REMEMBER!

You do not need to know the models of climate change or the work of the IPCC but you do need to know how to evaluate the quality of evidence.

DID YOU KNOW?

Petrol, used in car engines, comes from a fossil fuel. It burns in oxygen to make carbon dioxide. Increased carbon dioxide levels have been linked with climate change.

Global climate change

Learning objectives:

- describe four potential effects of global climate change
- discuss the scale and risk of global climate change
- discuss the environmental implications of climate change.

KEY WORDS

average global temperature
distribution
food producing capacity
erosion

There are many different causes of climate change, most of which are natural causes. Examples are changes in radiation from the Sun, plate tectonic movements and volcanic eruptions. There are certain human activities that are thought to be causes of recent climate change. If humans cause a change, it is called an 'anthropogenic change'.

DID YOU KNOW?

The natural causes of increasing average global temperature are called 'forcings'.

The scale of global climate change

The major cause of global climate change is an increase in the **average global temperature**.

As we have seen before, the increase in the amounts of greenhouse gases is thought to cause an increase in average temperature.

Human activities producing greenhouse gases add to the natural causes of the rise in average temperature.

These activities are burning fossil fuels, deforestation, increased agriculture and rubbish decomposition.

The production of extra carbon dioxide and methane is thought to lead to increased global temperatures. In the media this is called 'global warming.'

The potential effects of global climate change by warming include:

- sea level rise, which may cause flooding and increased coastal **erosion**
- more frequent and severe storms
- changes in the amount, timing and distribution of rainfall.

Figure 6.101 Areas at sea level are more at risk of flooding as sea levels rise

Figure 6.102 Storms become more severe and more frequent

1 **Describe which human activities may add to 'global warming'.**

2 **Explain two things that might happen if global warming continues.**

Figure 6.103 Distribution of rainfall may change

The risks of global climate change

Scientists are concerned that if the average global temperature rises by 2 °C there will be irreversible changes taking place caused by a change in climate. Climate is not a short-term weather pattern change that follows cycles, but a long-term change in the whole weather system. A large number of scientists contribute evidence to the Intergovernmental Panel on Climate Change (IPCC) so that the risks of climate change can be assessed. No government can do this on their own, as it is a global issue.

The scientists use a number of indicators to demonstrate how the climate is changing. The most sensitive indicator is the retreat of glaciers.

Glaciers are mountain reserves of fresh water that are frozen. The mass of water now held in glaciers continues to reduce.

3 **Explain the difference between a weather pattern and a climate change.**

4 **Suggest why it is important that glaciers remain high in mass.**

Figure 6.104 Glaciers are sources of frozen fresh water that are diminishing

Environmental concerns of climate change

There will be big changes in the environment while global warming continues. The impact will affect different regions in different ways. In some regions the impacts are already changing lives – for example, for the people of the Arctic region where ice sheets are retreating.

The main effects are predicted to be:

- temperature stress for humans and wildlife (some areas will become too hot for people to make a living)
- water stress for humans and wildlife (fresh water supplies will reduce in some regions)
- changes in the **food producing capacity** of some regions (the production of wheat and maize has already been affected)
- changes to the **distribution** of wildlife species (migration patterns are changing).

Scientists are working on complex models to try to determine what may happen, as there are many interrelated factors to take into account.

5 **Explain how each of the potential social effects of climate change listed in this topic are related to the physical changes that are taking place.**

REMEMBER!

You are not expected to remember all the details that are involved in the complex issue of global climate change, but you do need to be able to discuss the possible causes and implications.

Carbon footprint and its reduction

Learning objectives:

- explain that the carbon footprint can be reduced by reducing emissions of carbon dioxide and methane
- describe how emissions of carbon dioxide can be reduced
- describe how emissions of methane can be reduced.

KEY WORDS

alternative energy
carbon capture
carbon off-setting
carbon neutrality

There are two responses to increasing average global temperature – reduce it or adapt to it. Adapting to it means coping with the social changes and ways of living. Reducing it means reducing greenhouse gases. The reduction of carbon dioxide and methane emissions is called a mitigation action so that we do not have to adapt.

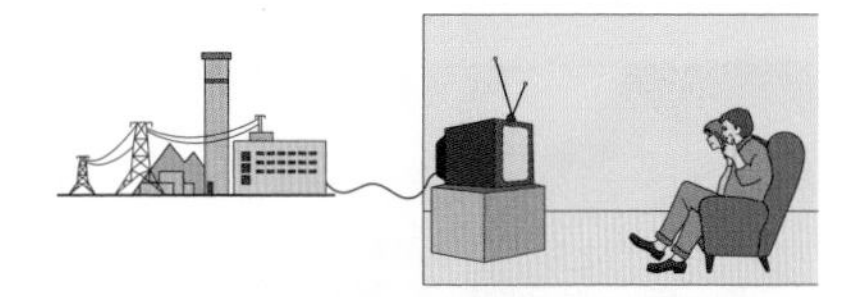

Figure 6.105 We do not burn fuel when watching TV – or do we?

What is a carbon footprint?

The carbon footprint is the total amount of carbon dioxide and other greenhouse gases emitted over the full life cycle of a product, service or event.

We can often see what we are doing when burning a fuel that gives out CO_2. Examples are when we ride in a car or bus. At other times we may not notice. For instance, when we turn on the light or electric shower we do not always realise that the electricity we are using was probably generated by burning fossil fuels. We also do not take into account how much energy was used in making the car or the shower in the first place, and so how much carbon fuel was used to make it.

There is now an intergovernmental agreement, called the Kyoto Protocol, that, in essence, states that we should all reduce our carbon footprint.

Figure 6.106 Alternative energy sources: using solar energy

1. **Explain how playing a computer game has a carbon footprint.**
2. **Explain why governments agree that carbon footprints need reducing.**

Figure 6.107 Alternative energy sources: using nuclear energy

Reducing the personal carbon footprint

Actions to reduce the our personal carbon footprint include:

- increased use of **alternative energy** supplies and hydrogen fuel cells
- energy conservation and energy efficiency in our homes
- energy efficiency by driving cars that use less fuel (i.e. have higher mpg figures)

3. **Explain how using solar energy reduces CO_2 emissions.**
4. **Explain whether using solar energy panels reduces the carbon footprint.**

Figure 6.108 Alternative energy sources: using wind energy

5 **Look at Figure 6.109. Explain why the methods with labels in blue are energy conservation methods and why the methods with labels in green are energy efficiency methods.**

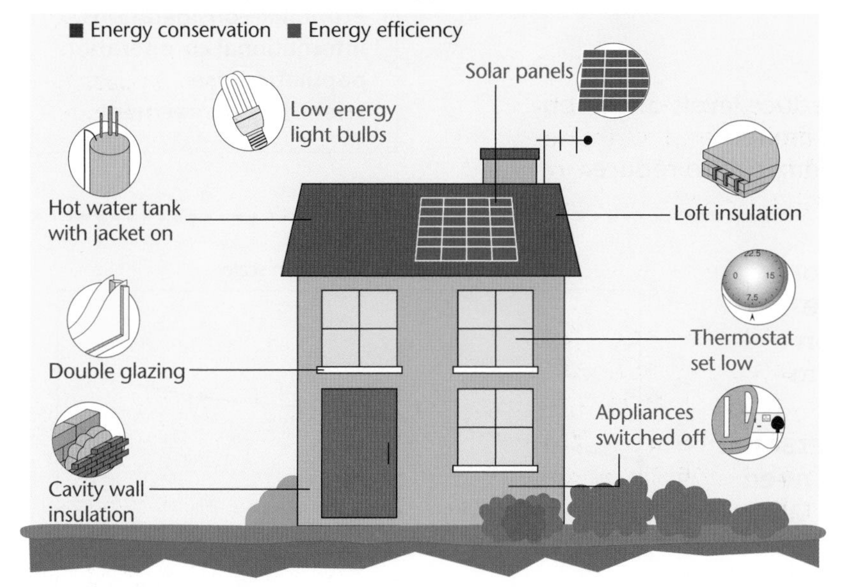

Figure 6.109 Using less energy by installing more insulation or using energy more efficiently

Governments reducing the carbon footprint

It is important that we reduce our own personal carbon footprint but there are some measures that governments or large companies or organisations will need to implement as we cannot do this on our own. These include:

- **carbon capture** and storage – taking the emissions of CO_2 from a power station and depositing the CO_2 into an underground geological formation so that it will not enter the atmosphere. This is a new idea and so will need investment.
- carbon taxes and licences – a system where a polluter pays a tax if they are emitting greenhouse gases. This is to give an incentive to the user to reduce CO_2 emissions. Another way is to give a licence or a permit to many users and put a cap on the total amount of CO_2 that can be emitted overall. The users then trade the permits amongst each other, so that some can emit more than others if they have bought their permits.
- **carbon off-setting**, including increasing the carbon sink through tree planting and reforestation. Carbon off-setting can take place on a small local scale or a large scale such as the European Union Emission Trading scheme. This trading market grew out of the agreements reached through the Kyoto Protocol.
- **carbon neutrality** – zero net release. This is a scheme where people and organisations aim to take out of the atmosphere as much CO_2 as they put into it, for example by planting trees that use the equivalent CO_2 quantity as their emission from their electricity use. The idea is to produce a zero carbon footprint.

DID YOU KNOW?

This idea could have been tried by storing the CO_2 in the ocean, but this would have made the acidification of the oceans worse.

REMEMBER!

You do not need to know all the details of this topic but you need to know where to find current information and be able to discuss it.

6 **Explain two different ways by which forestation can help reduce the carbon footprint.**

Limitations on carbon footprint reduction

Learning objectives:

- give reasons why actions to reduce levels of carbon dioxide and methane may be limited
- give reasons why methane is difficult to reduce.

KEY WORDS

economic considerations
international co-operation
population rise
scientific disagreement

We have seen that the reduction of carbon dioxide and methane emissions is called a mitigation action so that we do not have to adapt. We have also seen that although we can reduce our personal carbon footprint some actions need organisation by governments. Why does this not always work?

Difficulties in reducing methane levels

Methane is another greenhouse gas that has risen considerably in the last decades alongside the **rising population**. This is because as the population has risen:

- more grazing animals have been farmed to produce more meat, causing more methane waste. (This is an especially important factor where deforestation has also occurred to make way for grazing lands).
- more wet cultivation fields for rice growing have developed
- more rubbish landfill sites have opened.

These all cause methane levels to rise.

1 Explain why the trend in population growth correlates with the trend in methane increase.

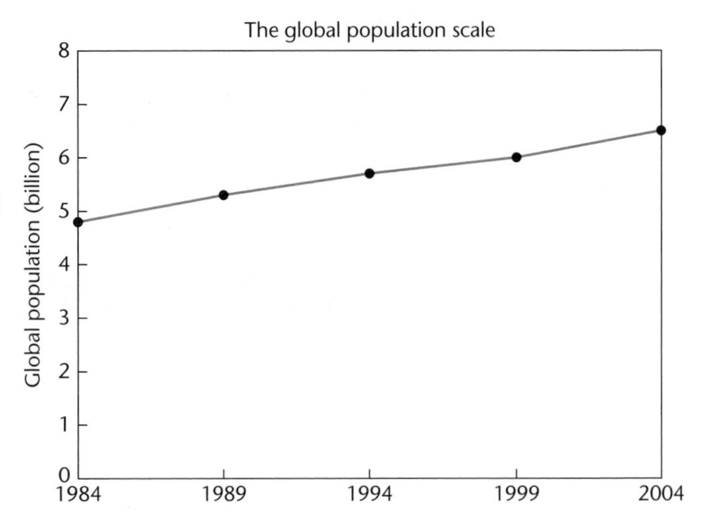

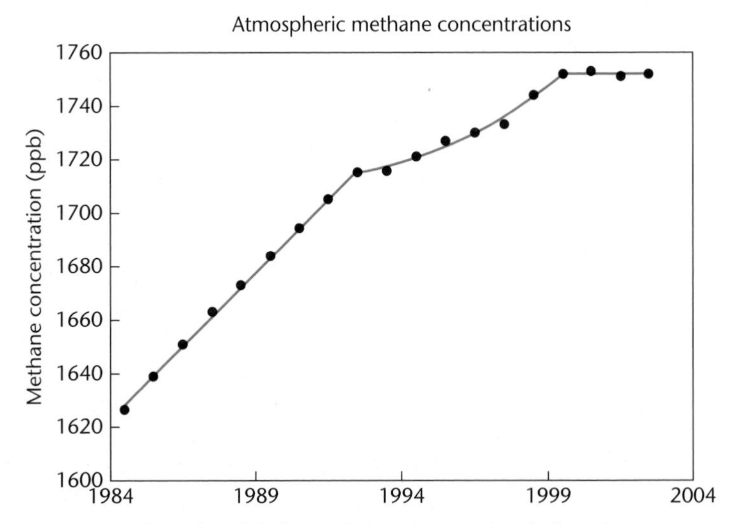

Figure 6.110 As the global population rises the level of methane rises

Trying to reduce the carbon footprint

Although there is some agreement that the carbon footprint has to be reduced it is often seen as someone else's problem. We do not always make a success of reduction because of:

- lifestyle changes needed. Everybody needs to change their need for fuel in order to make a difference and it is not always easy or comfortable to do that. If everybody used a lot less fossil fuels or used different sources of energy that

KEY SKILLS

When looking at two graph plots, look for pattern changes and correlations.

would make a huge difference, but it has to happen right across the world, particularly in the developed nations, especially those with big populations.

- lack of public information and education. This happens because scientists do not always explain their evidence in a way that is easy to understand by people who do not have the technical language. Unfortunately, the scientific models are often complex and cannot be easily simplified. It also needs people to be more aware that they need to find out information and be pro-active in looking.

2 Suggest five steps people could take to change their lifestyle to reduce CO_2 emissions.

3 Suggest how public information could be improved so that evidence about global warming and the need for carbon footprint reduction could be debated more fully.

Other problems in reductions

There are also some major international problems with trying to reduce the carbon footprint which are not easily solved. These include:

- **scientific disagreement** over the causes and consequences of global climate change. This disagreement is being challenged and reduced by the development of peer-reviewing of evidence by large numbers of scientists who are experts in a wide range of related fields, being drawn together by the IPCC. The bias of organisations who benefit from selling fossil fuels is also being scrutinised.
- **economic considerations**. There is disagreement over a model that tries to predict the cost of trying to reduce the CO_2 concentration. The model is to reduce levels from 450 ppm, which governments have accepted as a target, to 350 ppm by changing the technology we use. This is because the lower figure is said to be the CO_2 level we should really be aiming for but it will cause instability. There are a wide range of opinions including one that says there would be widespread unemployment and another that says in the long term there would be better employment prospects. Another says it will cost more whilst another says that emissions can be eliminated at no extra cost. This uncertainty does not help decision-making.
- incomplete **international cooperation**. Not all countries agree with the vast majority who have signed agreements on carbon reduction. There are 192 signatories to the Kyoto Protocol, but Canada has withdrawn and the USA has not ratified its signature. 26 states have accepted the new Doha Amendment with targets for reduction up to 2020.

DID YOU KNOW?

There is a target figure for the level of CO_2 that governments agree as a global target. It is measured in parts per million, ppm, not as a percentage.

4 Suggest three reasons why information about the need to reduce the carbon footprint could be *biased*.

5 It is often difficult to decide the ownership of emissions contributing to a carbon footprint. For instance, a large sports stadium may claim that it has a very small carbon footprint when empty. Evaluate this claim.

Atmospheric pollutants from fuels

Learning objectives:

- describe how carbon monoxide, soot, sulfur dioxide and oxides of nitrogen are produced by burning fuels
- predict the products of combustion of a fuel knowing the composition of the fuel
- predict the products of combustion of a fuel knowing the conditions in which it is used.

KEY WORDS

sulfur dioxide
oxides of nitrogen
particulates
hydrocarbons

Burning a fuel produces energy, which is what we want. However, burning a fuel can also produce unwanted products. If a fuel contains sulfur, then acid rain can result. If a fuel is burned in a shortage of oxygen then soot (carbon particles) may be produced, which is dirty and bad for human health. Burning fuels is a major source of pollution in the atmosphere.

DID YOU KNOW?

Carbon dioxide is not included in this group as, although it is a greenhouse gas and levels need to be reduced, it is not a harmful gas.

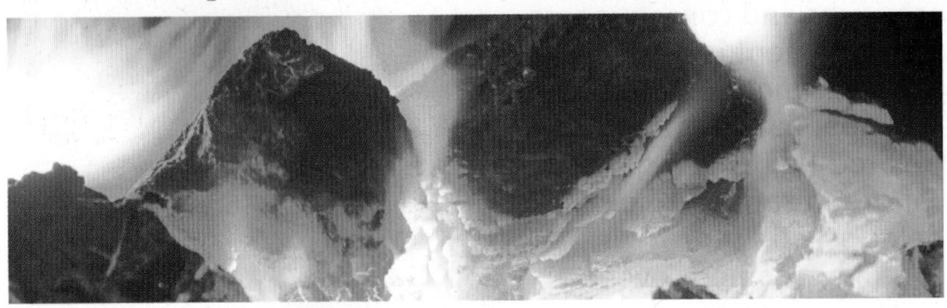

Figure 6.111 There are regulations in many urban areas on the types of fuels you can burn so that the air in cities is kept clean

Common atmospheric pollutants

The combustion of fuels is a major source of atmospheric pollutants.

Most fuels, including coal, contain carbon and other elements such as hydrogen and may also contain some sulfur. The sulfur combines with oxygen to make **sulfur dioxide**.

When a fuel is burned the gases given off will include carbon dioxide and water vapour. However, they may also include unwanted gases such as carbon monoxide, sulfur dioxide and **oxides of nitrogen**.

Burning fuels may also release solid particles and unburned **hydrocarbons**. These form **particulates** in the atmosphere.

1. **Write down three unwanted pollutants that are released when a fuel is burned.**
2. **Explain the term 'incomplete combustion'.**

Gaseous pollutants

Carbon monoxide, sulfur dioxide and oxides of nitrogen are the gases that are pollutants.

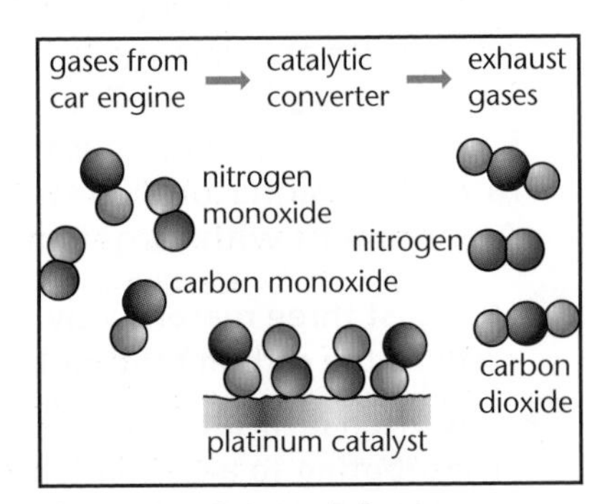

Figure 6.112 A catalytic converter

The environmental effects, how they are caused and how they can be removed or reduced are given in the table:

Gaseous pollutant	Carbon monoxide (CO)	Sulfur dioxide (SO_2)	Oxides of nitrogen (NO_x)
how they are caused	incomplete combustion in a shortage of oxygen	burning coal or petrol containing sulfur in oxygen	from N_2 and O_2 at very high temperatures such as in an engine
how they can be reduced	converted to CO_2 by a catalytic converter	removed from power stations by capturing with limestone	converted to N_2 by a catalytic converter

Figure 6.113 Sulfur dioxide is removed and captured at source in this power station. Note that the clouds from the cooling towers are droplets of water not smoke.

$$S + O_2 \rightarrow SO_2$$

$$C_7H_{16} + 8O_2 \rightarrow CO_2 + 6CO + 8H_2O \quad \text{Incomplete combustion}$$

$$C_7H_{16} + 11O_2 \rightarrow 7CO_2 + 8H_2O \quad \text{Complete combustion}$$

REMEMBER!

You do not need to remember these equations but you do need to be able to predict the products of combustion.

3. **Describe how some pollutants can be removed before they enter the atmosphere.**
4. **Use the equations above to explain why CO can be formed when fuels are burned.**
5. **The pollutant nitrogen monoxide is initially formed in engines. Give a balanced chemical equation for its formation.**

Pollution by particulates

Solid particles and unburned hydrocarbons form particulate matter (PM) in the atmosphere, and are measured by their diameter. The size of the particles ranges in diameter and some are more penetrating in the lungs than others. Particles with a diameter of less than 10 micrometers (PM_{10}) can get past the filtering mechanism in the nose and go into the airways to the lungs. With diameters of less than 2.5 micrometers ($PM_{2.5}$) they can penetrate the alveoli. These particles come from unburned fuel and cause health problems. They need to be reduced.

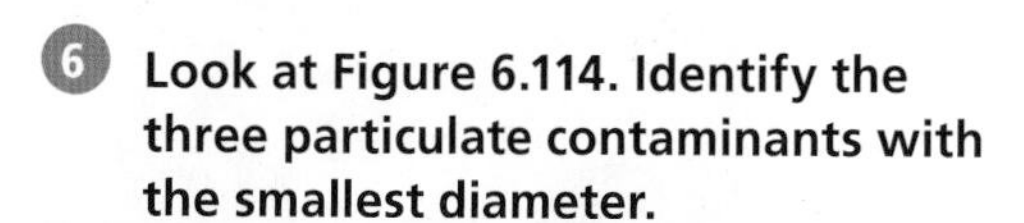

6. **Look at Figure 6.114. Identify the three particulate contaminants with the smallest diameter.**
7. **Use Figure 6.114 to explain why pollution from vehicles is a problem.**
8. **Suggest how particulate emissions from cars might be reduced.**

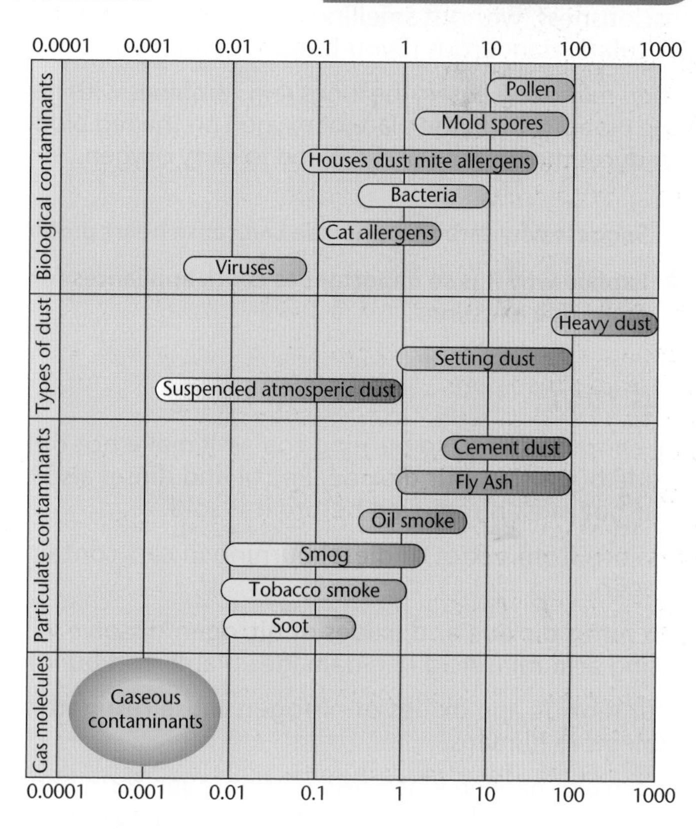

Figure 6.114 Diameters of airborne particles. The measurements are in micrometers (1000 micrometers = 1 mm).

Properties and effects of atmospheric pollutants

Learning objectives:

- describe and explain the problems caused by increased amounts of oxides of carbon, sulfur and nitrogen as pollutants in the air
- describe and explain the effects of acid rain
- evaluate the role of particulates in damaging human health.

KEY WORDS

acid rain
global dimming
particulates
toxicity

Gases from burning fossil fuels pollute the atmosphere, but that isn't the only problem. Solid particles can also be released, which are referred to as particulate matter (PM). However, is burning fossils fuels the only source of these PMs? Are there natural causes too?

Toxic gases

Carbon monoxide is a **toxic** gas made during the incomplete combustion of carbon. It is colourless and does not smell and so is not easily detected. It is often called the silent killer as it makes people feel mildly unwell at first. Then they lose consciousness, without smelling any gas, and can die. It is particularly dangerous if you breath it when you are asleep.

Carbon monoxide enters the lungs and combines with haemoglobin in the blood. It takes the place of oxygen on the red blood cells and so reduces the capacity of the blood to carry oxygen.

Figure 6.115 Regularly checking boiler emissions for carbon monoxide is important

1. **Suggest why carbon monoxide can cause heart problems.**
2. **Explain why it is so important to check appliances for carbon monoxide emissions.**

DID YOU KNOW?

Nitrogen dioxide is NO_2 but there are other oxides of nitrogen in exhaust gases so they are grouped together as NO_x. The common name is then NOX gases.

Acid rain

Coal contains sulfur, so burning coal will make not only carbon dioxide but also sulfur dioxide. Petrol and diesel also contain sulfur but most is removed before it is sold.

Emissions from petrol or diesel burning in cars contain oxides of nitrogen.

Both sulfur dioxide and oxides of nitrogen dissolve in water causing **acid rain**. Acid rain damages plants and buildings.

Sulfur dioxide and oxides of nitrogen cause respiratory problems in humans.

This can also happen when oxides of nitrogen form photochemical smog, when sunlight interacts with the polluting gases.

Figure 6.116 Cars need to have an annual check for the levels of NO_x gases they emit

3. **What is the cause of acid rain?**
4. **Explain how acid rain causes damage.**

Figure 6.117 Forests in the path of acid rain can be severely damaged if they are grown on certain types of soil.

Particulates

Particulates cause **global dimming**, reducing the amount of sunlight that reaches the Earth's surface. This can not only happen with burning fossil fuels but also through natural processes such as volcanic eruptions.

Figure 6.118 Dust and other particulates from natural causes or burning fossil fuels can cause global dimming

Particulates cause health problems for humans because of damage to the lungs. The

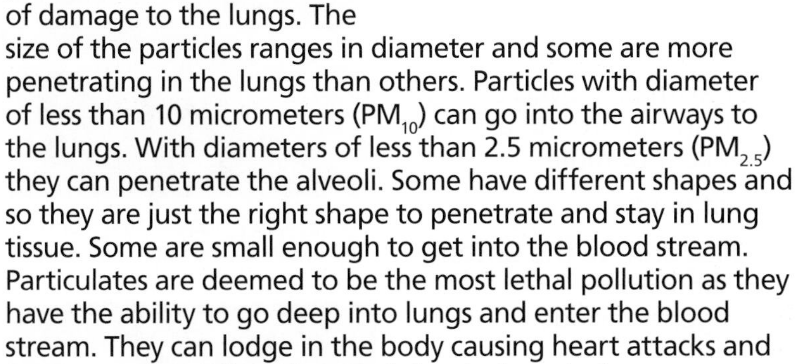

size of the particles ranges in diameter and some are more penetrating in the lungs than others. Particles with diameter of less than 10 micrometers (PM_{10}) can go into the airways to the lungs. With diameters of less than 2.5 micrometers ($PM_{2.5}$) they can penetrate the alveoli. Some have different shapes and so they are just the right shape to penetrate and stay in lung tissue. Some are small enough to get into the blood stream. Particulates are deemed to be the most lethal pollution as they have the ability to go deep into lungs and enter the blood stream. They can lodge in the body causing heart attacks and DNA mutations. There is no safe level of particulate matter.

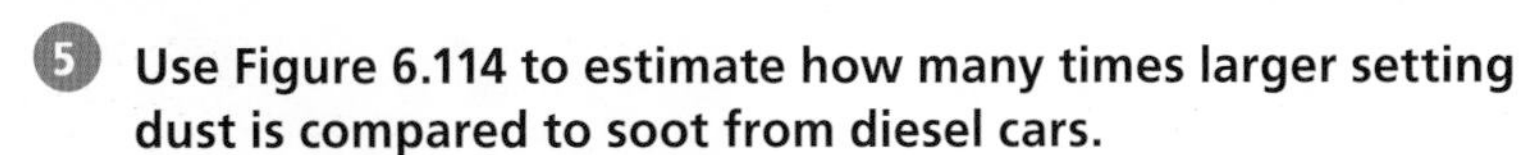

5 **Use Figure 6.114 to estimate how many times larger setting dust is compared to soot from diesel cars.**

6 **Global dimming is caused by particulates in the Earth's atmosphere.**

a **Suggest how global dimming may affect the Earth's climate and global warming.**

b **Governments have put measures in place to reduce particulate emissions. Explain the impact this may have on the climate.**

KEY INFORMATION

Remember rain is already slightly acidic as it dissolves CO_2 from the atmosphere. Sulfur dioxide and oxides of nitrogen make the rain *more* acidic to levels where damage is done.

Figure 6.119 The damaging effects of acid rain

Potable water

Learning objectives:

- distinguish between potable water and pure water
- describe the differences in treatment of ground water and salty water
- give reasons for the steps used to produce potable water.

KEY WORDS

potable
sedimentation
desalination
reverse osmosis

Safe drinking water is essential for life. Where does it come from and how is it made safe? For humans, the appropriate quality drinking water should have sufficiently low levels of dissolved salts and microbes. How do you get safe water if it does not rain?

Producing potable water

Water that is safe to drink is called **potable** water. In a chemical sense potable water is *not pure* water because it contains *dissolved substances*, which are needed by the body.

There are several methods used to produce potable water. The method chosen depends on available supplies of water and local conditions.

Traditionally, settlements grew up on the sides of river banks or lakes because the source of water was *abundant*. Other settlements relied on water from *groundwater* that collected in *aquifers*. They extracted it from *wells*.

In the United Kingdom, as the population grew in cities, *reservoirs* were built to provide constant fresh drinking water for everyone.

The water is provided by *rain*. This water, with low levels of *dissolved substances (fresh water)* collects in the ground and in lakes and rivers and is then treated to make it potable.

There are three main stages in water treatment:

- **sedimentation** of particles so that solids drop to the bottom
- *filtration* of very fine particles using sand
- *sterilising* to kill microbes. Sterilising agents include chlorine, ozone and ultraviolet light.

1. **Explain the stages of how rain becomes drinking water.**
2. **Suggest why the sterilisation stage of water purification is carried out last.**
3. **Some data for a drinking water sample is shown below. Explain whether the water is potable.**

	Bacteria per 100 ml	Lead / µg per ml	Nitrate / mg per *l*
Sample of drinking water	2	2	15
Maximum allowed	0	10	50

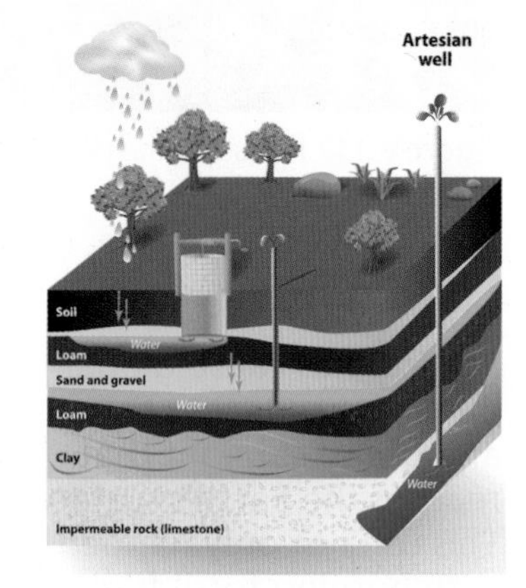

Figure 6.120 Groundwater is held in aquifers and is reached from the surface through artesian wells

Figure 6.121 Huge reserves of water are held to provide drinking water

Figure 6.122 A sedimentation tank at a water-treatment works

Potable water from seawater

Seawater has so many substances dissolved in it that it is undrinkable. However, if supplies of fresh water are limited, **desalination** of salty water or seawater is needed to remove the dissolved substances.

Desalination can be done by *distillation* or by processes that use *membranes* such as **reverse osmosis**. These processes require large amounts of energy, so it is very expensive. It is only used when there is not enough fresh water.

Many countries with little fresh water rely on desalination processes using distillation or reverse osmosis. Spain is the largest operator of desalination plants in Europe.

4 Suggest why Spain operates desalination plants whereas the UK does not.

5 Explain which stage in distillation makes the process so costly.

DID YOU KNOW?

Remember all these steps cost money so it is important to conserve water. In hot months the water resources tend to have less in them so sometimes bans on using a hosepipe are announced. This is done to conserve water so that everybody has access to clean drinking water. Water is now often metered to remind people not to waste this precious resource.

Potable water for all

Water is a *renewable* resource. However, that does not mean that the supply is endless. If there is not enough rain in the winter, reservoirs do not fill up properly for the rest of the year. In the UK today, more and more homes are being built, which increases the demand for water. Industry and agriculture also use huge amounts. Producing potable water does have costs. It takes energy to pump and to purify it. If this energy is sourced from fossil fuels this adds to the increases in greenhouse gases and in turn to increases in global warming.

This then causes stress to *glaciers* that are a source of river production that provides us with fresh water. To minimise that stress we need to consider water *conservation*.

Figure 6.123 A Spanish desalination plant producing potable water using reverse osmosis

6 Suggest three ways in which domestic water can be conserved.

7 Suggest three ways in which industrial or agricultural water can be conserved.

8 The bar chart shows water consumption per capita for various regions per day.

a Estimate the ratio of water consumption between North America and sub-Saharan Africa.

b Suggest why sub-Saharan Africa has such a low consumption of water per capita.

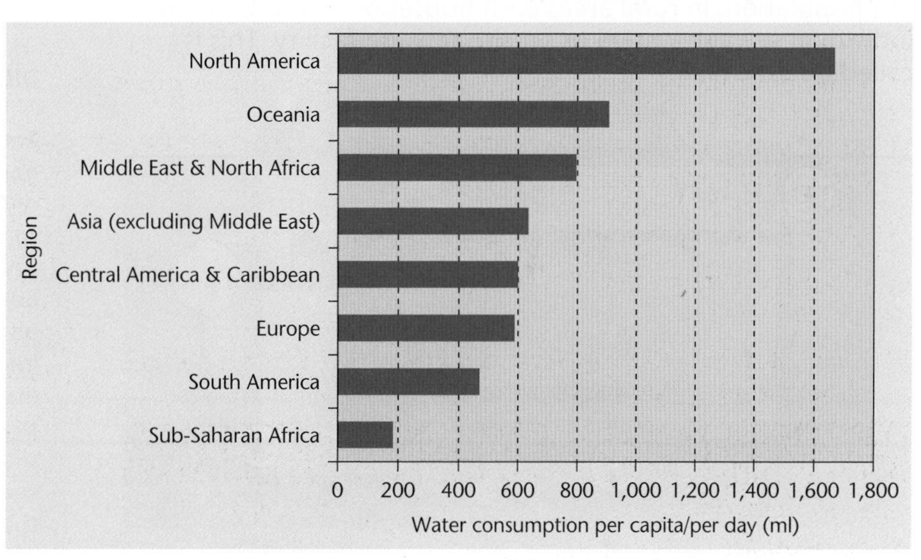

Waste water treatment

KEY WORDS

sewage
anaerobic
aerobic
sedimentation

Learning objectives:

- explain how waste water is treated
- describe how sewage is treated
- compare the ease of treating waste, ground and salt water.

Water is a precious commodity that we tend to take for granted in the UK as we have so much rain. The natural water cycle works well for us most of the time but what happens when urban living, industrial development, intensive agriculture or dry weather put pressure on this natural resource?

Figure 6.124 Sewage pouring into a river. In the UK, sewage is treated to make sure it doesn't contaminate the waterways.

Water cycles

There is no new water on the planet; it is ancient water that is continually recycled.

The water cycle is a natural cycle, discussed on the previous pages, where water from the oceans evaporates, condenses to forms clouds, moves over high ground and precipitates water back to the ground to rivers, lakes and aquifers.

For thousands of years we have lived with this natural cycle, drinking water from wells, rivers and lakes and excreting waste into the ground where the solids are broken down by bacteria and the water finds its way back to groundwater or rivers or evaporates to form clouds.

With growing populations, ways of treating **sewage** are needed as well as ways of providing clean fresh water to everyone.

Systems for dealing with sewage depend on the density of the population. In rural areas each household has their own individual sewage system or on-site sewage facility. This is called a *septic tank*.

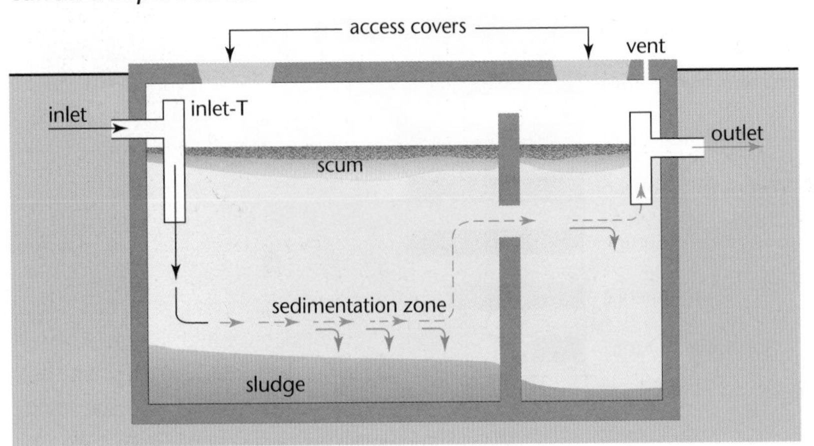

Figure 6.125 How a septic tank works

DID YOU KNOW?

The Great Stink of 1858 galvanised the Members of Parliament into action to pass laws allowing new sewage systems to be developed that did not put untreated sewage straight into the River Thames, literally straight under their noses: Go to www.skepticalscience.com and search for 'changing minds the great stink'.

Septic tanks allow **anaerobic** bacteria to develop that treat the sewage by decomposing it. The treated water leaks out through finger drains on to the land and the remaining sludge collects in the tank. Periodically the tank needs to be emptied.

1. **Describe how water from the sea ends up in aquifers.**
2. **Explain why some rural communities need septic tanks.**

Figure 6.126 Industrial use of water is increasing

The urban water cycle

Urban lifestyles and industrial processes produce large amounts of waste water that require treatment before being released into the environment. This can be from domestic washing machines, dishwashers and showers. It can be from industrial processes such as cleaning cycles and solvent usage in manufacturing processes and from cooling systems.

The *industrial waste water* may contain pollutants such as organic matter and harmful chemicals that require extra treatment before it can go back into the fresh water cycle.

Agriculture is by far the biggest user of water. *Agricultural waste water* can cause problems to ecosystems by having too much nutrient in it. This can lead to *eutrophication*.

This is why sewage and agricultural waste water require treatment to ensure the removal of organic matter and harmful microbes.

Sewage treatment includes:

- *screening* and grit removal
- **sedimentation** to produce sewage sludge and *effluent*
- *anaerobic digestion* of sewage sludge
- **aerobic** *biological* treatment of effluent.

DID YOU KNOW?

Clean water saves more lives than medicines do. That is why, after earlier disasters and issues in developing countries, relief organisations concentrate on providing clean water supplies.

3. **Suggest how sedimentation works and predict one limitation.**
4. **Explain why urban lifestyles need organised waste water management.**

The water footprint

Just as there is the monitoring of the carbon footprint, there is an organisation encouraging the monitoring of the water footprint.

International bodies describe water in categories. Green water is water from precipitation, blue water is in rivers, lakes and aquifers and grey water is waste water from agricultural, industrial and domestic use. It is grey water that needs treatment to go back into the water cycle.

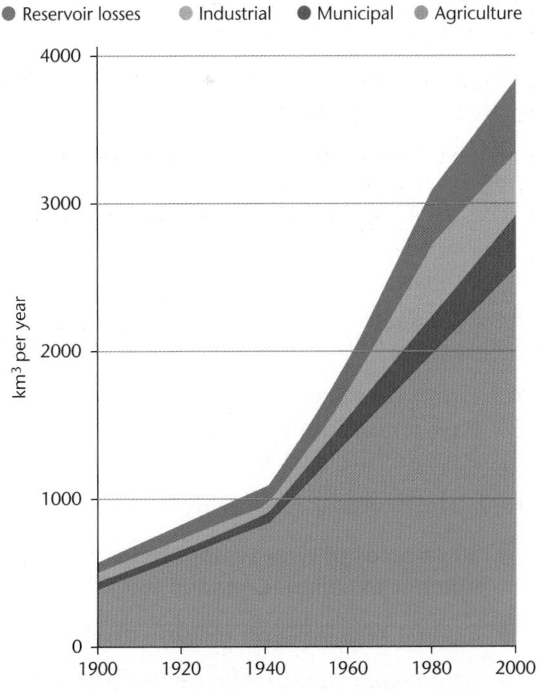

Figure 6.127 Graph of water usage between 1900 and 2000.

5. **Suggest why Australia may be most affected by changes in water footprint.**
6. **Compare the relative ease of obtaining potable water from groundwater, saltwater and waste water.**

PRACTICAL

Analysis and purification of water samples from different sources, including pH, dissolved solids and distillation

Learning objectives:

- describe how safety is managed, apparatus is used and accurate measurements are made
- recognise when sampling techniques need to be used and made representative
- evaluate methods and suggest possible improvements and further investigations.

KEY WORDS

distillation
evaporation
condensation
purity
boiling point

Fresh drinking water is essential for everyone. Water needs to be made safe which means testing and treatment. River water can be filtered and bacteria removed or sea-water can be distilled.

These pages are designed to help you think about aspects of the investigation rather than to guide you through it step by step.

Analysing water samples

Different skills are needed to analyse water and carry out a **distillation**, including applying sampling techniques, manual dexterity and the safe use of heating devices.

One way of selecting samples is by layered or section sampling. If 30 samples of river water are to be tested, they must be taken equally in each section over the whole stretch.

Imagine a new town was being planned which needs potable water. A river runs nearby. Your job is to test the water to find the best place to site a water treatment plant. The pH needs to be 6.5–9.5

The river section is 1 km long. It can be divided into three sections:

A. Open countryside over chalk (0.5 km)
B. Past a large dairy farm (0.2 km)
C. Past an old lead mine (0.3 km)

Think about these questions:

1. **30 samples of river water are tested altogether. Calculate the number of samples needed for each selection.**
2. **Explain how will you test the pH of the samples.**
3. **Other than 'what is the pH?', which other questions will you ask about the sites that the samples are taken from?**

DID YOU KNOW?

There are two steps to getting the most accurate results of chemical analysis:
i) collecting samples
ii) analysing samples.
Errors need to be reduced in both steps.

Distilling sea water

Some countries do not have access to fresh water, only sea-water. This has dissolved salts in it that need to be removed. This is done by distillation. Figure 6.126 shows the apparatus for the standard technique.

Think about these questions:

4. **Explain why the water in the flask is heated.**
5. **Why is a condenser attached to the arm of the heating flask?**
6. **Describe how distillation works.**
7. **Explain the purpose of the thermometer.**
8. **Why does the salt not evaporate?**
9. **Explain why the cold water flows from the *bottom* of the condenser out into the sink from the *top* of the condenser.**

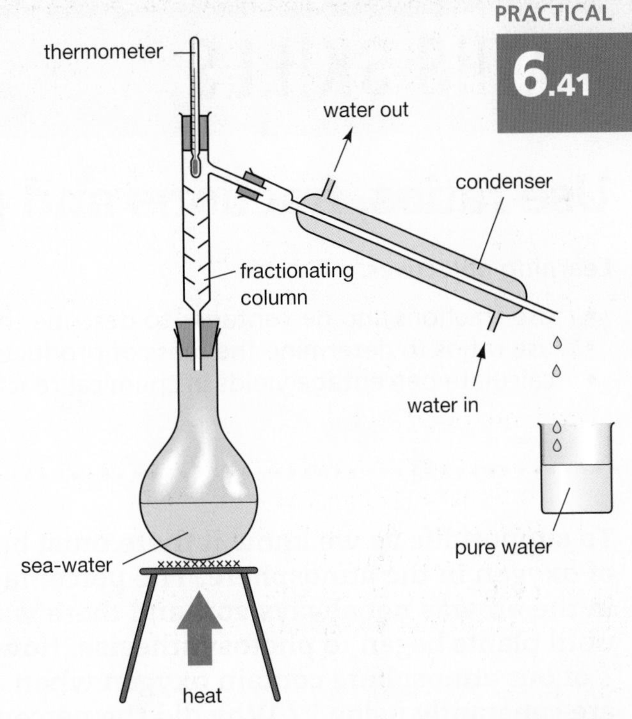

Figure 6.128

Analysing the results

When three samples are gathered, they are tested for **purity** by finding the **boiling point**. Jo and Akira tested their three samples:

	1st sample	2nd sample	3rd sample
Boiling point before distillation in °C	102.1	102.0	101.9
Boiling point after distillation in °C	101.4	101.6	100.8

10. **Explain the results.**

The way the equipment was used, and how techniques can be improved, should be considered.

11. **Suggest ways to improve the sampling of river water above. For example, after answering question 3, would all the sites need to be included in the sampling?**
12. **Suggest ways to improve the measurement of pH or the distillation technique shown.**
13. **Suggest what other investigations could be done to analyse drinking water samples.**
14. **Taking averages is common in science.**
    - **a** **Calculate an average of the three sample boiling points both before and after the distillation.**
    - **b** **Explain whether separate or average boiling points are more useful.**
    - **c** **The pH for 50 samples of river water at different locations was measured. Suggest whether an average or separate values would be more useful.**

REMEMBER!

The water in the cooler jacket of the condenser does not mix with the water going through the centre of the condenser.

KEY INFORMATION

A simplified equation for section sampling could be: Sample size for each section = size of whole sample × size of section

MATHS SKILLS

Use ratios, fractions and percentages

Learning objectives:

- use fractions and percentages to describe the composition of mixtures
- use ratios to determine the mass of products expected
- calculate percentage yields in chemical reactions.

KEY WORDS

fractions
percentages
approximately
segment

To support life as we know it there must be a proportion of oxygen in the atmosphere. The percentage of oxygen in the air was not always 20% and there was no oxygen until plants began to photosynthesise. How does a $\frac{1}{5}$ of our atmosphere contain oxygen when animals are constantly using it? Why did the percentage composition of the atmosphere change?

The composition of air

Air is a mixture of gases. It contains nitrogen and oxygen and there are other gases present such as carbon dioxide, water vapour and noble gases like argon in small amounts.

If the other gases are taken out of the diagram it will show that the air is **approximately** $\frac{4}{5}$ nitrogen and $\frac{1}{5}$ oxygen. Approximately means 'almost'. If the circle is divided into 5 equal parts, 4 parts represent nitrogen and 1 part represents oxygen. The **fractions** are $\frac{4}{5}$ and $\frac{1}{5}$, respectively.

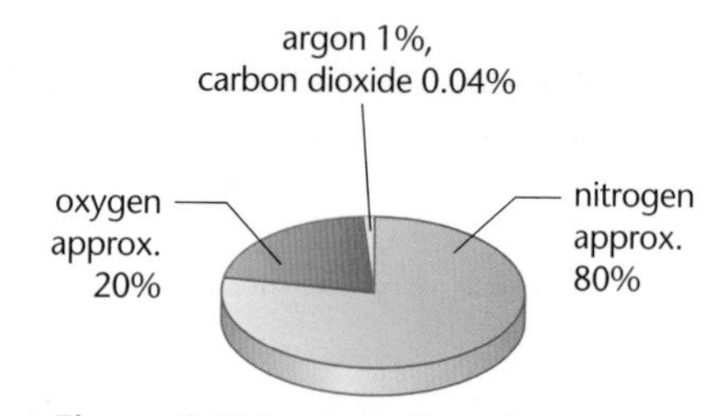

Figure 6.129 proportion of gases

If the circle is now divided into 10 equal parts (so double the number of parts) then 8 represent nitrogen and 2 represent oxygen. The fractions are $\frac{8}{10}$ and $\frac{1}{10}$, respectively.

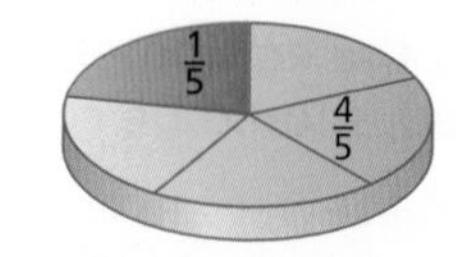

Figure 6.130 proportion of gases

Now imagine that each of the ten **segments** were divided into 10. There would be 100 segments in the circle. 80 of these would represent nitrogen and 20 segments would represent oxygen. The fractions are $\frac{80}{100}$ and $\frac{20}{100}$, respectively.

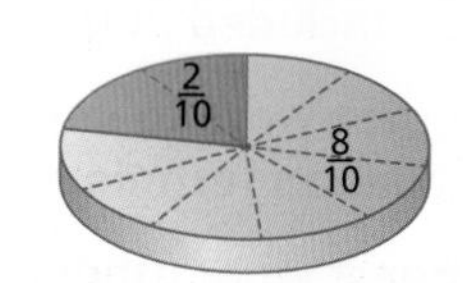

Figure 6.131 proportion of gases

Fractions representing $\frac{1}{100}$ are **percentages**. So $\frac{80}{100}$ is 80% and $\frac{20}{100}$ is 20%.

Now as fractions, percentages and a ratio the summary is:

$\frac{4}{5}$	$\frac{8}{10}$	$\frac{80}{100}$	80%	Ratio 4:1
$\frac{1}{5}$	$\frac{2}{10}$	$\frac{20}{100}$	20%	

REMEMBER!

If you are drawing a pie chart, you draw the segments in order from largest to smallest clockwise.

1. **The atmosphere of Mars is approximately $\frac{19}{20}$ CO_2, $\frac{1}{20}$ other gases. What is: a the percentage of gases?; b the ratio of gases**

2. **If an ancient atmosphere had been 3:1 methane to ammonia, what would be the percentage of each gas?**

Ratios of substances

In a chemical reaction an equation will tell us how many molecules will react in the process. Look at these two reactions:

$NaOH + HCl \rightarrow NaCl + H_2O$	$NaOH + H_2SO_4 \rightarrow Na_2SO_4 + H_2O$
This is balanced and mass is conserved.	This is not balanced and mass is not conserved. It needs to be: $2NaOH + H_2SO_4 \rightarrow Na_2SO_4 + 2H_2O$
Ratio is 1:1 [1 NaOH to 1 HCl]	Ratio is 2:1 [2 NaOH to 1 H_2SO_4]

3 What is the ratio of reactants in this reaction? $Mg + 2HCl \rightarrow MgCl_2 + H_2$

4 What is the ratio of reactants in this reaction? $3Fe + 2O_2 \rightarrow Fe_3O_4$

5 Determine the ratio of H_2O:O_2 in

$4NH_3 + 5O_2 \rightarrow 4NO + 6H_2O$

6 N_2 reacts with O_2 in the ratio 1:2. A single product is formed. Give the balanced chemical equation for the reaction.

Calculating the percentage yield

If there are two ways to make a substance it is really useful to know which method will produce the highest yield (amount of product) – but why use percentages and not fractions?

You may use one method where you need to filter a solution to get a solid product and lose a lot of it on the filter paper. You were expecting to make 12 g and only made 9 g. Your yield would be $\frac{9}{12}$ of what you were expecting. Another method may yield 20 g when you were expecting to achieve 24 g. Your yield would be $\frac{20}{24}$. How do you compare them easily? In this case you can change the fraction to the same, lowest denominator – that is to twelfths.

	Reaction 1 Method A	Reaction 1 Method B	Reaction 2
Yield	$\frac{9}{12}$	$\frac{20}{24}$	
Change to same lowest denominator	$\frac{9}{12}$	$\frac{10}{12}$	
You can see that method B produces more.	$\frac{9}{12}$	$\mathbf{\frac{10}{12}}$	
What if it is not easy to change the fractions?	$\frac{9}{12}$	$\frac{10}{12}$	$\frac{5}{7}$
In order to compare the fractions, a common denominator is needed. In this case, the lowest common multiple is 84	$\frac{63}{84}$	$\frac{70}{84}$	$\frac{60}{84}$
It is much easier to make all the fractions into percentages for comparison. Use a calculator, divide the numerator by the denominator and multiply the result by 100. Round your answer up or down to the nearest whole number for convenience in this case.	$\frac{9}{12} \times 100 =$ 75%	$\frac{10}{12} \times 100 =$ **83%**	$\frac{5}{10} \times 100$ = 50%

So if you want to make comparisons in yields you can use the formula:

$$\text{Percentage yield} = \frac{\text{mass of product actually made} \times 100}{\text{mass of product expected}}$$

7 Calculate the percentage yield if you obtained 17.2 g and were expecting to get 26 g.

8 What mass did you make if your percentage yield was 80% and you expected 32 g.

9 0.58 g of product was made. 5.37 g of product was expected in theory. Work out the percentage yield.

10 27.0 g of product was produced. This was a yield of 31%. Calculate the mass of product expected.

DID YOU KNOW?

The percentage yield is a big consideration in making an industrial chemical.

Check your progress

You should be able to:

state examples of natural products that are supplemented or replaced by agricultural and synthetic products	→	distinguish between finite and renewable resources from given information	→	extract and interpret information about resources from charts, graphs and tables
describe the process of phytomining	→	describe the process of bioleaching	→	evaluate alternative biological methods of metal extraction
describe the components of a Life Cycle Assessment (LCA)	→	interpret LCAs of materials or products from information	→	carry out a simple comparative LCA for shopping bags
describe the composition of common alloys	→	interpret the composition of other alloys from data	→	evaluate the uses of other alloys
compare quantitatively the physical properties of materials	→	compare properties of glass and clay ceramics, polymers, composites and metals	→	explain how the properties of materials are related to their uses and select appropriate materials
apply the principles of dynamic equilibrium to the Haber process	→	explain the trade-off between rate of production and position of equilibrium	→	explain how the commercially used conditions for the Haber process are related to the availability and cost of raw materials and energy supplies, control of equilibrium position and rate
describe why crude oil is a finite resource	→	identify the hydrocarbons in the series of alkanes	→	explain the structures and formulae of alkanes
describe the difference between an alkane and an alkene	→	draw displayed structural formulae of the first four members of the alkenes	→	explain why alkenes are called unsaturated molecules
explain how a voltage can be produced by metals in an electrolyte	→	evaluate the uses of cells	→	interpret data for the relative reactivity of different metals
discuss the scale of global climate change	→	discuss the risk of climate change	→	discuss the environmental implications of climate change
describe how emissions of carbon dioxide can be reduced	→	describe how emissions of methane can be reduced	→	give reasons why actions on reductions may be limited
explain how waste water is treated	→	describe how sewage is treated	→	compare the ease of treating waste, ground and salt water

Worked example

1. **Fertilisers are added to soil to provide three essential elements. Identify the essential elements.**

   a NKNa **b** (NPK) **c** NPS **d** NKS

The answer NPK is correct

2. **Draw the apparatus needed to make a fertiliser *solution* from acid and alkali. Do not include crystallisation.**

1. Use a measuring cylinder to pour alkali into a conical flask.

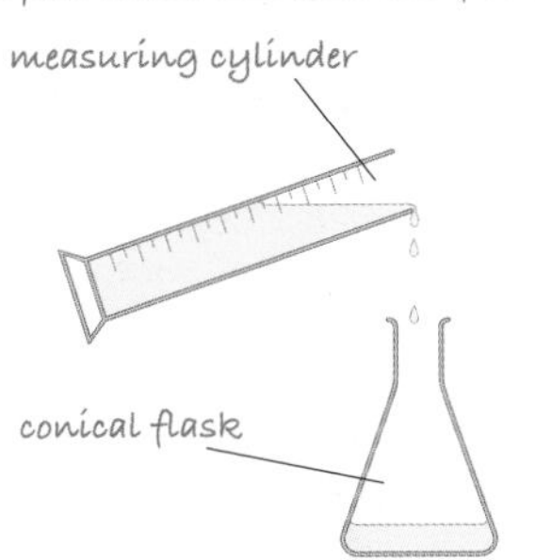

2. Add acid to the alkali until it is neutral.

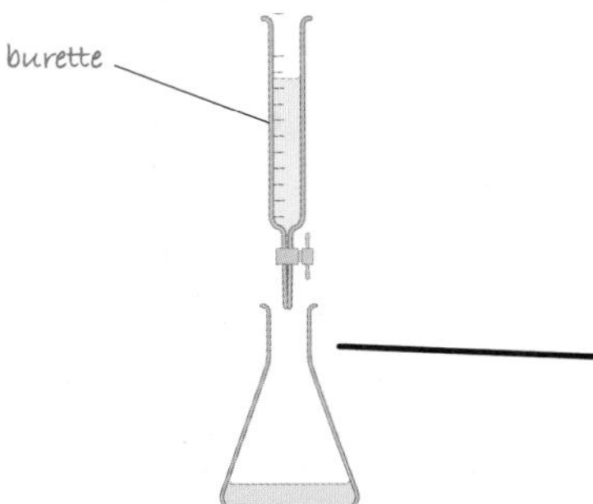

The apparatus for a titration is correct.

3. **Fertilisers are needed to improve crop growth. Give reasons why they are needed.**

They provide more crops for a growing population.

A better answer would include the fact that natural deposits of fertilisers are running out and not always in the place needed.

4. **Give two *different* ways that corrosion of iron can be prevented.**

The iron can be painted or covered with grease.

This is partly correct. A better answer would include another type of prevention, e.g. sacrificial protection.

5. **Evaluate which metal would be best for making the following objects. a lighting circuit b coiled spring for holding a toy c a decorated wrist band d heating coil for a hot drinks cup**

Alloy	Electrical conductivity S/m	Hardness Mohs	MP °C	Thermal conductivity W/mK	Tensile strength N/m²
A	5.9×10^7	3	1062	380	220
B	4.6×10^6	1.5	318	35	17

a A b A c B d A

The answer is partially correct but not complete. The student identifies the best alloys for each object but has not explained why they chose them (not evaluated). The choice must be justified with a reason.

End of chapter questions

Getting started

1. Which of the following is not true about the Haber process? 1 Mark

   a ammonia is produced

   b high pressure is used

   c nitrogen and hydrogen are not recycled if they do not react

   d a catalyst is used

2. Which of the following is a sustainable resource? 1 Mark

   a Managed forest **b** Oil **c** Gold **d** Coal

3. Identify the alloy.

   a copper **b** bronze **c** nickel **d** magnesium 1 Mark

4. Which stage is used to destroy microbes when treating water?

   a filtration **b** sedimentation **c** chlorination **d** distillation 1 Mark

5. Identify the alkene.

   a C_3H_8 **b** C_3H_6 **c** C_3H_8O **d** $C_3H_6O_2$ 1 Mark

6. Monomer X is an alkene. State the type of polymerisation it undergoes and draw a short section of the polymer. 2 Marks

7. Two students are concerned about using resources wisely. They discuss whether to make a new object using wood from a managed forest or plastic from crude oil. They choose wood. Suggest why. 2 Marks

Going further

8. Suggest what should happen to waste glass to reduce the use of resources. Explain your answer. 1 Mark

9. Using carbon fibre as an example, explain what is meant by a composite material. 1 Mark

10. Cracking is carried out on fractions produced from crude oil. 2 Marks

    a Explain the purpose of cracking.
    b Write a balanced equation for the cracking of heptane, C_7H_{16}.

11. Describe one experiment you would do to show that salt water and air makes iron rust quicker than water and air. Predict the outcome. 4 Marks

12. When three new materials, a metal, glass and ceramic, are compared the electrical conductivity is different.

    a is 7 units **b** is 0.00003 units **c** is 0.0000005 units

    Which is a metal? Give a reason for your answer. 4 Marks

More challenging

13. How is potable water is made from seawater? 1 Mark
14. Why water sterilised with chlorine before being distributed to customers? 1 Mark
15. Explain how galvanising iron prevents it rusting. 2 Marks
16. Describe the structure of one strand of DNA. 2 Marks
17. The lifetime of methane in the atmosphere is much shorter than carbon dioxide although it has a greater global warming potential. Explain how these factors might affect global warming. 2 Marks

Most Demanding

18. Explain why a compromise temperature is used in the Haber process. 2 Marks
19. Explain how the iron catalyst in the Haber process works and suggest how it affects the yield of ammonia. 2 Marks
20. Substance D is a hydrocarbon containing six carbons. When heated with a catalyst it forms E which has two carbons and one other compound, F. E reacts to form G. When sodium is added to G, the gas evolved pops when a lighted splint is inserted.
    - **a** Identify D, E, F and G, writing equations where relevant.
    - **b** Describe any additional reactions that E undergoes. 4 Marks

37 Marks

key
atomic number
symbol
name
relative atomic mass

(1)	(2)											(3)	(4)	(5)	(6)	(7)	(0)
1	2	3	4	5	6	7	8	9	10	11	12	13	14	15	16	17	18
1 **H** hydrogen 1.0																	2 **He** helium 4.0
3 **Li** lithium 6.9	4 **Be** beryllium 9.0											5 **B** boron 10.8	6 **C** carbon 12.0	7 **N** nitrogen 14.0	8 **O** oxygen 16.0	9 **F** fluorine 19.0	10 **Ne** neon 20.2
11 **Na** sodium 23.0	12 **Mg** magnesium 24.3											13 **Al** aluminium 27.0	14 **Si** silicon 28.1	15 **P** phosphorus 31.0	16 **S** sulfur 32.1	17 **Cl** chlorine 35.5	18 **Ar** argon 39.9
19 **K** potassium 39.1	20 **Ca** calcium 40.1	21 **Sc** scandium 45.0	22 **Ti** titanium 47.9	23 **V** vanadium 50.9	24 **Cr** chromium 52.0	25 **Mn** manganese 54.9	26 **Fe** iron 55.8	27 **Co** cobalt 58.9	28 **Ni** nickel 58.7	29 **Cu** copper 63.5	30 **Zn** zinc 65.4	31 **Ga** gallium 69.7	32 **Ge** germanium 72.6	33 **As** arsenic 74.9	34 **Se** selenium 79.0	35 **Br** bromine 79.9	36 **Kr** krypton 83.8
37 **Rb** rubidium 85.5	38 **Sr** strontium 87.6	39 **Y** yttrium 88.9	40 **Zr** zirconium 91.2	41 **Nb** niobium 92.9	42 **Mo** molybdenum 95.9	43 **Tc** technetium	44 **Ru** ruthenium 101.1	45 **Rh** rhodium 102.9	46 **Pd** palladium 106.4	47 **Ag** silver 107.9	48 **Cd** cadmium 112.4	49 **In** indium 114.8	50 **Sn** tin 118.7	51 **Sb** antimony 121.8	52 **Te** tellurium 127.6	53 **I** iodine 126.9	54 **Xe** xenon 131.3
55 **Cs** caesium 132.9	56 **Ba** barium 137.3	57-71 lanthanides	72 **Hf** hafnium 178.5	73 **Ta** tantalum 180.9	74 **W** tungsten 183.8	75 **Re** rhenium 186.2	76 **Os** osmium 190.2	77 **Ir** iridium 192.2	78 **Pt** platinum 195.1	79 **Au** gold 197.0	80 **Hg** mercury 200.6	81 **Tl** thallium 204.4	82 **Pb** lead 207.2	83 **Bi** bismuth 209.0	84 **Po** polonium	85 **At** astatine	86 **Rn** radon
87 **Fr** francium	88 **Ra** radium	89-103 actinides	104 **Rf** rutherfordium	105 **Db** dubnium	106 **Sg** seaborgium	107 **Bh** bohrium	108 **Hs** hassium	109 **Mt** meitnerium	110 **Ds** darmstadtium	111 **Rg** roentgenium	112 **Cn** copernicium		114 **Fl** flerovium		116 **Lv** livermorium		

Glossary

#

3D model model of a chemical using spheres (not flat circles) as atoms or ions

A

absorb take in a liquid

acid solution with a pH of less than 7

acid rain rain water which is made more acidic by pollutant gases

activation energy the energy needed to start a chemical reaction

active ingredient the ingredient in a formulation that provides the chemical reaction needed but which is bulked out with other substances

actual yield the yield measured by mass that is actually obtained in a reaction

aerobic processes (usually respiration) using oxygen

algae small green organism that grows in ponds or stagnant water

alkalis substances which produce OH^- ions in water

alkanes a family of hydrocarbons found in crude oil with single covalent bonds, e.g. methane

alkenes a family of hydrocarbons with with at least one double C–C bond

alloys a mixture of two or more metals designed to change the properties of the main metal

Alternative energy energy obtained from sources other than burning fossil fuels for example from the sun, wind turbines, tides

aluminium an element that is a metal of low density and has an atomic number of 13

amino acids small molecules from which proteins are built

anaerobic processes (usually respiration) without oxygen

analyse identify and measure the chemical constituents of a substance

anode electrode with a positive charge

approximately almost (usually with reference to numbers/quantities)

argon an element that is a non-metal, a noble gas in Group 0 and has an atomic number of 18

atmosphere the mixture of gases making the layer of gas around a planet. The current Earth's atmosphere contains nitrogen, oxygen, carbon dioxide, water vapour and other traces

atmospheric pressure the pressure exerted by the gases of the atmosphere (this varies at different heights above sea level)

atom economy a way of measuring the amount of atoms that are wasted or lost when a chemical is made

atomic number the number of protons found in the nucleus of an atom

atomic structure the number of protons, neutrons and electrons in atoms

Average global temperature the mean of temperatures taken at the surface of the Earth different points of latitude. Currently it is approximately 14˚C

Avogadro number this number is 6.02×10^{23}. It is the number of particles or units of 1 mole of any substance, where 1 mole is the formula mass measured in grams

axes the horizontal and vertical lines on which a scale is drawn so that points are plotted to establish a relationship between them

B

balanced equation chemical equation where the number of atoms on each side of the equation balance each other

barium chloride testing chemical for sulfates in water

base reacts with an acid to form a salt

bauxite an ore of aluminium consisting of aluminium oxide

battery two or more electrical cells joined together

bioleaching process that uses bacteria to leach metals

boiling point temperature at which the bulk of a liquid turns to vapour

bond breaking the process of uncoupling bonds between atoms. Energy is taken in during the process

bond making the process of coupling bonds between atoms. Energy is given out during the process

brass an alloy of copper and zinc

bromine an element that is a non-metal, a liquid halogen in Group 7 and has an atomic number of 35

bronze an alloy of copper, tin and other metals

burette a piece of apparatus for measuring variable amounts of liquid with accuracy. It is a long thin tube with a tap with a graduated scale down its length

by-product a product of a chemical reaction that is not the required product

C

carbon anode the positive electrode in a electrolysis that is made of carbon, so is usually inert

carbon capture the process of capturing carbon dioxide so that it does not contribute to the amount of greenhouse gases

carbon dioxide (CO_2) a greenhouse gas which is emitted into the atmosphere as a by-product of combustion

carbon neutrality the process of growing enough trees that take in carbon dioxide to offset the production of carbon dioxide by burning fossil fuels

carbon off-setting the process of trading one nations carbon dioxide output with another nation that produces less in order to keep to agreed global reduction limits

carbonate a negative ion (anion) made of one carbon atom and three oxygen atoms that together carry a charge of –2. The formula is CO_3^{2-}

catalysis the process of making a reaction go faster by adding a catalyst

catalyst a chemical that speeds up a reaction but is not changed or used up by the reaction

catalytic action the action of a catalyst in making a reaction go faster by lowering the activation energy

cathode the negative electrode in a circuit or battery

cell a single unit that is capable of producing electricity, Batteries are made up of two of more cells

cellulose a natural polymer found in plants for example grass

changes of state the processes of changing a substance from solid to liquid and from liquid to gas, and also the reverse of these processes

charge(s) a property of matter charge exists in two forms, positive and negative, which attract each other

chemical amount an amount of substance of a chemical, usually expressed in moles, which is equivalent to the formula mass of the chemical measured in grams

chlorine an element that is a non-metal, a gaseous halogen in Group 7 and has an atomic number of 17

chromatogram is a visible record showing separated substances that have travelled set distances. The distance an unknown substance travels is compared with the distance a standard substance travels

chromatography a method for splitting up a substance to identify compounds and check for purity

chromium a metal of the transition metal block atomic number 24

cobalt a metal of the transition metal block atomic number 27

collision the meeting of two particles moving with a certain amount of energy either causing a successful collision for a reaction to proceed or an unsuccessful collision where there is no reaction

composites are made of two materials, a matrix or binder surrounding and binding together fibres or fragments of the other material, which is called the reinforcement

compromise not the best solution for either of two situations but the best for both together. Usually applied to conditions of a reaction where, for example, a low temperature may give a high yield but also makes the reaction too slow. A compromise temperature will give less yield but at a quicker rate

compound two or more elements which are chemically joined together, e.g. H_2O

concentration the amount of chemical dissolved in a certain volume of solution

concentrate to remove the solvent to produce a more concentrated solution

condensation the product of turning gas to liquid

condensing the process of turning gas to liquid

conservation of mass the total mass of reactants equals the total mass of products formed

Construction lines lines drawn when calculating a tangent to a curve at a specific point. The lines are sections of the *x* and *y* scale

correlation the relationship between two variable, if one increases as the other increases, there is a positive correlation, or if one increases but there is no relation to the behaviour of the other variable then there is no correlation

corrosion the unwanted oxidation of metals to make the metal oxide

counteract go against as in Le Chatelier's principle where an equilibrium may be altered to counteract the effect of a rise in temperature

covalent bonds bonds between atoms where some of the electrons are shared

cracking the process of breaking down larger hydrocarbons into more useful molecules

crimson colour of dark red used to describe the flame produced by burning calcium compounds

crude oil a viscous dark liquid drilled out of the ground or from under the sea. It is a mixture of hydrocarbons which are separated by fractional distillation to produce useful fuels and other products

cryolite common name for sodium hexafluoroaluminate, Na_3AlF_6, which is the substance used to lower the temperature of the melted electrolyte used to obtain aluminium by electrolysis

crystallisation the process of cooling a warm saturated solution of a substance to produce large crystals if cooled slowly or small crystal if cooled quickly

cylindrical description of an object in 3D having a base and a top as circles of the same radius where all points on one circle edge are joined vertically to the identical position on the other circle edge to give a curved surface

D

Decimal point the point separating whole numbers from the value indicating the amount of 1/10ths or 1/100ths represented

Decimals the place value indicating amount of 1/10ths or 1/100ths represented

Decimetre cubed a unit of measurement of volume representing 10 cm × 10 cm × 10 cm written as 1 dm^3 which is equal to 1000 cm^3 or 1 litre

deforestation the destruction of forests without replanting. This provides less opportunity for carbon dioxide to be taken in by plant leaves

delocalised electrons electrons which are free to move away through a collection of ions – as in a metal

density the density of a substance is found by dividing its mass by its volume

deoxyribonucleic acid the chemical containing genetic information made of two strands coiled in a double helix. Each strand consists of a backbone of a sugar and a phosphate group on which there are a random sequence of 4 bases. Each base is paired with a complementary base on the other strand

desalination the process of making potable (fit for drinking) water from sea water by distillation or reverse osmosis

diameter the length of the largest line across a circle, the line goes through the midpoint

diamond a substance used for jewellery and drilling. They are made from carbon atoms held together by covalent bonds acting in all directions making it the hardest natural substance

dinitrogen tetroxide an oxide of nitrogen with the formula N_2O_4

directional bonds covalent bonds made by the sharing of electrons that occur in a set direction for example in water, ammonia, methane, diamond

Directly proportional a mathematical term for describing the relationship between two variables when both increase (or decrease) together

discharged given off. Usually used for electrolysis reactions where substances are 'discharged' at electrodes

displacement reaction chemical reaction where one element displaces or 'pushes out' another element from a compound

distillation the process of evaporation followed by condensation

distort the process of adding atoms of another metal to a main metal in order for the layer of atoms to be out of line so that they do not slip easily over one another, making the metal harder

DNA bases four chemicals that are found in DNA, they make up the base sequence and are given the letters A, T, G and C

dot and cross diagram a diagram representing the number of electrons in the outside shell of bonding atoms or ions

ductile able to be pulled into a wire

E

economic considerations factors taken into account when turning a chemical process into a manufacturing process on a large scale , for example a high pressure may produce a high yield but the cost of maintaining a high pressure may be too expensive

electrical conductors materials which let electricity pass through them

electrical non-conductor a substance that does not conduct electricity, which is an electrical insulator for example wood or plastic

electrode ions are discharged at the electrodes during electrolysis

electrode reaction a reaction during electrolysis that occurs at a positive electrode (where electrons are passed from the ion to the electrode) or a negative electrode (where electrons are passed from the electrode to the ion)

electrolysis is the process of passing direct current through a solution or melted ionic compound to move the ions apart and so break the compound down

electrolyte a liquid that conducts electricity and breaks down during electrolysis

electron arrangement how the electrons fill the shells around the nucleus for example 2,8,1 means 2 electrons in the 1st shell, 8 in the 2nd, 1 in the 3rd

electron gain gaining of electrons to form negative ions electron loss losing of electrons to form positive ions electron shells the orbit around the nucleus likely to contain the electron

electron loss the removing of electrons from the outer shell of an atom during a chemical process. Electron loss is a process of oxidation

electron shell an area around the nucleus which will probably hold electrons. It is not a physical 'hard ring' but a model of an area where electrons may be 'found'. An energy level

electron transfer the process of electron movement from one atom or ion to another

electronic a pattern or process referring to electrons

electronic structure the number of electrons in sequence that occupy the shells, e.g. the 11 electrons of sodium are in sequence 2.8.1

electrons small particles within an atom that orbit the nucleus (they have a negative charge)

electroplating the process of coating the negative electrode of an electrochemical cell with metal

electrostatic the non-moving charges that reside on particles

electrostatic attraction attraction between opposite charges, e.g. between Na^+ and Cl^-

elements substances made out of only one type of atom

empirical formula the simplest formula that a larger formula can be reduced to. For example the empirical formula of C_8H_{16} is CH_2

endothermic reaction chemical reaction which takes in heat

energy the ability to 'do work'

energy levels the level of energy that an electron occupies or the level of energy that moving particles have for collision in a reaction or the level noted on the y axis of an energy level diagram or profile

energy transfer the transfer of energy out (an exothermic reaction) or of energy in (an endothermic reaction)

enzymes biological catalysts that increase the speed of a chemical reaction

equation a mathematical or chemical statement where one side equals another

equilibrium when the forwards and backwards reactions are occurring at the same rate in a closed system

equilibrium position in a reversible reaction where the forward reaction and backward reaction are set up together producing products and reforming reactants by a set amount (which can be altered by changing conditions)

Ernest Rutherford a Nobel Laureate scientist working in Manchester and Cambridge who proposed that the atom had a nucleus with electrons surrounding it and that the atom was mostly empty space

erosion the process of rock breaking down

ethandiol the compound containing 2 carbon atoms and 2 alcohol functional groups, $C_2H_4(OH)_2$

evaporation when a liquid changes to a gas, it evaporates

evolve produce or give off (especially in relation to a gas)

excess the reactant that is present in an amount more than another reactant and is more than is needed

exhaust gases gases discharged into the atmosphere as a result of combustion of fuels

exothermic reaction chemical reaction in which heat is given out

extracting taking from, obtaining from (usually using another chemical or electrolysis)

F

fertilisers chemicals made to enhance the growth of crops, usually containing high levels of nitrogen or phosphorus or potassium or all three in set ratios

filtration the process of filtering river or ground water to purify it for drinking water

finite resource a resource that was made millions of years ago and cannot be made again and will eventually run out, for example crude oil

flame test test where a chemical burns in a Bunsen flame with a characteristic colour – tests for metal ions

force of attraction the force with which positive and negative ions attract each other or that molecules attract each other

formulation a mixture that has been designed as a useful product

fraction a number that is expressed as a number of parts of the total number of parts for example $\frac{2}{3}$ means 2 parts out of a total of 3

fractional distillation crude oil is separated into fraction using this process of distillation where fraction of different boiling points distil off at different times

free electrons electrons that are not bound to an atom or ion but can move away from the atom to be 'delocalised'

frequency the amount of times a process occurs, for example the number of time per sec that a 'collision' of particles takes place

fuel cell cells supplied by an external source of fuel (e.g. hydrogen) and oxygen or air. The fuel is oxidised electrochemically within the fuel cell to produce a potential difference

fullerenes cage-like carbon molecules containing many carbon atoms, e.g. buckyballs

G

galvanising the process of coating with zinc to prevent rusting

gas the state of a chemical where all the particles are moving rapidly and randomly

gas syringe a piece of apparatus made of glass with a plunger inside a scaled tube for collecting gas

gas volume the volume of a gas measured in cm^3 or dm^3. At room temperature and pressure the volume of 1 mole of gas occupies approximately 24 dm^3

gas-liquid chromatography a process for separating mixtures of chemical using gas as the mobile phase over liquid as the stationary phase

Geiger and Marsden experiment an experiment using the scattering of alpha particles on gold leaf that showed that some particles were deflected by a concentrated charge leading to the idea of an atom with a nucleus

gelatinous a jelly like substance for example the coloured precipitates formed by metal hydroxides

giant covalent structures a large regular three-dimensional covalently bonded structure containing more that one non-metal element

giant lattice the large structure made when large numbers of positive and negative ions are attracted together to make a regular pattern in 3D

global dimming the process of blocking out the sun with polluting gases, resulting in a decrease of temperature

glycine an amino acid with the simplest amino acid structure

gradient rate of change of two quantities on a graph; change in *y* / change in *x*

graph a representation of data between two variables (eg temperature and time) by plotting both data points against two scales, one vertical and one horizontal

graphene the strongest substance made, consisting of sheets hexagonal rings of carbon atoms joined together that are 1 atom thick

graphite a type of carbon made of rings of atoms in layers

greenhouse effect The effect noted when ultraviolet ray from the sun hit the Earths surface and are re-radiated as infra-red rays which are then trapped by gases in the atmosphere such as carbon dioxide and methane, causing energy to be retained and the temperature to increase (like in a greenhouse)

group a vertical column on the Periodic Table that contains atoms with similar electronic configurations, especially in the outer shell, for example Group 1 contains lithium, sodium and potassium, all of which have 1 electron in the outer shell

H

Haber process industrial process for making ammonia

half equations are used to describe oxidation or reduction using electron movement

halides a compound of two elements, one of which must be a halogen

halogen an element in group 7 of the periodic table

helium second element in periodic table. It has the lowest boiling point

hexanedioic acid a dicarboxylic acid $(CH_2)_4(COOH)_2$ used in the manufacture of nylon 6.6

Hydrocarbons compounds containing only hydrogen and carbon

hydrochloric acid a strong acid made by dissolving hydrogen chloride in water

hydrogen an element that is a colourless, odourless, highly flammable gas with an atomic number of 1

hydroxide a metal compound containing the hydroxide ion (OH^-)

hydroxyl ion an anion with the formula OH^-

hyperaccumulators plants that absorbs heavy metals

I

ignite to catch fire

impure not pure, contains unwanted matter

indicator shows visually when the pH of a solution has reached a certain value, e.g. the end point of an acid alkali titration

inert unreactive

insoluble does not dissolve in a solvent usually water

intermolecular force force between molecules

intramolecular forces within a molecule

iodine–131 radioactive isotope of iodine used in diagnosis and treatment of thyroid cancer

ion migration the movement of ions through an electrolyte

ionic bond a chemical bond between two ions of opposite charges

ionic equations show only the ions that take place in a chemical reaction

ions charged particles (can be positive or negative)

iron (III) oxide a compound of iron and oxygen. It occurs naturally as haematite

isotopes atoms with the same number of protons but different numbers of neutrons

J

J. J. Thompson suggested the plum pudding model of an atom

James Chadwick discovered the neutron

John Dalton suggested that atoms were small particles like billiard balls

L

Le Chatelier's principle If a system is at equilibrium and a change is made to any of the conditions, then the system responds to counteract the change

Life cycle assessments (LCAs) are carried out to assess the environmental impact of products in each of several stages

limewater calcium hydroxide particles in water – this clear liquid turns milky in the presence of carbon dioxide

limiting reactant the reactant that is used completely in a chemical reaction

limited resource a resource such as metal ores that do not renew themselves at a sufficient rate to supply demand for them

liquid petroleum gas propane or butane or a mixture of both propane and butane

lustrous shiny

lysis to split apart

M

magnitude size of something

malleable substances that can be hammered or pressed into different shapes

manganese a transition metal

mass describes the amount of something; it is measured in kilograms (kg)

measuring cylinder a narrow cylindrical container with a scale used for measuring liquids

metal ions atom of a metal that has lost electrons and has become positively charged

metallic relating to or being a metal

metalloids the line of elements separating the metals from the non-metals are called the metalloids

metals solid substances that are usually lustrous, conduct electricity and form ions by losing electrons

mixture a substance consisting of two or more elements or compounds not chemically combined together

mobile phase in chromatography this is the phase that moves

molar mass the mass in grams of one mole of a substance

molar volume the volume occupied by 1 mole of a substance

mole a unit for a standard amount of a substance. One mole of any substance contains the same number of particles, atoms, molecules or ions as one mole of any other substance

molecule two or more atoms which have been chemically combined

Monomer a small molecule that can be polymerised to make a much larger molecule (polymer)

N

nanometre units used to measure very small things (one billionth of a metre)

nanoparticles very small particles on the nanoscale

nanotube carbon atoms formed into a very tiny tube

negative ion an ion made by an atom gaining electrons

neon one of the noble gases

neutral a neutral substance has a pH of 7

neutralisation a reaction that takes place when an acid and base react

neutrons small particle which does not have a charge found in the nucleus of an atom

nickel a transition metal with an atomic number of 28

Niels Bohr proposed that electrons orbit the nucleus in fixed orbits called shells

nitrogen a colourless gas that makes up about 78% of the air by volume

nucleus central part of an atom that contains protons and neutrons

O

optimum conditions the conditions for a reaction that give the best results

outer shell the shell containing electrons that is furthest from the nucleus

oxidation a chemical reaction that is either the addition of oxygen, or the removal of hydrogen or the loss of electron

oxide a compound of two elements, one of which is oxygen

oxides of nitrogen compounds containing oxygen and nitrogen only

oxygen a colourless gas that makes up about 21% of the air by volume

P

particle a minute amount of matter

patterns another word for trends. A regular way in which properties change

peptide a compound containing two or more amino acids joined by an amide link (–OC–NH–)

percentage yield comparing the amount of useful product made to the amount expected

period a row in the periodic table

periodic a regular way in which properties change because of electronic structure and noted across the periodic table

pH a number which shows the acidity or alkalinity of a substance dissolved in water related to the number of hydrogen ions in the solution

phosphorus a non-metal with symbol P It is found in Group 5 of the Periodic Table

photosynthesis the process whereby green plants synthesise carbon dioxide and water into carbohydrates

phytomining process that uses plants to extract metals

pipette a glass tube used for transferring measured amounts of liquid, usually 10 cm^3, 25 cm^3, 50 cm^3 or 100 cm^3

polyester a condensation polymer made from a dicarboxylic and a diol. E.g. Terylene

polymer very large molecules with atoms linked to other atoms by covalent bonds

polymerisation the formation of a polymer from monomers

positive ion an ion made by an atom losing electrons

potable water water that is safe to drink

potassium a reactive metal, symbol K, in Group 1 of the Periodic Table

precipitate solid formed in a solution during a chemical reaction

preferential discharge when an anion or cation is discharge in preference to other anions or cations present

pressure the ratio of force to area

product molecules produced at the end of a chemical reaction

product mass the amount of product formed

profile see reaction pathway

protein a biological polymer of amino acids joined by amide groups see peptide

protons small positive particles found in the nucleus of an atom

purity a pure substance is a single element or compound. Pure substances have a sharp melting point and a sharp boiling point

R

random having no regular pattern

radiation a stream of particles (usually alpha or beta particles) from a radioactive source

radius a straight line from the centre to the circumference of a circle or sphere

rate of reaction the rate of disappearance of a reactant or the rate of appearance of a product (sometimes referred to as speed of a reaction)

ratio the relative sizes of two or more different amounts

reactant mass the mass of a reactant. It is usually converted to moles by diving the value by it relative formula mass

reactants chemicals which are reacting together in a chemical reaction

reaction pathway usually a diagram which shows the energy changes that occur during a chemical reaction

reactivity see relative reactivity

rechargeable a battery that can be recharged over and over again

recycling see reuse of resources

reduction the process of electron gain

reduction with carbon removes oxygen from an oxide to formed the element. Haematite (iron(III) oxide) is reduced by carbon in the Blast furnace to form iron

reinforcement increase the strength of. Reinforced concrete contains steel bars to increase its strength

relative compare with a standard measure (usually 1) of something else

relative atomic mass the average mass of an atom compared to 1/12 of a carbon–12 atom

relative formula mass the sum of the relative atomic masses in a compound

relative reactivity describes how one element (a metal) is more reactive than another. The series formed is called a reactivity series

repeating unit see monomer

reuse of resources the recovery and processing of materials after they have been used in order to reuse them

reverse osmosis a method of obtaining pure water from water containing salts such as sea water

reversible reaction a reaction that can go forwards or backwards depending on the conditions

R_f value is the ratio $\left(\frac{\text{distance moved by a compound}}{\text{distance moved by the solvent}}\right)$ It is used in chromatography to identify compounds. It always the same for a particular compound if the chromatography has been performed in the same way

room temperature and pressure abbreviated to r.t.p. 25 °C and 1 atmosphere pressure

S

sacrificial protection using a more reactive metal to protect another metal

salt a compound formed by replacing one or more hydrogen ions in an acid by a metallic ion or an ammonium ion

'Sea' of electrons occurs in a metal where the outer electrons are free to move between the metal ions as they are 'delocalised'

sediment matter that settles to the bottom of a liquid

sedimentation a process during water purification where small solid particles are allowed to settle

segment: parts of a circle when divided round by a line from its radius

separation technique a method used to separate mixtures. For example, filtration or distillation

sewage effluent waste from humans that needs treatment before released back into waterways

silicon dioxide a white solid with the formula SiO_2

silver nitrate a chemical used for testing halide ions in water

single bond a bond formed when one pair of electrons is shared between two atoms

slope a line that is not horizontal or vertical but drawn in 2D at an angle from the horizontal

solute the substance which dissolves in a solvent to form a solution

solution when a solute dissolves in a solvent, a solution forms

solvent the liquid used to dissolve a solute

solvent front how far the solvent travels during chromatography

spectroscopy the spectra produced when matter is subjected to electromagnetic radiation

square bracket a pair of brackets in the form [ ], used in the formula of polymers or in dot and cross diagrams

stable electronic structure an achieved structure where the outer electron shell of an atom is full

standard form a way of writing a very large number with one number before the decimal point, multiplied by a power of 10

stationary phase the phase in chromatography that does not move. In paper chromatography it is the paper

steel an alloy or iron and carbon

steeper the slope of a line in relation to another slope, used in context of identifying from a graph whether a reaction is going at a greater or lesser rate than another reaction

strength (of an acid) strong acids ionise completely in water. Weak acids partially ionise

sulfates a salt formed from sulfuric acid. It is represented by SO_4^{2-}

sulfur dioxide a colourless, pungent gas with the formula SO_2

supply and demand the amount of a substance that is available and the amount of the substance required, used in context of cracking hydrocarbons

surface area to volume ratio the area of the outside of an object divided by its volume, used in the context of splitting large lumps of chemical to powder to make a greater surface area

surroundings the matter surrounding the site of a chemical reaction into which or from which energy is being transferred

sustainability can be used without completely being used up and being replaced or recycled

symbol a way of representing elements. It consists of one or two letters

T

tangent a line that touches a curve, used to measure the rate of reaction at a specific point by dividing the measured variable on the y axis by time on the x axis

temperature a measure of hotness or coldness of matter

tensile strength measures how easy it is to break a material

tetrahedral bonding a molecule which has an atom in the centre and four atoms at the corners of a tetrahedron

tetrahedron an object with four faces, also used to describe a model of a carbon atom with four bonds joined to other atoms

titration a method to measure the amount of acid or alkali that react with each other

theoretical yield the amount of product that we would expect to make from a reaction

thermal conductivity a measure of how well a material allows heat to flow through it

thermal decomposition the breaking down of a compound into two or more products on heating

thermosetting describes a plastic that cannot be remelted

thermosoftening a plastic which softens on heating but hardens again on cooling

toxic metals how poisonous metals such as lead are

toxicity how poisonous a substance is

transfer move from one place to another

transition element an element in the middle section of the periodic table, between the group 1 and 2 block and the group 3 to group 0 block

trend showing the direction in which something is changing

two dimensional a diagram or drawing showing only length and breadth but not depth of an object

U

universal indicator a mixture of indicators that change colour over a pH range.

unreactive does not chemically react

V

viscosity the resistance of liquid to flow

volatile an easily evaporated liquid

volcanic product produced by a volcano

volume the amount of space that a substance occupies The S.I. unit for volume is the cubic metre (m^3). The cubic centimetre (cm^3) is also used

W

wavelength (λ) distance between two wave peaks

Y

yellow the colour of Universal Indicator in a solution with a pH of 6 (slightly acidic), or the colour of the flame of sodium compounds

yield the amount of product formed in a reaction

Index